Applied Economics in the Digital Era

"Professor Gary Madden was one of the major scholars in ICT economics. His deep knowledge of theory and methods allowed him to tackle a broad set of domains and problems. Those who knew him could only admire his productivity and commitment. The articles in this book reflect not only his extensive network of close colleagues but also the scope and nature of his scientific interests. If possible, Madden would certainly have continued his research on many highly relevant topics covered in this book such as ICT vs. GDP, adoption of e-commerce, two-sided platforms, regulation of Internet giants, and network neutrality."
—Heikki Hämmäinen, *Professor of Networking Technology, Aalto University*

"This book was edited to honour the work of Gary Madden, a pioneer in the field of ICTs and Forecasting. It takes us through the different technological revolutions of the human species leading to the emergence of the current cyber-civilization, where forecasting the evolution of consumers' needs and tastes becomes the key driver of economic success. In this setting, privacy and dominance in the information market become the key relevant issues faced by todays' civilization, leading to interesting current policy questions such as asking whether the acquisition of Whatsapp by Facebook should have been opposed by antitrust authorities.

All the chapters provide illuminating contributions on the different aspects of modeling and forecasting key questions arising in ICT markets, covering related and essential themes from machine learning, net neutrality, to multisided markets and platform dominance, jointly providing a key essential reading, for scholars, students and practitioners."
—Emanuele Giovannetti, *Professor of Economics, Anglia Ruskin University*

"Gary Madden's interests ranged widely across the fast-developing world of ICT economics. This book of essays, in recognition of his contributions, by some of the leading scholars in the area captures the breadth and depth of current research. Papers discuss both regulatory issues as they apply to the internet and the adoption of new technologies, but also analyze the fast developing world of online activity. All scholars in the area will find something new and interesting and the broad view of research methods and perspectives in applied econometrics and forecasting will appeal to practitioners."
—Robert Fildes, *Distinguished Professor of Management Science, Lancaster University Management School*

"When Gary Madden spent his sabbatical at Columbia, he made quite an impression as a scholar and person. This Festschrift captures the wide spectrum of his intellectual interests and research skills. It covers new methods of analysis as well as several topics ranging across the area of information and communications technology (ICT). This commendable volume should be of interest and significance those new to the field as well as to professionals exploring new techniques."
—Eli Noam, *Director, Columbia Institute for Tele-Information*

"This book of readings dedicated to the memory of Gary Madden reflects his broad interests by some of his students and co-authors who worked with him. Those of us who knew Madden recognize that this is a small, but significant, offering of his research interests. That said, the book is useful for both experienced practitioners or those just starting their research career. For the latter, they can get a sense of the field beginning with Faulhaber's overview on the information and communications technology (ICT) sector. It covers the latest on antitrust issues for the sector. For the practitioners, it opens up new methodologies; for example, Taylor shows new methods to determine the impact of price changes throughout the economy. Insightful and a delight to read."
—Jonathan Liebenau, *Associate Professor of Technology Management, London School of Economics*

James Alleman · Paul N. Rappoport ·
Mohsen Hamoudia
Editors

Applied Economics in the Digital Era

Essays in Honor of Gary Madden

Editors
James Alleman
College of Engineering and Applied Science
University of Colorado Boulder
Boulder, CO, USA

Paul N. Rappoport
Temple University
Wyncote, PA, USA

Mohsen Hamoudia
Orange Business Services
Saint Denis, France

ISBN 978-3-030-40600-4   ISBN 978-3-030-40601-1   (eBook)
https://doi.org/10.1007/978-3-030-40601-1

© The Editor(s) (if applicable) and The Author(s), under exclusive license to Springer Nature Switzerland AG 2020
This work is subject to copyright. All rights are solely and exclusively licensed by the Publisher, whether the whole or part of the material is concerned, specifically the rights of translation, reprinting, reuse of illustrations, recitation, broadcasting, reproduction on microfilms or in any other physical way, and transmission or information storage and retrieval, electronic adaptation, computer software, or by similar or dissimilar methodology now known or hereafter developed.
The use of general descriptive names, registered names, trademarks, service marks, etc. in this publication does not imply, even in the absence of a specific statement, that such names are exempt from the relevant protective laws and regulations and therefore free for general use.
The publisher, the authors and the editors are safe to assume that the advice and information in this book are believed to be true and accurate at the date of publication. Neither the publisher nor the authors or the editors give a warranty, expressed or implied, with respect to the material contained herein or for any errors or omissions that may have been made. The publisher remains neutral with regard to jurisdictional claims in published maps and institutional affiliations.

This Palgrave Macmillan imprint is published by the registered company Springer Nature Switzerland AG
The registered company address is: Gewerbestrasse 11, 6330 Cham, Switzerland

Editors and Contributors

About the Editors

James Alleman is Professor Emeritus at the University of Colorado—Boulder. He was Visiting Professor CIMBA Italy, Paderno del Grappa, Italy in the Winter of 2017. In 2014, he was awarded a Senior Fellowship at Institut Barcelona d'Estudis Internacional (Barcelona, Spain) to research regulation of Information and Communications Technology (ICT) in Latin America. He was a Visiting Senior Scholar at IDATE in Montpellier, France in the fall of 2005 and continues his involvement in IDATE's scholarly activities. During the calendar years 2001 and 2002, he was a Visiting Professor in the Economics and Finance Division at Columbia Business School, Columbia University. He was a Senior Fellow and Director of Research at Columbia Institute of Tele-Information (CITI), Columbia Business School, Columbia University from 2001 to 2016.

Dr. Alleman was previously the Director of the International Center for Telecommunications Management at the University of Nebraska at Omaha, Director of Policy Research for GTE, and an economist for the International Telecommunication Union. He has conducted research in the area of telecommunications policy, with emphasis on pricing, costing, and regulation as well as on interconnection, international telephony settlements, communications in the infrastructure and related areas. More recently, he has been researching the application of real options valuation techniques to network industries and issues of income and wealth inequality. He provides litigation support in these areas.

Dr. Alleman was co-founder and former Chair of the International Telecommunications Society (ITS) and continues to serve on its Board. He is also on the Editorial Board of *Telecommunications Policy*.

Mohsen Hamoudia is Vice President—Head of Strategy and Market Intelligence of Large Accounts Division within Orange Business Services, Paris. He has held positions in France Telecom Group, *inter alia*, as Head of the Planning and Forecasting Division of France Telecom Long Distance (1997–2001) and Head of Strategic Marketing of Large Accounts (2007–2013). Dr. Hamoudia teaches Forecasting and Quantitative Techniques since 1989 at Toulouse Business School, ESDES Business School-Lyon and ISM (Institut Supérieur du Marketing), Paris. He is a Board member of the International Institute of Forecasters (IIF) and its past President (2012–2016). He is also a Board member of OrangeFab, the startups acceleration program of Orange.

Prior to France Telecom Group, Dr. Hamoudia was Head of Marketing for Electronic Payments System at Atos Company (1988–1993) and Traffic Forecasting Manager at Air Inter (1984–1988) and Researcher in the Software and IT Center of Toulouse (1979–1983). Dr. Hamoudia is published in a wide variety of international journals.

Gary Madden (1952–2017), for whom this volume is dedicated, was a Professor of Economics at Curtin University, Perth, Australia, and Director of the Communication Economics and Electronic Markets (CEEM) at Curtin. As demonstrated in this Festschrift, his research interests were broad in scope, but nevertheless he had a laser-like focus on the areas he studied. His primary focus was on empirical aspect of the information and communications technology (ICT) sector, network economics, productivity, and real options. His extensive publications are listed in the Appendix. He was a Board Member of the International Telecommunications Society and served as an advisor to many international agencies and the Australian Government. Dr. Madden was an Associate Editor for the International Journal of Management and Network Economics, for the Telecommunications Policy journal and was an Editorial Board Member of Communications & Strategies, and the

Preface

Professor Gary Madden was a "renaissance" economist. He worked in the areas of applied microeconomics, econometrics, forecasting, and policy research, primary in the Information and Communications Technology (ICT) sector. Gary's primary research area concerns the economic modeling of digital networks. Within this gambit, Gary's particular research fields encompass the economic impacts of disruptive technology such as 3G–4G mobile (including analysis of spectrum assignment mode) and fixed-line broadband, network externalities, blended competition and the internet, emerging technology forecasting, and welfare and technology-led economic growth (including digital divide issues). He was a true collaborator and had an endless drive to push the boundaries of research.

The essays in this Festschrift are based on his wide-ranging research by his friends and colleagues and, in two cases, articles he was working on prior to his untimely death. The contributors took up several themes on which Gary worked for several years as they brought some new developments and applications.

The Festschrift will provide inspiration to readers to broaden their depth in the field, discover unique techniques, to share ideas, and in general to provide inspiration to new researchers.

Boulder, USA	James Alleman
Wyncote, USA	Paul N. Rappoport
Saint Denis, France	Mohsen Hamoudia

Acknowledgments The editors would like to thank all the contributors to this Festschrift for their patience and flexibility with the editors. In particular, Chris Dippon and Andy Banerjee for encouraging and supporting the work and feedback on the articles.

Contents

1 Civilizations' Economic Drivers: Babylon to Bitstream   1
Gerald R. Faulhaber

Part I   Applied Econometrics

2 A Different Approach to Deriving Price and Income Elasticities: Applications to Telecommunications and Gasoline & Motor Oil   29
Lester D. Taylor

3 A Cross-Country Analysis of ICT Diffusion, Economic Growth, and Global Competitiveness   63
Aniruddha Banerjee, Paul N. Rappoport, and James Alleman

4 Adoption of E-Commerce by Individuals and Digital-Divide   103
Ángel Valarezo, Rafael López, and Teodosio Pérez Amaral

Part II Forecasting

5 Predictive Accuracy Tests for Prediction of Economic
   Growth Based on Broadband Infrastructure   137
   Walter Mayer, Gary Madden, and Xin Dang

6 Modelling Internet Diffusion Across Tourism Sectors   151
   Miriam Scaglione and Jamie Murphy

7 Exploring Drivers of Online Political Participation
   in the European Union   175
   María Rosalía Vicente and Hiroaki Suenaga

8 Machine Learning and Forecasting: A Review   193
   Petrus H. Potgieter

9 Stated Consumer Behavior with D-Efficient Choice
   Set Designs: The Case of Mobile Service Bundles   209
   Christian M. Dippon and Gary Madden

Part III Governance of the Internet

10 Platforms with Restrictive Licensing   231
    Christiaan Hogendorn

11 QoS Investment, Vertical Integration
    and Regulation of the Internet   249
    Edmond Baranes and Cuong Hung Vuong

12 Operations of Internet Platform Intermediaries   267
    Rob Frieden

13 Escalating Instability of Network Neutrality
    Policy in the U.S.   285
    Barbara A. Cherry

14 Multisided Markets and Platform Dominance   303
James Alleman, Edmond Baranes,
and Paul N. Rappoport

Conclusion   325

Appendix: Gary Madden's Publications   327

Index   335

Journal of Media Economics. Not only was he a prolific researcher, he freely helped others in their research including the authors in this book. He was a most decent human being and a good friend; we feel the pain of his untimely death. This volume is our lasting tribute to him.

Paul N. Rappoport is Professor Emeritus at Temple University, Fox School of Business, USA. Dr. Rappoport earned his Ph.D. in Economics from Ohio State University. He is an applied econometrician and a specialist in data mining and advanced visualization. He has served as a senior industry consultant for over 40 years. He is widely published in the areas of demand modeling, health economics, telecommunications and broadband analysis, forecasting, and price elasticity measurement and his research has been used to construct a wide range of commercial applications. Dr. Rappoport was the founder and CEO of PNR & Associates. There he led the National Telecommunications Demand Study (NTDS), created Bill Harvesting and the ReQuest survey. Dr. Rappoport was EVP and Chief Research Officer of Centris. There he led the development of business segmentation models. Most recently, Dr. Rappoport is a senior consultant to Johnson and Johnson where he mentors and works with the company's data scientists on studies ranging from Global Brand Protection to forecasts of medical products. Dr. Rappoport is the co-editor with James Alleman and Áine Marie Patricia Ní-Shúilleabháin of *Demand for Communication Services—Insights and Perspectives: Essays in Honor of Lester D. Taylor*, Springer-Verlag (2016).

Contributors

Aniruddha (Andy) Banerjee is an applied economist and statistician who specializes in applying advanced analytics to find business solutions. Over much of his career, he has conducted econometric analysis and forecasting to support the strategic, regulatory, and litigation aspects of network industries such as telecommunications, cable, media, and home security companies. Currently, as Director of Analytics at IHS Markit, Dr. Banerjee is applying recent advances in data science and survey methods to support a wide array of industries. Business solutions provided pertain to, among others, demand forecasting, consumer behavior and choice, pricing and price elasticity, marketing channel effectiveness, market segmentation, and economic impacts.

Dr. Banerjee was formerly Senior Vice President at SSRS, Senior Vice President at Centris Marketing Science, and Vice President at economic and strategic consulting firms, Analysis Group, Inc. and National Economic Research Associates. In these positions, he had lead responsibilities for advanced analytics, market research, strategy formulation, and regulatory and policy analysis. He has also held positions in AT&T's Market Analysis and Forecasting Division, Bell Communications Research's Regulatory Economics group, and BellSouth Telecommunications' Pricing Strategy and Econometrics group.

Dr. Banerjee is a thought leader and former university professor who has published several papers in peer-reviewed journals and academic books, co-authored a book *Mobile Virtual Network Operators: Blessing or Curse?* and presented his work extensively at national and international conferences. He is a member of the International Telecommunications Society and serves on its Board of Directors and Publications Committee. He holds a Ph.D. in Agricultural Economics from the Pennsylvania State University and an M.A. in Economics from the Delhi School of Economics.

Edmond Baranes is Professor of Economics at the University of Montpellier, France. His main research fields are industrial economics, competition policy, and regulation. He has especially conducted research in the area of telecommunications and internet policy, and more recently in digital economics. His researches include pricing and costing models, regulation issues, and antitrust case analysis. He has also provided litigation support in these areas.

Barbara A. Cherry is Professor at The Media School at Indiana University in Bloomington, Indiana, USA. Her research is primarily focused on the evaluation of deregulatory policies, governance structures, comparative analysis of infrastructure industries both domestically and internationally, and framing of analyses from a complexity theory perspective. Dr. Cherry's research reflects an interdisciplinary academic background integrated with policymaking experience in the telecommunications industry and government, having held positions of Senior Counsel in the Office of Strategic Planning and Policy Analysis at the Federal Communications Commission; Associate Professor and Associate Director of the James H. and Mary B. Quello Center for Telecommunication Management and Law at Michigan State University; Director of Public Policy Studies at Ameritech; and Regional Attorney for State Government Affairs at AT&T.

Dr. Cherry holds a Ph.D. in Communication Studies from Northwestern University, a J.D. from Harvard Law School, and an M.A. in Economics and Law from Harvard University while recipient of a National Science Foundation Fellowship in Economics, and a B.S. with Highest Honors in Economics from the University of Michigan. Dr. Cherry has published articles in scholarly social science journals and law reviews, including *Telecommunications Policy*, *Information Economics and Policy*, *The Information Society*, *International Journal of Communications*, *Federal Communications Law Journal*, and *CommLaw Conspectus*. Her books include *The Crisis of Telecommunications Carrier Liability*, and *Making Universal Service Policy: Enhancing the Process Through Multidisciplinary Evaluation* (coedited with S. Wildman and A. Hammond). Dr. Cherry is a Board Member of the International Telecommunications Society and a member of the Internal Board for the Ostrom Workshop at Indiana University.

Xin Dang is Professor in the Department of Mathematics at the University of Mississippi, USA. She received her Ph.D. degree in Statistics from the University of Texas at Dallas in 2005. Her research interests include robust and nonparametric statistics, statistical and numerical computing, bioinformatics, and Econometrics. In particular, she has focused on statistical learning, data depth and application, soft and rough sets, and robust procedure computation.

Dr. Dang has published more than 35 journal papers including *Journal of Multivariate Analysis*, *Statistics in Medicine*, *International Journal of Forecasting*, *IEEE Transactions on Pattern Analysis and Machine Intelligence*, *IEEE Transactions on Neural Networks and Learning Systems*, *Pattern Recognition*, *IEEE Transactions on Knowledge and Data Engineering*, *BMC Bioinformatics*, *Knowledge Based Systems*, *Journal of Nonparametric Statistics*, *Journal of Statistical Planning and Inference*, *Statistical Papers*, etc. She is a member of the Institute of Mathematics Statistics, American Statistics Association, International Chinese Statistical Association, and International Neural Network Society. She teaches various statistics classes to undergraduate and graduate students at the University of Mississippi.

Christian M. Dippon is Managing Director at NERA and a leading authority in complex litigation disputes and competition matters in the communications, internet, and high-tech sectors. He is also the Chair of

NERA's Global Energy, Environment, Communications & Infrastructure (EECI) Practice, where he leads over 100 experts in the areas of energy, communications, media, Internet, environment, auctions, transport, and water. Global Arbitration Review (2019) ranks Dr. Dippon among the world's leading commercial arbitration experts.

Dr. Dippon advises his clients in economic damages assessments, class certifications and damages, false advertising, antitrust matters, and regulatory and competition issues. He has extensive testimonial and litigation experience, including depositions, jury and bench trials in state and federal courts, domestic (AAA) and international arbitrations (UNCITRAL, ICC, ICSID), and submissions before international courts. He assists clients with a broad range of litigation disputes related to wireline, wireless, cable, media, Internet, consumer electronics, and the high-tech sector. Dr. Dippon also routinely testifies before the US and international regulatory authorities, including the Federal Communications Commission, the International Trade Commission, the Canadian Radio-television and Telecommunications Commission, and the Competition Bureau Canada.

Dr. Dippon has authored and edited several books as well as book chapters in anthologies and has written numerous articles on telecommunications competition and strategies. He also frequently lectures in these areas at industry conferences, continuing education programs for lawyers, and at universities. National and international newspapers and magazines, including the *Financial Times*, *Business Week*, *Forbes*, the *Chicago Tribune*, and the *Sydney Morning Herald*, have cited his work.

Dr. Dippon serves on NERA's Board of Directors, the Board of Directors of the International Telecommunications Society (ITS), and on the Editorial Board of *Telecommunications Policy*.

Gerald R. Faulhaber is Professor Emeritus of Business Economics and Public Policy at the Wharton School of the University of Pennsylvania, USA and Professor Emeritus of Penn Law School. He served as Chief Economist of the Federal Communications Commission in 2000–2001. His research is the microeconomics, management, and public policy aspects of technology and telecommunications firms. Professor Faulhaber's current research is wireless telecommunications, cybersecurity and the Internet of Things, public policy and the internet, and the political economy of regulation. He has also written on file sharing and music copyright, public safety radio, and network neutrality. He has expertise in Applied Microeconomics, Industrial Organization, Network Neutrality for the Internet, File Sharing and Fair Use Copyright, Regulation, and Spectrum Policy for Wireless Telecommunications.

Professor Faulhaber holds a Ph.D. (1975) and an M.A. from Princeton University (1974), an M.S. from New York University (1964) and an A.B. from Haverford College (1962).

Rob Frieden holds the Pioneers Chair and serves as Professor of Telecommunications and Law at Penn State University where he teaches courses in technology management, regulation, business, policy, and law. Professor Frieden has published four books, written over 100 articles in law and telecommunications policy journals and provides biannual updates for *All About Cable and Broadband* (Law Journal Press). He frequently provides insights on current topics in telecommunications law and policy for media, conference attendees, and consultancy clients throughout the world.

Professor Frieden serves as an educator, researcher, grant seeker, and consultant in the law, regulation, and business of broadband networks, cybersecurity, electronic commerce, intellectual property, the Internet of Things, privacy, regulatory reform, satellites, and spectrum management.

Professor Frieden holds a B.A., with distinction, from the University of Pennsylvania (1977) and a J.D. from the University of Virginia (1980).

Christiaan Hogendorn is Associate Professor of Economics at Wesleyan University, USA and co-editor-in-chief of *Information Economics and Policy*. His research focuses on market structure and competition in network and infrastructure industries. He is currently working on projects studying network neutrality, online news aggregators, and intermodal rail transport. He was formerly a Member of Technical Staff at Bell Laboratories, and he has been a visiting scholar at the Columbia Institute for Tele-Information (CITI), the Research Institute in Industrial Economics (IFN) in Stockholm, and the Hitachi Center for Technology and International Affairs at the Fletcher School, Tufts University. He received his Ph.D. in Applied Economics from the Wharton School, University of Pennsylvania and his B.A. in Economics from Swarthmore College.

Rafael López is Professor in the Department of Economic Analysis at Complutense University, Spain. He received a B.A. in Economics and a B.S. in Mathematics from Los Andes University (Bogotá, Colombia) and a Ph.D. in Economics from Pompeu Fabra University (Barcelona, Spain). He has published in international refereed journals such as *Telecommunications Policy* and *Tourism Economics*.

Walter Mayer is retired Professor of Economics at the University of Mississippi, USA. He earned a Ph.D. from the University of Florida. His research has been published in *Journal of Econometrics*, *Journal of Business and Economic Statistics*, *International Journal of Forecasting*, among other journals.

Jamie Murphy is Adjunct Professor with the University of Eastern Finland. His industry background includes European Marketing Manager for PowerBar and Greg Lemond Bicycles, Lead Academic for the Google Online Marketing Challenge. He received a Ph.D. from Florida State University. Dr. Murphy's industry and academic experience spans continents and includes hundreds of academic publications and presentations, as well as dozens of *New York Times* and *Wall Street Journal* articles. His research focus includes food waste, sustainability, MOOCs, robots, and the effective internet use for citizens, businesses, and governments.

Teodosio Pérez Amaral is Professor of Economics at the Universidad Complutense de Madrid, Spain. He has published in international refereed journals such as *Tourism Economics*, *Oxford Bulletin of Economics and Statistics*, *Econometric Theory*, *Telecommunications Policy*, *Information Economics and Policy*, and *Applied Economics*, among others.

Petrus H. Potgieter is Professor in the Department of Decision Sciences at the University of South Africa and founder of Smarter Analytics (Pty) Ltd., a consulting company that focuses on mathematical modeling solutions for modern business practice, including the application of cloud-based machine learning solutions. Dr. Potgieter's current active startup projects include with Yimi Africa for secure and certified transmission of documents and Phulukisa Health Solutions for algorithmic triage and e-medicine.

His academic interests are wide, ranging from the economics of telecommunications and media, including the modeling of complex pricing problems for digital goods, to problems of computational complexity and theoretical computability in general, especially exact computation, as well as the governance and development of common-pool resources such as blockchains. He is a participant in the European Union Horizons

2020 research project Computing with Infinite Data which aims to establish and test software infrastructure for the representation of and operations on mathematical object with infinitary nature such as real numbers (as opposed to rationals).

Petrus is a Board member of the International Telecommunications Society, to which he was introduced by the late Gary Madden.

Miriam Scaglione is Professor at the University of Applied Sciences Valais (HES-SO Valais) School of Management & Tourism. She is president of the Swiss Chapter of the International federation for IT and Travel & Tourism and a member of the International Association of Scientific Experts in Tourism and the International Institute of Forecasters.

She has a Ph.D. in Computer Sciences and M.Sc. in Statistics (*Université de Neuchâtel*), M.Sc. in Cognitive Sciences (EHSS, Paris); database certification from the Swiss Federal Institute of Technology in Lausanne (*EPFL*), Switzerland, and a B.Sc. in Scientific Computing (*Universidad de Buenos Aires*).

Dr. Scaglione's research interests include applied statistics, time series, and technology diffusion forecasting in tourism, e-tourism, telecommunications, and transportation. She leads tourism regional impact studies and *big data* projects that monitor visitor flows using mobile phone data and visitor guest cards. She has published more than 40 academic publications in peer-reviewed journals such as *Technovation, International Journal of Forecasting, Information Technology & Tourism, Erdkunde* and *International Journal of Hospitality Management*, as well as dozens of international congress proceedings. Since 2012 she has earned eight best paper awards in different international congresses.

Hiroaki Suenaga is Senior Lecturer in the School of Economics, Finance and Property at Curtin University. He received his Ph.D. degree in Agricultural and Resource Economics from the University of California, Davis. His fields of research interest are in energy economics, commodity markets, and applied micro-econometrics and time-series analysis. His research collaboration with Gary Madden in the areas of telecommunication policy has produced several journal articles on the diffusion and pricing of network technologies.

Lester D. Taylor is Professor Emeritus of Economics at the University of Arizona, USA where he taught for 36 years. He has a B.A. in economics and mathematics from the University of Iowa (1960) and a Ph.D. in economics from Harvard University (1963), and taught at Harvard and the University of Michigan prior to Arizona. He was a senior staff economist at the President's Council of Economic Advisers in Washington, DC, in 1964–1965, and spent 18 months in 1967–1968 with a group of Harvard economists as an advisor to the Ministry of Planning of the Government of Colombia. Since 1975, Dr. Taylor's research has had a heavy emphasis on the econometric estimation of price and income elasticities in the electricity and telephone industries and has consulted extensively with electricity and telecommunications companies worldwide. He is the author (or co-author) of more than 80 journal articles and a dozen books, including *Probability and Mathematical Statistics* (1974), *Telecommunications Demand in Theory and Practice* (1994), *Consumer Demand in the United States* (Third Ed., 2010), *Capital, Accumulation, and Money* (Second Ed., 2010), *The Internal Structure of Consumption* (2013), and *The Life and Art of Archie Boyd Teater* (2016).

Ángel Valarezo is Researcher at the Instituto Complutense de Análisis Económico, Universidad Complutense de Madrid, Spain. He received a Ph.D. in Economics from Universidad Complutense de Madrid (UCM), a Master in e-Commerce from UCM, a Master in Telecommunications Management from Universidad Politécnica de Madrid, and a B.A. in Business and Administration from UCM. His research areas of interest are: Digital and Telecommunications Economy, Quantitative Models of Adoption and Use of Digital Services, Digital Divide, Digitization, and Individual Digital Skills. He has a professional background as Digital Project Manager Consultant, Marketing Manager, and Digital Entrepreneurship.

María Rosalía Vicente is Associate Professor in the Department of Applied Economics at the University of Oviedo, Spain. She has a Ph.D. in Economics. Dr. Vicente's research interests focus on information and communication technologies (ICT), their socioeconomic effects and the use of big data for economic forecasting. She has published in several scientific journals such as *Applied Economics Letters, Economic Letters, Computers in Human Behaviour, Government Information Quarterly, Information & Management, The Information Society, Technological Forecasting and Social Change*, and *Telecommunications Policy*, among others. She has been visiting researcher at Massachusetts Institute of Technology, the Organization for Economic Cooperation and Development (OECD), and Curtin University. She was involved in

Cisco-Oxford Saïd Business School's project for the development of a broadband quality index. The results of this project were cited by *The Economist, The Financial Times, The New York Times, The International Herald Tribune*, and the BBC, among others. Dr. Vicente participated in the Seventh European Framework ICTNET: European Network for the Research on the Economic Impact of ICT, coordinated by OECD and with the participation of the ZEW, the Imperial College and the University of Parma.

Cuong Hung Vuong is Postdoctoral Researcher at the Montpellier Researche en Economie, University of Montpellier, France. He holds a Ph.D. in Economics from University of Paris 2 Panthéon-Assas, France. He worked for various universities and the public sector in Vietnam until 2018. His research articles have been published in *Journal of Regulatory Economics, Communications & Strategies, Intereconomics*. His current research is in the fields of applied microeconomics and industrial organization including telecommunication, internet, and open-source software.

List of Figures

Fig. 2.1	Iterative paths for gasoline & motor oil expenditures for 50% increase in price of gasoline & oil (*Source* Authors' computation)	51
Fig. 3.1	Information and Communications Technology (ICT) per 100 inhabitants (*Source* ITU World Telecommunications/ICT Indicators Database, http://www.itu.int/ict/statistics)	69
Fig. 3.2	Global Competitiveness Index (*Source* World Economic Forum Global Competitive Report)	70
Fig. 3.3	ICT Development Index (IDI) (*Source* ITU, https://www.itu.int/en/ITU-D/Statistics/Pages/publications/mis2017/methodology.aspx)	71
Fig. 4.1	Households with private broadband access at home, penetration rates of internet use, and e-commerce. As percentage of people aged 16–74 (2007–2017) (*Source* Authors' creation from *INE* (2018). Survey on Equipment and Use of Information and Communication Technologies in Households)	108
Fig. 4.2	Penetration rates of e-commerce in the EU-28 as percentage of people aged 16–74 (2008 and 2017) (*Source* Authors' creation from *Eurostat* (2018b). Survey on Equipment and Use of Information and Communication Technologies in Households)	109
Fig. 4.3	Evolution of individual penetration rates of e-commerce in Spain (2008–2017): Gender and Age (*Source* Authors' creation from *INE* [2018]. Survey on Equipment and Use of	

	Information and Communication Technologies in Households)	110
Fig. 4.4	Evolution of individual penetration rates of e-commerce in Spain (2008–2017): by Education and Digital Skills (*Source* Authors' creation from *INE* [2018]. Survey on Equipment and Use of Information and Communication Technologies in Households)	111
Fig. 4.5	Evolution of individual penetration rates of e-commerce in Spain (2008–2017): Income and Employment Situation (*Source* Authors' creation from *INE* [2018]. Survey on Equipment and Use of Information and Communication Technologies in Households)	112
Fig. 4.6	Decision process of the *internet* and e-commerce adoption (2008–2017) (*Source* Authors' creation based on processed data from *INE* [2018]. Survey on Equipment and Use of Information and Communication Technologies in Households)	124
Fig. 4.7	Odds ratios of e-commerce. Pool (2008–2017) (*Source* Authors' creation from estimation results, Table 4.2)	131
Fig. 4.8	Odds of e-commerce. Pool (2008–2017) (*Source* Authors' creation from estimation results, Table 4.2)	131
Fig. 5.1	Density estimation of two-step ahead prediction error for the models (*Source* Authors' computation)	147
Fig. 6.1	Cumulative bass distribution models with different q/p ratios; T^* is the peak rate of adoption, namely the inflexion point of the cumulative distribution (*Source* Authors' computation)	154
Fig. 6.2	Non-cumulative bass distribution models with different q/p ratios; T^* is the peak rate of adoption, namely the inflexion point of the cumulative distribution (*Source* Authors' computation)	155
Fig. 6.3	Principal components analysis of Bass parameters, α level of heterogeneity and final penetration rate across tourism sectors/countries (*Source* Authors' computation)	162
Fig. 6.4	Swiss Destination Managements Organizations (DMOs) (*Source* Authors' computation)	164
Fig. 6.5	Austrian DMOs (*Source* Authors' computation)	165
Fig. 6.6	German DMOs (*Source* Authors' computation)	165
Fig. 6.7	Swiss restaurants (*Source* Authors' computation)	166
Fig. 6.8	Swiss cable cars companies (*Source* Authors' computation)	166
Fig. 6.9	International hotel chains (*Source* Authors' computation)	167
Fig. 6.10	European tour operators (TOs) (*Source* Authors' computation)	167

Fig. 6.11	Malaysian hotels (*Source* Authors' computation)	168
Fig. 6.12	Swiss non-affiliated hotels (*Source* Authors' computation)	168
Fig. 6.13	Swiss affiliated hotels (*Source* Authors' computation)	169
Fig. 6.14	Swiss hotels (*Source* Authors' computation)	169
Fig. 6.15	Swiss *gasthouses* (*Source* Authors' computation)	170
Fig. 6.16	Swiss travel agencies (*Source* Authors' computation)	170
Fig. 11.1	The model's structure (*Source* Authors' design)	253
Fig. 11.2	The ISP's optimal qualities and social welfare under both regimes (*Source* Authors' computation)	263
Fig. 14.1	Population who use social media in the U.S. (*Source* Social media usage in the United States [https://www-statista-com.colorado.idm.oclc.org/study/40227/social-social-media-usage-in-the-united-states-statista-dossier/])	307
Fig. 14.2	Visitors to social network site in the U.S. (*Source* Social media usage in the United States [https://www-statista-com.colorado.idm.oclc.org/study/40227/social-social-media-usage-in-the-united-states-statista-dossier/])	307
Fig. 14.3	Facebook's advertising revenue (*Source* Social media usage in the United States [https://www-statista-com.colorado.idm.oclc.org/study/40227/social-social-media-usage-in-the-united-states-statista-dossier/])	308
Fig. 14.4	Search market revenue share, U.S. (*Source* Social media usage in the United States [https://www-statista-com.colorado.idm.oclc.org/study/40227/social-social-media-usage-in-the-united-states-statista-dossier/])	308
Fig. 14.5	Platform effect: demand for internet advertising (*Source* Authors' construction)	311
Fig. 14.6	Product effect: extra cost of product due to monopoly (*Source* Authors' construction)	313
Fig. 14.7	Extra cost to consumers due to the elimination of competition (*Source* Authors' construction)	314

List of Tables

Table 2.1	Means of intra-budget expenditures regression coefficients: 14 categories of U.S. consumption expenditure (1996 Q1–2005 Q4)	31
Table 2.2A	Intra-budget regressions: 18 categories of U.S. Consumption expenditure (2012 Q4; t-ratios in parentheses)	34
Table 2.2B	Intra-budget regressions: 18 categories of U.S. Consumption expenditure (2012 Q4; t-ratios in parentheses)	36
Table 2.3	Own- & cross-price elasticities: for a 5 and 10% increase in price of food, 14 categories of expenditure (total expenditure held constant)	39
Table 2.4	Own- & cross-price elasticities: for a 5 and 10% decrease in price of food, 14 categories of expenditure (total expenditure held constant)	40
Table 2.5	Effects of A ± 10% change in price of telecommunications: 18 categories of expenditure (total expenditure held constant)	45
Table 2.6	Estimated own- and cross-price elasticities: for 10% decrease/increase in price of telecommunications (from Table 2.5)	46
Table 2.7	Effects of ±50% change in price of gasoline & motor oil with total expenditure held constant: 18 categories of expenditure	47
Table 2.8	Estimated own- and cross-price elasticities for ±50% change in price of gasoline & oil (from Table 2.7)	48

Table 2.9	Effects of 50% decrease/+75% increase in price of gasoline & motor oil: 18 categories of expenditure (total expenditure held constant)	49
Table 2.10	Effects of 50% decrease/+75% increase in price of gasoline & motor oil: 18 categories of expenditure (total expenditure held constant)	50
Table 2.11	Simulated own- and cross-price elasticities with respect to 50% decrease/75% increase in price of gasoline & oil (18 categories of expenditure)	50
Table 2.12	Estimated own- and cross-price elasticities with respect to ±10% and ±50% changes in price of gasoline & motor oil: 18 categories of expenditure	52
Table 2.13A	Means of intra-budget expenditures regression coefficients: 18 categories of U.S. Consumption expenditure (2006 Q1–2012 Q4)	57
Table 2.13B	Means of intra-budget expenditures regression coefficients: 18 categories of U.S. Consumption expenditure (2006 Q1–2012 Q4)	58
Table 2.14	Changes in gasoline & motor oil expenditures from ±50% changes in gasoline & motor oil price: first 10 iterations of Expressions (2.15) and (2.18) (all households)	60
Table 3.1	Test of Granger-causality from IDI to GDP per capita, 2008–2017	78
Table 3.2	Test of Granger-causality from GDP per capita to IDI, 2008–2017	79
Table 3.3	Summary of Granger-causality findings (IDI to GDP per capita), 2008–2017	80
Table 3.4	Summary of Granger-causality findings (GDP per capita to IDI), 2008–2017	81
Table 3.5	Pairwise Pearson correlation among mobile internet penetration, fixed broadband penetration, and percentage of households with broadband access	81
Table 3.6	Test of Granger-causality from mobile internet penetration to GDP per capita, 2008–2017	82
Table 3.7	Summary of Granger-causality findings (mobile internet penetration to GDP per capita), 2008–2017	83
Table 3.8	Test of Granger-causality from fixed broadband penetration to GDP per capita, 2008–2017	84
Table 3.9	Summary of Granger-causality findings (fixed broadband penetration to GDP per capita), 2008–2017	85

Table 3.10	Test of Granger-causality from mobile internet penetration and fixed broadband access (jointly) to GDP per capita, 2008–2017	86
Table 3.11	Summary of Granger-causality findings (mobile internet penetration and fixed broadband penetration jointly to GDP per capita), 2008–2017	87
Table 3.12	Test of Granger-causality from percentage of households with broadband access to GDP per capita, 2008–2017	88
Table 3.13	Summary of Granger-causality findings (percentage of households with broadband access to GDP per capita), 2008–2017	89
Table 3.14	Test of Granger-causality from GDP per capita to mobile internet penetration, 2008–2017	90
Table 3.15	Summary of Granger-causality findings (GDP per capita to mobile internet penetration), 2008–2017	91
Table 3.16	Test of Granger-causality from GDP per capita to fixed broadband penetration, 2008–2017	92
Table 3.17	Summary of Granger-causality findings (GDP per capita to fixed broadband penetration), 2008–2017	93
Table 3.18	Test of Granger-causality from GDP per capita to percentage of households with broadband access, 2008–2017	94
Table 3.19	Summary of Granger-causality findings (GDP per capita to percentage of households with broadband access), 2008–2017	95
Table 4.1	Individual penetration rates of e-commerce in Spain (2008 and 2017)	112
Table 4.2	E-commerce	117
Table 4.3	Explanatory variables	117
Table 4.4	Models of adoption of e-commerce by individual internet users. Linear Probability Model (LPM), Logistic Regression Model (LRM), and Heckman Selection Model (HSM). Pool data (2008–2017)	118
Table 4.5	Models of adoption of e-commerce by individual internet users. Linear Probability Model (LPM) and Logistic Regression. Complete estimates for the interaction term Digital Skills × Age. Pooled data (2008–2017)	128
Table 4.6	Polychoric correlations among selected independent variables	129
Table 4.7	Model of internet use. Heckman first stage estimates. Probit Regression Model. Pooled data (2008–2017)	129

Table 5.1	Variable definitions and descriptive statistics	140
Table 5.2	Arellano-Bond GMM estimates of Log GDP per capita growth, 2008–2012	142
Table 5.3	AIC and BIC of both models with and without Luxembourg	146
Table 5.4	P-values of various tests for model forecasting accuracy with and without Luxembourg included	148
Table 6.1	Study sample	158
Table 6.2	Estimated Bass and G/SG α parameter with its standard deviations in (parentheses)	160
Table 6.3	Residual and forecasting error statistics	171
Table 7.1	Description of variables	183
Table 7.2	Descriptive statistics	184
Table 7.3	Results of the probit regression on online political participation	185
Table 9.1	Definition of service bundle attributes	217
Table 9.2	Service bundle attribute levels	217
Table 9.3	An illustrative choice set	218
Table 9.4	Selected sample and population mean comparisons	219
Table 9.5	Exploded logit estimates, pilot data	221
Table 9.6	Ngene optimization results	221
Table 9.7	Mixed exploded logit estimates	222
Table 9.8	Mixed exploded logit estimates, lognormal	222
Table 9.9	D-optimality comparisons	224
Table 9.10	Ngene optimization results with perfect foresight	225

CHAPTER 1

Civilizations' Economic Drivers: Babylon to Bitstream

Gerald R. Faulhaber

FIRST WAVE: CIVILIZATIONS AND ECONOMIC DRIVERS

Homo Sapiens has lived on our planet for over 200,000 years. What we call "civilization" has only been with us comparatively recently, from about 3500 BCE, less than 3% of our species lifetime.

Typically, historians mark the beginning of civilizations to events and developments in Mesopotamia.[1] What emerged in the valleys of the Tigris and Euphrates Rivers was the *organization of supra-family groups*

---

[1] Some experts argue that the evidence supports some form of civilization emerging 20,000 years earlier, in a cave in South Africa (Pappas 2012). Our use of the term "civilization" is more in accord with the later date of origin of 3500 BCE.

An earlier version of this material was originally presented at Cyber Civilization Research Conference, Keio University, Tokyo, Japan, on December 7, 2018. My thanks to David Farber and his research center at Keio University for his support of this work.

G. R. Faulhaber (✉)
University of Pennsylvania, Philadelphia, PA, USA
e-mail: faulhabe@wharton.upenn.edu

© The Author(s) 2020
J. Alleman et al. (eds.), *Applied Economics in the Digital Era*,
https://doi.org/10.1007/978-3-030-40601-1_1

to ensure the stability, encouragement, and security of the economy and society, which we use as a rough working definition of a civilized society.

What existed prior to civilizations? Humans were hunter-gatherers and/or nomadic herders, organized by family or tribes. These small organizations were sufficient to defend the lifestyle of their members; hunters and soldiers (generally men) defended the tribe, provided weapons and supplied the tribe with meat. Gatherers and maintainers (generally women) collected plant food, processed it for the tribe and provided utensils and implements to the tribe.

What is most important is that prior to civilization, farming was minimal to nonexistent. There was almost no effort put into growing plants or tending domestic animals (with the exception of herding). What was unique about Mesopotamia in 3500 BCE was the emergence of agriculture as the primary means of food production for the population. In fact, civilization was the price that humans paid to develop and maintain agriculture at the time. Agriculture dramatically changed the efficiency by which humans could feed themselves. Previously, humans had to spend close to 100% of their time gathering food. Farming permitted a much more efficient means of humans feeding themselves, so farming was a great technological innovation. But not without cost; agriculture demanded farmland, which demanded property rights[2] on a large scale to induce farmers to plant and harvest crops and invest in future crops. Without such protection, outsiders or indeed neighbors could raid the product of farms, reducing the incentives for farmers to farm. Farming also included irrigation (or water management generally), generally on a large scale. Managing the irrigation in the Tigris and Euphrates valleys required large-scale organized and ongoing efforts, to which all society's members had to agree (or at least abide by). Similarly, ancient Egypt's Nile Valley was subject to annual flooding. After such floods, property boundaries had to be reestablished, requiring a strong government (Baines 2011). Agriculture moved on from its origins in Mesopotamia and Egypt to East Asia, South Asia, and to Mediterranean Europe. Eventually, agriculture spread to Africa and North and South

---

[2] "Property rights" is meant to be broadly construed. These need not be private property rights; they could well be communal property with individual farmers' access to property guaranteed by law, or some other form of property management that ensured farmers' output would be controlled by the farmer, and any investment in the property would redound to the benefit of the farmer-investor.

America. In order to support the agricultural economies around the world, some form of civilization had to develop, along with agriculture. These civilizations were quite different; Babylonian civilization was much different from Chinese civilization, Indian much different from Mediterranean. But all civilizations had to accomplish the same ends: organizing a large population to ensure stability, encouragement, and security of the economy and society.

But any economy must be based on *economic drivers*: scarce resources that are absolutely essential for the economy to function. In the case of agriculture, there is one unique scarce resource, and that is *land*. Without land, agriculture is simply not possible. In Mesopotamia and Egypt, control of this scarce resource, land, was the basis not only of the economy but of the civilization. Although the civilizations of this first wave were different from each other, they were all based on the single scarce resource of land. Those who controlled the land controlled the society; as a result, all first wave civilizations shared these common traits:

- Steeply hierarchical
- Wealth and power determined by landownership
- Emperor/king, then landowning nobles, then fighting men, then craftspeople, then serfs
- Extreme income inequality

Succeeding agrarian civilizations were different, but all possessed these characteristics:

- Babylon: manage property rights in Tigris-Euphrates valley
- Egypt: pharaoh rules all; Nile flooding leads to problems with property rights control; irrigation
- China: Emperor's mandarins control
- Japan: decentralized shogunates control
- Britain: land-based aristocracy control
- US: land more equally distributed (except the Southern states)

These civilizations evolved slowly. The signing of the Magna Carta in thirteenth-century England shifted power between the king and nobles. England's Glorious Revolution of 1688, further dispersed power from the monarchy. Perhaps most dramatic, the American and French Revolutions of the late eighteenth century drastically shifted power away from monarchies toward a more popular distribution of power. But Europe and America were not the only loci of change: the Meiji Restoration in nineteenth-century Japan shifted power from the Tokugawa shogunates to the (restored) emperor and a central government.

While these changes have been celebrated, it must be borne in mind that all societies at this time continued to be dominated by the one scarce resource: land. This dominance stems from the economic base of all civilizations at this time: agriculture. Unless and until the economic driver were to change, all civilizations possessed the structure and properties outlined above; changes were minor.

Second Wave: Civilizations and Economic Drivers

The earliest indication that the key economic driver was about to change occurred in the middle of the eighteenth century in Great Britain. The economy began a shift to manufacturing: the production of goods with interchangeable parts using large powered factories. This is not to say that production of goods was not part of human history before; the earliest civilizations did produce goods for personal and military uses. However, production was piecemeal; parts were not interchangeable, the location of production was often the home of the craftsperson, and there was no industrial-level power. In the late eighteenth century, however, the shift to manufacturing began in earnest, embarking on what we were to call the Industrial Revolution. British production of armaments in this time period signaled an early example of what would become known as mass production. Another early landmark occurred in 1793 when Samuel Slater brought British manufacturing technology to the US and built the first cotton mill in North America in Pawtucket, Rhode Island (Gaur 2013). These early examples were merely precursors of manufacturing and production in Britain, the US, Continental Europe, and indeed around the developed world. At first, the power for these factories was provided by water, but as steam power developed in the first half of the nineteenth century, steam became the choice for industrial production.

A key to industrial production was size; no longer would the manufacture of products and services take place in the craftsman's home, but rather in a far larger facility with costly machinery and equipment. While each industry required different facilities and machinery, they all required lots of it, and it was always costly. Essential to industrial production, then is *capital*, funds to be invested in costly facilities in order to produce goods and services in the future. Capital would then be repaid, with interest and dividends from the revenues from the sale of the produced goods and services. But industrial production required investors to surrender their money today for a promise of future cash. And they were being asked to give up a great deal of money in order to invest in the great enterprises of the nineteenth century. One major industry that required large amounts of capital was railroads. Not only in Great Britain but more so in the US, where railroads spanned the continent. Telegraph, steel production, and shipbuilding were but a few of the industries demanding capital.

Capital, then, became the driver of the second wave of civilization. *Capital* was the *scarce resource* required to drive the Industrial Revolution.

The second wave grew exponentially in the second half of the nineteenth century, particularly in the US. After the Civil War, manufacturing technologies grew rapidly; the size and rate of growth of the US market opened enormous opportunities to expand US industrial production. This growth occurred during a short time, between 1870 and 1915. Railroads, chemicals, retail, oil, automobiles, and other manufacturers rapidly became the largest firms: Standard Oil, Carnegie, Ford, US Steel, Dupont, Sears Roebuck.

Much of the capital required for this rapid expansion of US industry came from Europe. Even though Europe itself was undergoing the Industrial Revolution, the opportunities in the US, due to its size and growth, were apparently more attractive. Additionally, Japan was also undergoing rapid industrialization after the Meiji Restoration in 1868. However, the growth of US firms during this period allowed them to dominate the world economy for the first two-thirds of the twentieth century.

To identify the driver of this second wave as capital in no way suggests that all or even most industrialized economies are or were capitalist economies, in the sense that privately owned capital had the commanding role in the economy. But in virtually all industrialized economies, capital was the scarce resource that drove the economy, no matter what the chosen social and economic structure. For example, the Soviet Union adopted a communist economic model in which capital investment played as important a role in the Soviet economy as it did in the US economy. In the mid-twentieth century, countries in Western Europe opted for a more socialist economic model, but still capital investment was a driving factor for these economies as well. Capital (as opposed to capitalism) is and was the economic driver.

Not all countries industrialized at this time; many countries did not industrialize until well into the twentieth century, and many countries have yet to do so. Not surprisingly, such countries maintained the structure of agricultural societies, and many still do today. Even countries that industrialized early have not shaken off their agricultural social structures; for example, Great Britain's landed aristocracy still maintains strong social standing and even some political power (the House of Lords in Parliament).

If capital is the new scarce resource, then what about land? What about agriculture? Clearly, agriculture is needed to feed the people, but the structure of agricultural production has changed in farming; in 2017, less than 1% of US households are farmers. As of 2010, about 44% (Trading Economics 2016) of US land is committed to farming (this includes grazing land); in the nineteenth century, this was about 80% of US land. However, the US farm output is far greater than it was in the nineteenth century. Technological change has increased farm productivity enormously. Simultaneously, it has reduced the scarcity of land. Today, land for farming purposes is no longer scarce; still necessary, but not the scarce resource of the past.

In less developed countries, agriculture still predominates, land is still scarce, and social and political structures remain as they have for centuries. In some countries, a mix of industrial production and agriculture obtains. For example, India is rapidly developing a vibrant service economy in its metropolitan areas, but the vast majority of its citizens live in rural areas and are actively involved in (low productivity) agriculture. It is likely that this transition will continue for many years before India fully attains an industrial-age economy and society.

The unprecedented growth of industrial economies, particularly in the US, had significant economic and social impacts. The founders/owners of these fast-growing firms garnered vast amounts of wealth, building multiple large dwellings and giving away substantial amounts to charity.[3] The employees of these companies were often migrants from farms (often from the Southern US) or other countries and were required to

---

[3] For example, Andrew Carnegie, who amassed a fortune in the steel industry, funded the building of over 2500 community libraries worldwide. Almost half the total community libraries in the US by 1919 were funded by Carnegie.

work long hours with little time off; wages were low. The result was great income inequality: a few rich and many very poor.

This was a period in which the US became the world leader in GDP. Starting from behind the pack in 1870, the US became the highest producing economy in the early twentieth century (Wikipedia 2019a). It was also a period of great internal strife related to the dominance of capitalists over workers and the development of unions. Workers in some industries (such as mining) took up arms to oppose what they alleged was unfair treatment by factory owners. Lives were lost and in the early days of these conflicts, states called out the National Guard to defend factory/mine owners against union violence.

Over the first half of the twentieth century, the economic and political power of labor slowly increased, both in the US and other developed countries. A high point of union political strength was the passage of the Wagner Act in 1935, granting employees the right to organize in unions without interference, as well as to strike and maintain picket lines. By the 1950s, most unions were organized in the AFL-CIO.[4] Similar actions occurred in Europe and Japan (Japanese Trade Union Confederation—Rengo).

The postwar era saw success by labor in closing the income gap between workers and management/capitalists. The period from 1937 to 1970 is called the Great Compression, in which middle-class incomes rose quite substantially while high-income earners lost ground (Wikipedia 2019b). While this trend did not last into the last quarter of the twentieth century, the trend toward equalization of income in the US marked a correction to the more extreme effects of the advent of the second wave of civilization: the capital driver.

However, after 1970 the rest of the world outpaced the US in investment, building industrial economies in East Asia and Western Europe to challenge American firms. This diaspora of capital investment which marked the relative decline of US predominance of the industrial world economy also marked the beginning of a change in the driver of civilizations. The fact that capital became much more available throughout the world reflected the fact that capital was no longer scarce. Firms in East Asia, for example, had been unable to raise capital earlier found it progressively easier to attract investment from around the world post-1970.

---

[4] The Teamsters was an effective union for US truckers which stayed independent of the AFL-CIO.

Again, this is not to say that capital has become unimportant. Capital remains an essential component of virtually all production. This is also not to say that manufacturing is not an important part of the US economy. Manufacturing today generates 12% of the US real GDP (about constant since 1947) (Chien and Morris 2017), its output is the second largest (after China) (Kiprop 2019). However, capital is no longer the scarce resource, whose command enabled the US to dominate the world economy in the first two-thirds of the twentieth century.

THIRD WAVE: CIVILIZATIONS AND ECONOMIC DRIVERS

If capital is no longer the scarce resource driving our civilization, then what is the new scarce resource driving our economy and civilization?

The economy has become consumption-driven and technology-driven. Both consumption and technology have always been important in our capital-driven economy, but neither has been a scarce resource. But increasingly, these factors have become dominant in the economy, and both depend upon a new *scarce resource: information*.

In an economy in which production is easy and transportation super-quick, what matters is what people want, how they want it, and when they want it. What is it that can be produced? How it can be produced? When can it be produced? How can we get it to those who are willing to pay as quickly as they need it? How much are people willing to pay? How much will products and services cost? How can we upgrade/improve products and services? Firms that can manage these questions are the firms which will succeed in this third wave, and to succeed, they need *information*, which forms the basis of success in this economy: *cyber-civilization*.[5]

Success in the new cyber-civilization depends not on simply making things, but rather on making the *right* things, for the *right* customers, in the form and style they are willing to pay for, and at the *right* time. In the industrial wave, it was production that mattered most, and customers adjusted to the needs of the industry. In the information wave, production is the servant of consumption and must adjust to the needs of customers. Enabling this adjustment is what information is all about. The more we know about customers, their needs and desires, the better we

---

[5] Some have argued that the benefits of the information economy are exaggerated compared to the innovations of the nineteenth century (e.g., Gordon 2012).

can tailor the provision of goods and services to those needs and desires. The more we can forecast the evolving needs of customers, and the more we understand the technology for serving them, the more we can anticipate, indeed stay ahead of the market, which is critical to success in this cyber-civilization.

The change in the economy has been both substantial and quite rapid. The major players in the world economy have become much less dominant, even marginal. Firms such as General Motors (US), ThyssenKrupp (Germany), and Boeing (US) still employ large numbers of people, but both revenues and employees are substantially less over the last decade. (For employees at General Motors, see Wikipedia 2019c; for employees at Thyssen Krupp, see Wagner 2019a; for employees at Boeing, see Wagner 2019b). The new major players in today's economy are firms such as Microsoft, Google (now Alphabet), Facebook, and Amazon. Virtually all of these firms are in the business of gathering, analyzing, and selling information (including software). Additionally, none of these businesses have as many employees as the aforementioned "old" dominant firms.

An instructive, if extreme example of how the new information economy is overtaking the old production economy is from the field of photographic imaging. The US firm of Kodak had dominated the markets for cameras, film, and imaging from the end of the nineteenth century for over one hundred years, not only in the US but worldwide. Almost all commercial photo developing used Kodak film development equipment. At its peak around 1990, Kodak had over two-thirds worldwide market share and 145,000 employees (Hudson 2018).

The advent of cameras built into cell phones changed the business model for devices that capture visual images. However, there is still an active business in visual imaging. Today, that business is largely served by Instagram, founded in 2010 and now a subsidiary of Facebook, which permits uploading, sharing, and indexing of images. It currently has 450 employees (Wagner 2017), about 0.3% of Kodak's peak employment. With new technology and new information, the imaging business has radically changed, and Kodak, the globally dominant player, was unable to accommodate. In 2012, Kodak filed for bankruptcy.

This admittedly extreme example demonstrates three key principles: (i) how the mighty have fallen; even being a global market leader will not save a traditional firm; (ii) how tiny is its replacement; even with hundreds of millions of daily users, Instagram makes do with a tiny employment base; and (iii) most important, how quickly this happened. The shift from an agricultural economy to an industrial economy took just less than fifty years in the US, far more rapidly than in Great Britain or Japan. But this example suggests that the shift from an industrial economy to an information economy is occurring in far less time.

In this new cyber-civilization, value will be added by firms who provide information rather than by the firms that use it. For example, Google's Android operating system for cell phones will add more value than Samsung's actual phones. This is not to say that increasing the hardware capabilities of cell phones is not important; these advances will continue to be essential. However, the market will continue to be competitive and margins will thus be low. Driving the operating systems and software (apps) will be where the added value is realized, especially when targeted to relevant markets.

Cyber-Civilization: What Are the Consequences?

The deployment of information technology via software, cell phones, and the internet has been perceived as an (almost) unalloyed positive development. As the internet became available to first the developed world and then most everywhere globally, people could search for any information they desired, could buy retail goods from an almost unlimited inventory to be delivered to homes within days, could share photo images with the world, could communicate at a moment's notice with friends and family around the world. Until quite recently, most people in the developed world (and elsewhere) practically swooned over this development, and the firms responsible for it were viewed positively, even heroically. Even though these firms became quite large, they provided services unheard of before...*for free* or at low prices. Google search... free. Facebook access to friends...free. Microsoft search and operating systems...free. Amazon...low prices. What was not to like?

Apparently, plenty. The EU Antitrust authorities have levied a $5.1 billion fine against Google for alleged abuses of its power in the operating system for cell phones market (Satariano and Nicas 2018), not the first EU fine levied on Google; it has also opened an investigation concerning Amazon practices. In the past two years, public opinion seems to have turned against over big tech firms, such as Facebook and Microsoft (Bass 2018).

Why the concerns? The clearest reason for concern is *privacy*; the big tech firms collect detailed data about every user, and each instance of use

of its customers; it then makes use of this data for advertising and/or sales. The firm does not compensate its customers for its use of this data; it is assumed that the data is owned by the firm that collects it, not the customers from whom it is derived. Privacy may be a popular concept on the internet, but in most countries, there is no legal right to the control of data customers willingly share with a service provider.[6] In fact, the use and sale of this data make up a large fraction of the revenues of these firms; collecting, aggregating, and selling this data is how these firms are able to offer free services.

A second concern is *dominance*; the largest tech firms often have a dominant share of their market. Most obvious is Google, whose market share in the global search market is about 90% (87% in the US market and under 6% in China) (Clement 2019). Amazon has 50% of the US online retail market. However, online retail is only 10% of the total retail market. Facebook has 67% of the global social media, although its share is declining somewhat (Statcounter 2019). But there is little dispute that these firms are dominant in their markets. This appears to many to mimic large dominant firms in the industrial economy that would normally be subject to antitrust scrutiny and possible prosecution. Being large, even dominant, is not illegal, at least according to US antitrust law. If, however, any of these firms attained their dominant status by anticompetitive actions, or if they are abusing their dominant positions to disadvantage customers, then such scrutiny and possible prosecution may well be justified.[7] More likely, the information industry may well have returns to scale; if firms with large datasets have a competitive advantage over smaller firms in advertising, e.g., then large firms will grow larger because of this advantage, thus leading to dominance. Additionally, many (but not all) information firms exhibit network effects, in which more customers increase the value of their service to all other customers, an added cause of growth. Another reason for dominance is that it is the natural outcome to superior performance, and therefore should be rewarded, not punished. For example, in the early 2000s, Google was one of many players in the search engine market: Alta Vista, Yahoo, Excite, and Bing,

---

[6]The EU has adopted a measure that gives rights to customers concerning the use of their data; this measure "General Data Protection Regulation" was adopted by the EU in 2016 and was operational in 2018 (Wikipedia 2019d).

[7]The EU antitrust authorities appear to believe that such action is mandated (Satariano et al.).

among others. Perhaps Google emerged as the dominant search engine simply because it was more customer-focused rather than advertiser-focused, as many analysts have concluded (e.g., Heitzman 2017).

At the end of the nineteenth century, large firms threatened dominance of domestic markets. Many of these firms became large by purchasing a number of smaller firms, and thus gaining market power. Companies such as US Steel, Firestone, DuPont, General Motors, and Coca-Cola, to name a few, were perceived as dangerously dominant. The concern was the exercise of market power by keeping out competition, perhaps lowering innovation, and raising prices to customers. In 1890, the US Congress passed the Sherman Act, the first law governing the control of monopolies via antitrust actions. During the 1900s, the Federal government, operating through the Department of Justice, successfully brought several antitrust suits against existing firms. The first such action against a truly large firm was the Standard Oil case, which was resolved in 1911 with the requirement that Standard Oil be split into numerous companies, each of which would compete with one another.

The struggle to contain the dominance of large industrial firms through antitrust continued well through the twentieth century. There have been several landmark cases, including AT&T (1984) and Microsoft (2001). Antitrust laws were adopted in most OECD countries; recently the EU has been a rigorous enforcer of its antitrust legislation. Antitrust has also seen an evolution of economic thought by leading economists; the focus today of antitrust economics is to protect customers from monopoly pricing and (perhaps) reduced innovation. As opposed to earlier antitrust analysis, it is not designed to protect competitors of large firms, nor to punish firms for being large.

Antitrust has been an effective tool in containing abuse of market power in cases such as AT&T and Microsoft. Whether it has controlled most abuses of market power in the US economy is a matter of some debate. But where deployed, it has been effective in minimizing the use of market power to customers' disadvantage.

The third concern is *income inequality*; this is much less directly related to the large tech firms, but is a general social concern and does impinge on the information sector. The move from an industrial economy to an information-based economy has not only given rise to a well-paid elite of company founders, but also has resulted in substantial changes in middle-class jobs.

It is striking that a similar change in income inequality occurred at the end of the nineteenth century, with founders and CEOs of new industrial firms become quite wealthy. Today, we see similar changes (*Forbes Magazine* 2019); Founders and CEOs of new information firms have also done quite well: Jeff Bezos (Amazon) $112 Billion; Bill Gates (Microsoft) $90 Billion; Mark Zuckerberg (Facebook) $71 Billion; and Larry Page (Google/Alphabet) $48.8 Billion (the quoted numbers are Forbes' 2018 estimates). The similarity of this distribution of high-end wealth to the wealth distribution of the early industrial age, to men such as Carnegie, Mellon, and Rockefeller is striking. It is plausible that a change in the economic drivers of civilization leads to a concentration of great wealth among the leaders of the change.[8]

We previously noted that at the end of the nineteenth century, not only did the founders become wealthy, but the wages of the working class became lower, increasing the income inequality of US and worldwide economies. Unfortunately, the same increase in income inequality appears to be happening currently, as we segue to cyber-civilization.[9] Income stagnation and increasing inequality is evident in recent numbers: in 1971, the bottom 25% of the income distribution took home 6.2% of total income; in 2016, the bottom 25% took home 4.5%. In contrast, the top 10% of the income distribution took home 30.7% of total income in 1971; in 2016, the top 10% took home 47.6% of total income (Kuhn et al. 2018).

Has the emergence of cyber-civilization created increased income inequality? The evidence is anecdotal and meager at best. The substantial increase in income of the founders and CEOs of information firms is easy to measure and supports the hypothesis. The declining share of income captured by middle- and lower-income groups is clear from the data but there is little evidence linking this decline to the rise of information firms. A suggestive argument is that much of the output of the information industry is aimed at reducing costs of their corporate customers, which often involves replacing employees with repetitive jobs, typically lower middle-income workers, with robotic or computer-based inputs.

---

[8] An interesting note is that none of these individuals inherited their wealth; all of it was generated from the firms they founded and ran. It is reasonable to conclude that these individuals created their wealth through their own efforts and talent, rather than the good fortune of relatives.

[9] A popular work on the history of inequality is Piketty (2014).

This is likely to result in less jobs and lower wages for such workers, leading to increased income inequality. While this argument is plausible, there is no direct empirical evidence supporting it at this time.

What Is to Be Done?

These problems attendant to the development of the information economy and cyber-civilization bear a close resemblance to those problems attendant to the development of the industrial society a century ago. At the time, society evolved solutions to these problems: dominance and monopoly power gave rise to antitrust law, which continues to be used today to treat the abuse of monopoly power. The rise of income inequality gave rise to both labor unions and labor laws, which eventually (post-World War II) to substantially lessened inequality.[10]

However, the problems arising with the information economy are rather different than those of the earlier era, and are likely to require rather different solutions, which we discuss next.

What to do about Privacy? This problem is unique to the rise information age; nothing analogous occurred during the rise of the industrial age. It is a problem which is at the heart of the biggest of the new firms. The transaction which all customers make with the likes of Google, Facebook, etc. is that the firm give the customer a service that they value highly, in return for which the customer (implicitly) gives the firm the right to use the information incident to its transactions to the firm for its use or sale to others. While no individual's information is valuable, the aggregation of all customers' information is highly valuable to the firms themselves, in selling goods and services, and in selling advertising to other firms.

Traditionally, media such as newspapers and magazines have used advertising to support inexpensive customer prices for their information product. Today's information firms are now using customer transaction data to support their product, as well as advertising. Other firms collect customer transaction data as well, such as supermarkets who offer customers a discount for using the store's rewards card. Neither of these mechanisms are viewed as violating customer privacy. In the case of supermarket reward cards, customers are explicit in turning over

---

[10] In recent decades, the strength of unions has declined substantially.

data, whereas the collection of data by information platforms such as Google and Facebook is more implicit. In fact, information firms do give customers control over information sharing, but it finding where these controls are can be a quite a challenge: for example, searching for "privacy" in Windows 10 permits a user to make substantial changes in what data Microsoft is collecting. Similarly, in Google Gmail, selecting Settings, Advanced, Privacy also permits users to change what data is collected.

In the US, there are laws governing privacy; they do not appear to be well-suited for dealing with the privacy issues concerning information firms. For example, virtually all privacy laws require that an individual who claims their privacy has been violated to show actual damage from the violation (for example, the public revelation of gender orientation). It is unlikely that sharing transaction data with advertisers, for example, will cause economic or personal damage to any customer. So existing privacy laws are unlikely to deal with perceived problems with online privacy.

However, there is a view of many that online privacy is an important issue and must be dealt with. A recent publication by Senator Marco Rubio (Rubio 2019) addresses the problems and suggests how legislation might approach these concerns. As discussed in the footnote above, the EU has adopted the General Data Protection Regulation (GDPR) in 2016, an extensive and quite detailed regulation concerning the management of customer data by information firms.[11] Additionally, California has adopted a privacy regulation quite similar to the EU's GDPR. Japan's communications ministry already prohibits information firms from using the content of customers' e-mail and other communications data, and has indicated its intent to extend this prohibition to all information firms operating in Japan (*Japan Times* 2019).

Clearly, online privacy appears to be a concern. What is it we can do about it?

It would appear that the first order of business is to be specific about what problem we are trying to solve and why? If an information firm collects data from customers concerning their transactions with that firm, possibly to sell that data to others, who is being hurt, either personally or economically? If we believe that, say, Google should not be able to use or

---

[11] Matoo and Meltzer (2018) contains a discussion of the components of GDPR and the difficulties concerning trade with other countries with different privacy rules.

sell customer transaction data, then do we also believe that supermarkets should not be permitted to collect and use customer transaction data via rewards cards? If not, what is the difference between the two actions?

Next, presuming we have good reason to attempt to control the capture of customer transaction data by information firms, what are reasonable ways to accomplish this? The EU solution is to control via regulation what data firms can collect and what they can do with it. GDPR is a mandated top-down solution identical for all firms and all situations. Neither firms nor customers have any option but to follow the rules.

Another option might be to permit customers to make a choice about what data to share and under what conditions. For example, a law might require firms to give options to customers about which data can be shared, possibly including some form of payment. Some customers might be happy to share their data with firms in return for free or favorable service from the firm. Others might be happy to share some but not all their data, and receive less favorable service. Others might wish to share nothing with the firm, but would be willing to pay for the service provided. In this mode, customers and firms can negotiate over data sharing and terms of service to a mutually satisfactory solution. Not all customers are the same, so a single mandated rule may not be the best solution.

If this more flexible approach is used, then customers must have a clear and open mechanism to make their "opt-in" choice, and not require customers to dig through layers of menus in order to make their choices. For example, if the default choice is no sharing, then information companies would have incentive to make it easy for customers to share in return for favorable service. Only then would "opt-in" be a truly viable option for customers. Should this be the preferred policy, then something like GDPR could be the default option, with customers able to opt-in using an open and easy means of sharing what they wished under terms and conditions they understand and agree to.

There may well be cases in which the use of a customer's transaction information by a firm actually does create damages for the customer. For example, releasing data to a customer's employer that shows that this individual was a regular visitor to politically radical websites could well cause that individual to not be considered for promotion or even lose their job, thereby causing damage to that individual. Clearly, any privacy law should make recovery of damages in such cases relatively easy and straightforward.

What to do about Dominance? Dominance is an issue of which economists, policymakers, indeed most citizens are aware, as the problems are quite similar to market power issues in the industrial economy. Firms may attain dominance and market power via several means, and once endowed with market power may use it to enhance that market power using anticompetitive strategies, and ultimately to disadvantage customers. Economists, policymakers, and others have recently been quite concerned about the dominance of big information firms. Recently, the Administration nominated William Barr to head the Department of Justice, home of the Antitrust Division. During his confirmation hearings in the US Senate Judiciary Committee, several senators quizzed Barr closely on his views concerning large information firms and antitrust enforcement. He replied, "I don't think big is necessarily bad but I think a lot of people wonder how such huge behemoths that now exist in Silicon Valley have taken shape under the nose of the antitrust enforcers… I want to find out more about that dynamic."

In US antitrust law, being big is not necessarily bad. A firm can become big because it produces a better service or product than its competitors and so customers prefer them. Many would argue that Google bested Yahoo! and Alta Vista in the search engine market fair and square; no anticompetitive tactics. Their large market share, in this view, is a well-deserved reward for serving customers' needs better than competitors. The presence of Bing as a minor player in this market today, backed by large and well-skilled Microsoft, suggests that Google continues to earn its share via superior performance. Note that there is nothing that would prevent a competitor (such as Bing) from capturing market share from Google, if Google's performance advantage eroded with time, or they faced a new and superior competitor.

However, if a large firm attempts to become larger through mergers and acquisitions, it is likely to face strong resistance to such transactions from the antitrust authorities (Antitrust Division of the Department of Justice, and the Federal Trade Commission). In this case, the firm would be seeking to buy market share rather than earning it through organic growth, and would likely be restrained from such transactions by the antitrust authorities. An interesting example is the acquisition of Instagram by Facebook in 2012, when both companies were still fairly small. At the time, this did not arouse concerns with the antitrust authorities and the acquisition went through without challenge. It is likely that if the same acquisition occurred today, the acquisition would be challenged by the antitrust authorities, as Instagram is a critical competitor to Facebook in the under-25 market.[12] Perhaps more interesting is the failure of US antitrust authorities to intervene in the 2014 acquisition of WhatsApp by Facebook; WhatsApp was a competitor to Facebook, and yet no antitrust action was forthcoming.

Another means by which firms can grow is if the market is endowed with network effects, as discussed above. This now-popular term refers to a business in which the more customers it has, the more attractive it is to customers, both existing and new. There are three ways in which this can occur. (i) Facebook is a good example of the classic network effect, as it involves interactions among its customers; hence, a customer will find this service more attractive the more are other customers subscribe to Facebook. All things being equal, large firms are thus likely to get bigger, eventually resulting in large market share. (ii) A business may possess two-sided network effects, in which there are two sets of customers, and each set values the service more highly if the other set of customers is larger. For example, eBay hosts both buyers and sellers; sellers prefer a firm with lots of buyers, and buyers prefer a firm with lots of sellers. Again, large firms will be more attractive to all customers and are thus likely to get bigger. (iii) Information firms may possess a third form of network effect, in that large firms have a greater set of customers from whom to gather transaction data, which allows them (and their advertisers) to better predict customer's behavior. They are thus more attractive to customers and are likely to get bigger.

None of these methods of attaining large market share involves taking advantage of customers or engaging in any anticompetitive behavior. It would appear that most information markets exhibit some form of network effects, and we thus expect these markets to be dominated by a single firm, as a natural course of economic evolution. Breaking up such firms is likely to be only a short-run fix; if the market possesses network effects, it will evolve again to one (or a few) dominant firms.

There are other ways that firms can seek to increase market share that violate antitrust laws. For example, a large firm (such as Microsoft) can impose constraints on its customers (such as Dell Computer) that limit or even forbid that customer from using a competitive product. A large firm could require its customers to buy another product from the

---

[12] In fact, some have argued Facebook and Instagram should be broken up now (Wu 2018).

large firm in an attempt to extend its market power to a related market ("tying"). However, US antitrust law requires that any such action must be shown to damage customers; it is not enough, in fact it is irrelevant, whether its actions hurt a competitor. In fact, hurting competitors is the natural outcome of competition for best-performing firms, and is by no means illegal in the US.

The EU has a somewhat different basis for antitrust. For example, in a case against Google, it found that it had required any mobile phone producer offering Google's Android operating system had to offer the complete Google Play Store (an app that sells apps to users) and to prominently display the Play Store on each user's phone. This action was taken as an antitrust violation. It is clear how this might damage other firms (such as competitive mobile phone operating systems) but completely unclear how this could damage customers, and so it is unlikely it would have been pursued in the US.

In fact, it is less than clear exactly what *customer* problem[13] would be solved by aggressive antitrust action by the US government. With Google and Facebook offering customers free service and Amazon offering low-cost retail with superior delivery service,[14] it is difficult to identify how customers are being hurt. If there are concerns about privacy, then recasting privacy laws to provide protection to customers appears to be superior to exercising antitrust law against dominant firms (as privacy issues affect firms of all sizes).

Should large information firms in the US undertake anticompetitive actions that demonstrably harm the customer or the competitive market, then antitrust action should be swift and sure. We cannot assume this will not happen in the future, but it is difficult to see customer harms in today's market. However, careful and rigorous merger analysis of this industry, perhaps lacking in recent years, should be undertaken to ensure competition is not "bought up" by large firms. Keep in mind that

---

[13] Most information firms serve "two-sided" markets, in which both retail customers and, e.g., advertisers access each other through the firms' platform. This suggests that antitrust authorities should not only look at the impact of these firms on retail customers but also the impact on advertisers. Traditionally, there has been no antitrust actions based on harms to advertisers, so we do not consider it here. (See Alleman et al. 2020, this volume, for an alternative view.)

[14] It is difficult to characterize Amazon as "dominant"; Amazon's 2017 US retail sales was $102 B, while Walmart's was $374 B.

the basis of antitrust in US law is the protection of customers from the abuse of market power, and that antitrust action should only be justified upon evidence of damage to customers or to the competitive market... not to competitors. Large firms have opportunities for anticompetitive actions not available to smaller firms and antitrust officials should be particularly vigilant of large firms. But size alone without demonstrable damage to customers or competitive markets is no reason for antitrust prosecution.[15]

What to do about Income Inequality? In no way is the growth of income inequality the product of the large information firms. Rather, the growth in information firms and the growth of income inequality both arise from the same phenomenon: the rise of cyber-civilization. "Fixing" large information firms will do little to help redress income inequality. Not a privacy law, and not antitrust.

However, it is a problem, perhaps the most important problem, that the transition to cyber-civilization has given us, and most of the solutions that helped work in the industrial revolution are unlikely to work here. It helps first to fully understand the scale and scope of income inequality, and there is no better source than David Autor's excellent presentation (Autor 2019). He characterizes occupations in three buckets: technical/professional and managerial, office/clerical and skilled production, and service and unskilled/semiskilled production, in decreasing order of wages. He notes that (i) employment in the top bucket has grown substantially over the last few decades and wages have also increased; (ii) the middle bucket has decreased, and most employees have moved to the bottom bucket; and (iii) employment in the bottom bucket has increased. Further, these buckets roughly correspond to education levels: the top bucket is populated mainly be college grads or post-grads, the middle bucket by high-school grads and some college; while the bottom bucket is high-school grads and high-school dropouts. The increase in the bottom bucket can be explained partly as providing services such as health care and other home and retail services to the top bucket of managers and professionals. He notes that this shift in job distribution is due

---

[15] The history of antitrust in the US includes actions against firms based on their size or based on impacts on competitors. During the 1980s, the "Chicago School" of antitrust revised implementation of antitrust law based solely on potential customer harm or harms to competitive markets, eschewing antitrust actions base on size or on competitor impact. The defining paper is Bork (1979).

to (i) increased automation (primarily) and (ii) shifts of production overseas (secondarily), of which the first is likely to continue into the future.

Managerial and professional jobs are primarily occupied by employees with post-college education, whose income has almost doubled since 1980, and employees with college education, whose income has increased about 50% during that time. Employees with high-school or less education have seen their income fall during this time (for men) or stagnated (for women). Meanwhile, college and post-college graduates have increased from 12% in 1960 to 39% currently, while high-school or less has decreased from 75 to 33% over the same time period.

It is surprising that increasing the supply of college-educated workers leads to an increase in their income; normally, we might expect the reverse. It can only be explained if the number of jobs requiring a college degree increased even more rapidly, which would appear to be the case as we transition to cyber-civilization.[16]

Clearly, the transition to the information economy is driving the demand for a higher-skilled, better educated workforce. This is not new; during the period of rapid industrialization in the US, large firms looked for literate and numerate employees, but had some difficulty finding them. Most potential job-seekers had been raised on farms, with relatively little education. But during this period, policymakers developed a massive (and costly) program to provide public education, for free, to every child in the US, from grades K to 12, thus ensuring a much more educated workforce to populate the factories of the time. While this was more than sufficient for the second wave of civilization, it is not sufficient for the third wave. Individuals who see to their own college education are doing well in this new world, while the high-school-educated are doing poorly.

Many have called to bring manufacturing back from overseas to the US, in order to provide employment to the high-school educated, as it did for many years. But this does not take into account the massive changes in manufacturing, primarily due to information technology. In the days of Ford and General Motors, the business was based on a long assembly line manned by perhaps a thousand workers, each fitting their part into the whole of the car as it worked its way down the line; perfect for high-school educated or less. Today, however, these same products

---

[16] Clearly, the converse is true for high-school or less-educated workers.

are manufactured in assembly lines primarily (if not exclusively) using robots. The assembly line and its robots are run by perhaps 50 employees controlling the robots with computers; these employees have at least a two-year degree from a community college. Result: fewer, better educated employees. It is noteworthy that this switch to robotics is taking place not only in the US and other developed countries, but in China as well, thought by most to be the home of lots of cheap labor.

We can learn lessons from the successful policies of the second wave to point the way forward for the third wave.[17] Again, an educated workforce will help enormously. Recall that universal K-12 education was a successful policy then. Currently, the US's community and technical colleges are an invaluable resource in place to help further educate the existing and future workforce to meet the needs of the information economy. States can do (at least) three things to help our population adjust to this changing world: (i) improve support for community and technical colleges, including a liberal scholarship program for potential students; enable unemployment agencies to counsel beneficiaries on possible career-improving courses available locally; (ii) a national job bank available to all unemployment agencies to inform those seeking unemployment benefits of the existence of jobs that beneficiaries could qualify for, anywhere in the country; and (iii) assistance in moving households for beneficiaries willing to move to obtain a new job.

As the new economy develops, new jobs will require skilled and educated workers, including those willing to move their homes to take advantage of those jobs when and where they appear. While moving home is a major cost for anyone, recall that the US was largely populated by families that "upped stakes" in their country of origin to emigrate to the US in search of new opportunities; we are a nation of movers. States should put in place programs that make it easy for workers to upgrade their skills (as in (i) above), learn of employment opportunities (as in (ii) above) and to make the physical move easier (as in (iii) above).

Can labor laws help workers in the new "gig" economy? In the second wave, labor laws primarily affected workers and employers in large, long-lived companies, helping workers share in these (almost) permanent firms. Much of the third wave economy will likely be small start-ups based on new technology that would be discouraged

---

[17]The literature of policies to deal with income inequality is quite extensive; a good review article is Alleman and Liebenau (2016).

by the cost of strict labor laws, thus interfering with the development of new technology. Strictly interpreted labor laws are not likely to be helpful, either to employers or employees in the new information economy. As the economy matures, this may change, and labor laws may well be helpful, but this does not appear to be the case presently.

Is some form of taxation and redistribution a possible answer? The state could opt to solve this problem by giving workers cash subsidies to maintain their lifestyle. It is likely that most workers would far prefer to earn their living by working rather than depend on the dole, so this is likely to be an unpopular option, both with payers and receivers. However, a currently popular option being considered is a Universal Basic Income (UBI), in which every US adult receives a basic income from the state, whether needed or not. It is argued that UBI would eliminate poverty, encourage more entrepreneurship and better child care, although it would be quite expensive. Lowry (2018)[18] presents a short exposition of the arguments for UBI. Whatever the merits of the idea, it is not an adaptation to the new information economy. However, if taxpayers find its considerable cost worthwhile, this could be at least a partial solution to income inequality.

Conclusion

Economic drivers determine the social, political, and economic structure of civilizations, ever since the very first arose in Mesopotamia in 3500 BCE. When the economic driver, or scarce resource, changes, the civilization changes and its entire power structure changes as well. History teaches us that these changes are wrenching and give rise to extreme social tensions and often violence. Our economy is now in the middle of such a transformation, to an information economy: cyber-civilization. The social, political, and economic stresses are now emerging, and we can learn lessons from the past change of economic driver. Our lead recommendation for coping with these emerging stresses focus on primarily educating the workforce to be ready for the new economy, and reducing the cost of finding and accepting new employment in this economy.

---

[18] Lowry, Annie, "Why the US Should Provide Universal Basic Income," *The Atlantic*, August 1, 2018, at https://www.theatlantic.com/video/index/567739/universal-basic-income/.

Unfortunately, there is little evidence that today's policymakers understand the stresses that are now emerging and are prepared to take on what is required to resolve them.

REFERENCES

Alleman, James, and Jonathan Liebenau. 2016. Disparities in Income & Wealth: Policy Options. Presentation at Western Economics Association International, Portland, OR, June 30. https://www.colorado.edu/faculty/alleman/papers/Disparities_in_Income.

Autor, David. 2019. Work of the Past, Work of the Future (slides), Richard T. Ely Lecture, American Economic Association, Atlanta, GA, January 4, at https://economics.mit.edu/faculty/dautor/video [17 February 2019].

Baines, John. 2011. The Story of the Nile. BBC. http://www.bbc.co.uk/history/ancient/egyptians/nile_01.shtml [17 February 2019].

Bass, Alexandra. 2018. The Giants Everyone Loves to Hate. *The Economist*, the World in 2019 December, p. 114. https://worldin2019.economist.com/bashingtechgiants.

Bork, Richard. 1979. The Chicago School of Antitrust Analysis. *University of Pennsylvania Law Review* 127: 925–948. https://scholarship.law.upenn.edu/cgi/viewcontent.cgi?article=4863&context=penn_law_review.

Chien, YiLi, and Paul Morris. 2017. Is Manufacturing Really Declining? *Federal Reserve Bank of St. Louis*, April 11. https://www.stlouisfed.org/on-the-economy/2017/april/us-manufacturing-really-declining.

Clement, J. 2019. Share of Desktop Search Market Originating from Google in 2018. *Statista*, August 9. https://www.statista.com/statistics/220534/googles-share-of-search-market-in-selected-countries/.

Forbes Magazine. 2019. The World's Billionaires. *Forbes Magazine*, January 12, at https://www.forbes.com/billionaires/list/#version:static.

Gaur, Aakanksha. 2013. Samuel Slater, American Industrialist. *Encyclopedia Britannica*. https://www.britannica.com/biography/Samuel-Slater.

Gordon, Robert. 2012. Is US Economic Growth Over? Faltering Innovation Confronts the Six Headwinds. NBER Paper 18315, at https://www.nber.org/papers/w18315.

Heitzman, Adam. 2017. How Google Came to Dominate Search and What the Future Holds. Forbes Agency Council (2017), at https://www.forbes.com/sites/forbesagencycouncil/2017/06/05/how-google-came-to-dominate-search-and-what-the-future-holds/#588430633872.

Hudson, Andrew. 2018. The Rise & Fall of Kodak. *Photosecrets*, April 20. https://www.photosecrets.com/the-rise-and-fall-of-kodak.

Japan Times. 2019. Japan Seeks to Ban Global Tech Giants from Using Content Contained in Emails and Other Data, January 19. https://www.japantimes.co.jp/news/2019/01/19/national/japan-ban-global-tech-giants-using-content-contained-domestic-email-data/?appsule=1&idfa=345AD11F-06FF-4308-B97F-69BE5AC9BC2A#.XEOI1OTsZPZ.

Kiprop, Victor. 2019. 10 Countries with the Highest Industrial Outputs in the World. WorldAtlas, August 27. https://www.worldatlas.com/articles/10-countries-with-the-highest-industrial-outputs-in-the-world.html.

Kuhn, Morris, Moritz Schularick, and Ulricke Steins. 2018. Income and Wealth Inequality in America, 1949–2016. Opportunity and Inclusive Growth Institute, Federal Reserve Bank of Minneapolis, Institute Working Paper 9, June. https://www.minneapolisfed.org/institute/working-papers-institute/iwp9.pdf.

Lowry, Annie. 2018. Why the US Should Provide Universal Basic Income. *The Atlantic*, August 1. https://www.theatlantic.com/video/index/567739/universal-basic-income/.

Matoo, Aditya, and Joshua Meltzer. 2018. Resolving the Conflict Between Privacy and Digital Trade. Center for Economic and Policy Research, Vox, May 23. https://voxeu.org/article/resolving-conflict-between-privacy-and-digital-trade.

Pappas, Stephanie. 2012. Start Date for Human Civilization Moved Back 20,000 Years or So. *Christian Science Monitor*, July 30. https://www.csmonitor.com/Science/2012/0730/Start-date-for-human-civilization-moved-back-20-000-years-or-so.

Piketty, Thomas. 2014. *Capital in the Twenty-First Century*. Cambridge, MA: Belknap Press of Harvard University Press.

Rubio, Marco. 2019. Congress Needs to Address Consumer Data Privacy in a Responsible and Modern Manner. *The Hill*, January 16. https://thehill.com/blogs/congress-blog/technology/425557-congress-needs-to-address-consumer-data-privacy-in-a.

Satariano, Adam, and Jack Nicas. 2018. E.U. Fines Google $5.1 Billion in Android Antitrust Case. *New York Times*, July 18. https://www.nytimes.com/2018/07/18/technology/google-eu-android-fine.html.

Statcounter. 2019. Social Media Statistics Worldwide. http://gs.statcounter.com/social-media-stats.

Trading Economics. 2016. United States—Agricultural Land. https://tradingeconomics.com/united-states/agricultural-land-percent-of-land-area-wb-data.html.

Wagner, I. 2019a. Change in Employment Figures from Boeing. *Statista*, February 12. https://www.statista.com/statistics/268992/change-in-employment-figures-from-boeing/.

Wagner, I. 2019b. Number of Thyssenkrupp Employees. *Statista*, May 10. https://www.statista.com/statistics/274222/number-of-thyssenkrupp-employees/.

Wagner, Kurt. 2017. Inside Instagram's Reinvention. Recode, January 23. https://www.recode.net/2017/1/23/14205686/instagram-product-launch-feature-kevin-systrom-weil.

Wikipedia. 2019a. List of Regions by Past GDP, July 3. https://en.wikipedia.org/wiki/List_of_regions_by_past_GDP_(PPP).

Wikipedia. 2019b. Income Inequality in the United States, September 26. https://en.wikipedia.org/wiki/Income_inequality_in_the_United_States#The_Great_Compression,_1937%E2%80%931967.

Wikipedia. 2019c. General Motors, October 8. https://en.wikipedia.org/wiki/General_Motors.

Wikipedia. 2019d. General Data Protection Regulation, October 1. https://en.wikipedia.org/wiki/General_Data_Protection_Regulation.

Wu, Tim. 2018. The Case for Breaking Up Facebook and Instagram. *Washington Post*, September 28. https://www.washingtonpost.com/outlook/2018/09/28/case-breaking-up-facebook-instagram/?utm_term=.000a076f159f.

PART I

Applied Econometrics

CHAPTER 2

A Different Approach to Deriving Price and Income Elasticities: Applications to Telecommunications and Gasoline & Motor Oil

Lester D. Taylor

INTRODUCTION

Ordinarily, when we consider the information that is given in a household budget survey, we do so in terms of expenditures for different goods and services and how these relate to income, prices, and socio-demographic factors such as age, family size, and education. Allocation of expenditures amongst different categories of consumption is seen as being determined by tastes and preferences acting in conjunction with a constraint imposed by prices and income. The parameters thus obtained are obviously useful in analyzing the impact on consumption resulting from changes in income and prices (should the latter be available), but income and price elasticities, in themselves, say little about the internal structure of consumption spending. How expenditures for housing, transportation, and personal care—to pick three standard categories of consumption

L. D. Taylor (✉)
University of Arizona, Tucson, AZ, USA

spending—are related to expenditures for food, for example, has, to the best of my knowledge, never been a direct focus of empirical study. Might there be information lurking in these relationships that provide insight into consumer behavior that complements (or even goes beyond) that given by conventional price and income elasticities?[1]

In Taylor (2013),[2] an approach to the analysis of household consumption behavior is developed in which direct relationships amongst different categories of consumption become both the focus and engine of analysis. For an exhaustive set of expenditure categories, expenditures in a category are related, in turn, in a sequence of regression equations, to expenditures in each of the other categories. In *Internal Structure*, this was done for 14 categories of expenditure for each of 40 quarters of data (1996 through 2005) from the ongoing BLS Survey of Consumer Expenditure. Mean values from the 40 sets of estimates are then taken as input for a matrix of "intra-budget" coefficients, seen as representing the conjunctive effects on household expenditures of household tastes and preferences together with the constraints imposed by prices and income. The matrix of intra-budget coefficients thus obtained can be used in a variety of exercises to estimate what, in effect, are general-equilibrium consequences for all categories of consumption that arise from increases or decreases in expenditures in a particular category or from reallocation of expenditures from one category of consumption to another.[3] For reference and illustration, the matrix of intra-budget coefficients for the 14 categories of expenditure analyzed in *Internal Structure* is reproduced (Table 2.1).

---

[1] The traditional approach in demand analysis to questions concerning internal relationships of consumption has been in terms of the functional structure of the utility function, how utilities (or, more specifically, marginal utilities) of individual categories of consumption (whether singly or in groups) might be related to utilities of other categories. In great part, the motivation for such enquiries has been adjunctive to estimation, as demand functions derived from utility functions in which separability is postulated are generally much easier to estimate than demand functions derived from utility functions in which separability is absent.

[2] Hereafter, *Internal Structure*.

[3] The validity of the whole approach obviously depends upon the matrix of intra-budget coefficients representing stable distributional characteristics of households' expenditure decisions. Chapter 3 of *Internal Structure* presents 77a battery of exercises attesting that this assumption can be taken at face value.

Table 2.1 Means of intra-budget expenditures regression coefficients: 14 categories of U.S. consumption expenditure (1996 Q1–2005 Q4)

Coefficients	Food	Alcbev	Housing	Apparel	Trans	Health	EEnterm
Food	–	0.0251	0.5561	0.081	0.3378	0.0662	0.1058
Alcoholic beverages	0.5526	–	1.024	0.1618	0.4045	−0.1798	0.4259
Housing	0.0692	0.0058	–	0.0422	0.1809	0.0163	0.0869
Apparel	0.2318	0.0165	0.5669	–	0.2153	0.0001	0.1358
Transportation	0.0123	0.0006	0.0279	0.0038	–	0.0034	0.0153
Health	0.0741	−0.0092	0.0852	0.0145	0.1080	–	0.0319
Entertainment	0.0480	0.0079	0.1681	0.0322	0.1500	0.0139	–
Personal care	1.0484	0.0584	2.7116	1.0965	1.2542	0.6811	0.2417
Reading	0.6835	0.1430	2.9127	0.5761	0.0970	1.2433	1.0551
Education	0.0536	0.0011	0.1539	0.0170	0.0526	0.0055	0.0015
Tobacco	0.3067	0.1208	−0.4507	0.0027	0.4380	−0.0331	0.0457
Miscellaneous	0.0180	0.0046	0.2252	0.0303	0.0570	0.0471	0.0167
Contributions	0.0190	0.0014	0.1137	0.0111	0.0464	0.0346	0.0242
Insurance	0.0856	0.0083	0.4041	0.0178	0.2077	0.0007	0.0529
Intercept	732	−4.4	1154	−72	573	321	−54

Coefficients	Perscare	Reading	Educ	Tobacco	Misc	Contrib	Ins
Food	0.0120	0.0045	0.0641	0.0137	0.0047	0.0322	0.2321
Alcoholic beverages	0.0139	0.0214	0.0307	0.1220	0.0746	0.0836	0.4259
Housing	0.0692	0.0058	–	0.0422	0.1809	0.0163	0.0869
Apparel	0.2318	0.0165	0.5669	–	0.2153	0.0001	0.1358
Transportation	0.0123	0.0006	0.0279	0.0038	–	0.0034	0.0153
Health	0.0741	−0.0092	0.0852	0.0145	0.1080	–	0.0319
Entertainment	0.0480	0.0079	0.1681	0.0322	0.1500	0.0139	–
Personal care	1.0484	0.0584	2.7116	1.0965	1.2542	0.6811	0.2417
Reading	0.6835	0.1430	2.9127	0.5761	0.0970	1.2433	1.0551
Education	0.0536	0.0011	0.1539	0.0170	0.0526	0.0055	0.0015
Tobacco	0.3067	0.1208	−0.4507	0.0027	0.4380	−0.0331	0.0457
Miscellaneous	0.0180	0.0046	0.2252	0.0303	0.0570	0.0471	0.0167
Contributions	0.0190	0.0014	0.1137	0.0111	0.0464	0.0346	0.0242
Insurance	0.0856	0.0083	0.4041	0.0178	0.2077	0.0007	0.0529
Intercept	22	6.80	−81	71	−15	−66	102

Source Author's computation

In *Internal Structure*, questions of the following sort are investigated. Suppose that, from a point of equilibrium, there is an increase (for whatever reason) of $50 in expenditures for food. In terms of the matrix of intra-budget coefficients, this will perturb the equilibrium expenditures for the other categories, which will then feedback on expenditures for food, which will once again disturb other expenditures, and so on and so forth. In the new equilibrium (assuming that one is reached), how is the increase of $50 in food expenditures ultimately diffused amongst the other categories? Alternatively, one might ask what will be the effect on equilibrium expenditures when an increase in expenditures in one category is offset by a like decrease in another category, an increase, say, of $50 for housing coupled with a decrease of $50 for apparel? In addition, questions involving effects arising from price changes can also be investigated. Such is the focus of the present exercise, specifically on the calculation of own- and cross-price elasticities for two particular categories of consumption, telecommunications and gasoline & motor oil.

The format of the paper is as follows. The next two sections provide a description of the basic framework of the analysis together with derivation of the formulae used in the calculation of own-and cross-price elasticities. The next section illustrative elasticities for telecommunications and gasoline & oil. The last section finishes with summary and conclusions.

The Basic Framework[4]

For notation, let $y_1, \ldots, y_n$ denote an exhaustive n-category breakdown of expenditures, and

$$y_i = \alpha_i + \sum \beta_{ij} y_j + u_{ij} \quad i, j = 1, \ldots, n, \tag{2.1}$$

represent the least-squares regression equation of expenditures in the ith category on expenditures in the other $n - 1$ categories. Equation (2.1) has been estimated, for each of the 18 categories to be analyzed, for each of 28 BLS Quarterly Consumer Expenditure Surveys from 2006 Q1

---

[4]A much more detailed discussion of this framework is given in Chapters 2–4 of *Internal Structure* [also in Taylor (2017a)]. In Taylor (2017b), after-tax income is included in Expression (2.1) as an additional predictor in order to allow for straightforward calculation of income and total-expenditure effects. As shown there, price elasticities are little affected.

through 2012 Q4.[5] For illustration, the estimated coefficients from the 18 regression equations for 2012 Q4 are tabulated in Tables 2.2A and 2.2B.[6]

Next, let $y=(y_1, ..., y_n)$ denote a (column) vector of expenditures and let B be the $n \times (n+1)$ matrix:

$$B = [z, A], \qquad (2.2)$$

where z denotes the (column) vector of intercepts (as in the last row of Table 2.1) and A denotes the transpose of the $(n \times n)$ sub-matrix of "intra-budget" coefficients from the table. Hence:

$$y = z + Ay, \qquad (2.3)$$

or

$$[I - A]y = z, \qquad (2.4)$$

so that

$$y = [I - A]^{-1}z. \qquad (2.5)$$

Expressions (2.3) and (2.5) are referred to as the *structural* and *reduced-form* representations of the model, respectively. Both representations are used extensively in the exercises in Chapters 4 and 5 of *Internal Structure*, depending upon whether expenditures are assumed to be *exogenous* or *endogenous*. "*Exogenous*" in this context refers to a change in z, while "*endogenous*" refers to a change in y. However, since only exogenous changes are the focus in the present exercise, details regarding endogenous changes need not be gone into.[7]

---

[5] Unlike in *Internal Structure*, the present effort utilizes 18 categories of expenditure rather than 14. Food is disaggregated to food consumed at home and food consumed outside of the home, water and telephone are disengaged from housing, and gasoline & oil from transportation.

[6] Coefficient estimates are in rows, equations in columns; *t*-ratios are in parentheses, with 6715 degrees of freedom. The -1s on the diagonal reflect the fact that expenditure in that category is the dependent variable. Definitions of the expenditure categories are given in the Appendix.

[7] It might seem that new equilibria are independent of how changes in expenditure come about, but this is not the case, for equilibrium depends upon where and how changes in expenditure occur and at what point (and how) a total-expenditure budget constraint is imposed. In particular, exogenous changes [as should be clear from consideration of Expression (2.5)] leads to new equilibria that are different from the original. With endogenous changes, however, since z is not changed, the system will eventually return to the same equilibrium as before. Chapters 4 and 5 of *Internal Structure* provide discussion and examples.

Table 2.2A Intra-budget regressions: 18 categories of U.S. Consumption expenditure (2012 Q4; t-ratios in parentheses)

Coefficients

Category	Intercept	Fdhome	Fdaway	Alcbev	Housl	Telephone	Water	Apparel	Transl	Gasoil
Fdhome	604.96	−1	0.0627	−0.0781	0.0328	0.4550	0.5022	0.0430	0.0040	0.2428
	(36.04)		(5.37)	(−2.12)	(10.14)	(11.54)	(8.79)	(2.93)	(2.17)	(15.07)
Fdaway	49.67	0.0683	−1	0.6102	0.0272	0.1476	0.0282	0.0857	0.0061	0.0996
	(2.60)	(5.37)		(16.18)	(8.01)	(3.55)	(0.47)	(5.60)	(3.21)	(5.84)
Alcbev	15.01	−0.0086	0.0617	−1	0.0040	−0.0416	0.0315	0.0255	0.0011	0.0038
	(2.47)	(−2.12)	(16.18)		(3.65)	(−3.15)	(1.65)	(5.23)	(1.78)	(0.70)
Housl*	975.88	0.4605	0.3496	0.5027	−1	1.0466	1.9212	0.5276	0.0223	0.0283
	(14.42)	(10.14)	(8.01)	(3.65)		(7.04)	(8.99)	(9.66)	(3.25)	(0.46)
Telephone	109.30	0.0428	0.0127	−0.0355	0.0070	−1	0.1164	0.0165	−0.0001	0.0940
	(19.99)	(11.54)	(3.56)	(−3.15)	(7.04)		(6.63)	(3.66)	(−0.11)	(19.21)
Water	29.93	0.0227	0.0012	0.0129	0.0062	0.0560	−1	−0.0109	0.0005	0.0213
	(7.70)	(8.79)	(0.47)	(1.65)	(8.98)	(6.63)		(−3.50)	(1.22)	(6.14)
Apparel	−31.24	0.0297	0.0544	0.1600	0.0260	0.1211	−0.1670	−1	0.0027	0.0309
	(−2.05)	(2.93)	(5.60)	(5.23)	(9.66)	(3.66)	(−3.50)		(1.75)	(2.27)
Transl**	58.15	0.1768	0.250	0.4372	0.0708	−0.0297	0.4684	0.1715	−1	0.5849
	(0.48)	(2.17)	(3.21)	(1.78)	(3.25)	(−0.11)	(1.22)	(1.75)		(5.36)
Gasoil	182.86	0.1352	0.0509	0.0193	0.0011	0.5559	0.2625	0.0249	0.0073	−1
	(13.54)	(15.07)	(5.84)	(0.70)	(0.46)	(19.21)	(6.14)	(2.27)	(5.36)	
Health	258.26	0.0775	0.0262	−0.0717	0.0167	0.2306	0.8293	−0.0185	0.0036	0.0442
	(9.06)	(4.06)	(1.43)	(−1.25)	(3.29)	(3.71)	(9.29)	(−0.81)	(1.27)	(1.73)
Entertn	−166.73	0.0411	0.1512	0.2625	0.0467	−0.1474	0.0927	0.0850	0.0070	0.1018
	(−4.24)	(5.40)	(6.04)	(3.33)	(6.70)	(−1.73)	(0.75)	(2.70)	(1.79)	(2.90)
Percare	−2.22	0.0041	0.0080	0.0330	0.0059	0.0268	0.0475	0.0250	0.0007	0.0019
	(−0.82)	(2.25)	(4.63)	(6.11)	(12.32)	(4.57)	(5.64)	(11.67)	(2.40)	(0.77)
Reading	2.30	0.0024	0.0018	0.0141	0.0016	−0.0042	0.0134	0.0046	0.0005	−0.0060
	(1.39)	(2.19)	(1.74)	(4.26)	(5.43)	(−1.18)	(2.59)	(3.50)	(2.77)	(−4.09)

(continued)

Table 2.2A (continued)

Coefficients

Category	Intercept	Fdhome	Fdaway	Alcbev	Hous1	Telephone	Water	Apparel	Trans1	Gasoil
Educ	−326.17	0.0595	0.3031	−0.0610	0.1094	0.1488	−0.3569	0.0331	0.0086	0.0148
	(−6.76)	(1.84)	(9.88)	(−0.63)	(12.86)	(1.42)	(−2.36)	(0.85)	(1.78)	(0.34)
Tobacco	51.47	0.0169	−0.0090	0.0908	−0.0042	0.0510	−0.0194	−0.0051	−0.0004	0.0238
	(9.17)	(4.50)	(−2.50)	(8.05)	(−4.20)	(4.16)	(−1.10)	(−1.13)	(−0.70)	(4.73)
Misc	−70.64	0.0873	0.0379	0.0859	0.0048	−0.0140	−0.0475	0.0006	0.0003	0.0342
	(−2.34)	(4.35)	(1.97)	(1.42)	(0.89)	(−0.21)	(−0.50)	(0.02)	(0.11)	(1.27)
Contrib	−137.21	−0.0955	0.2335	−0.1778	0.0532	0.1632	−0.1450	0.1469	0.0020	−0.0049
	(−2.31)	(−2.41)	(6.17)	(−1.49)	(5.03)	(1.26)	(−0.78)	(3.08)	(0.34)	(−0.09)
Perins	−208.02	0.1218	0.3156	0.4015	0.1337	0.6607	0.3436	0.1653	0.0072	0.3552
	(−4.46)	(3.92)	(10.67)	(4.29)	(16.40)	(6.53)	(2.35)	(4.42)	(1.55)	(8.56)

Source Author's computation

Table 2.2B Intra-budget regressions: 18 categories of U.S. Consumption expenditure (2012 Q4; t-ratios in parentheses)

Coefficients										
Category	Health	Entertn	Percare	Reading	Educ	Tobacco	Misc	Contrib	Perins	R^2
Fdhome	0.0317 (4.06)	0.0307 (5.40)	0.1869 (2.25)	0.2970 (2.19)	0.0085 (1.84)	0.1786 (4.50)	0.0322 (4.35)	−0.0091 (−2.41)	0.0188 (3.92)	0.2586
Fdaway	0.0117 (1.43)	0.0358 (6.04)	0.4009 (4.63)	0.2459 (1.74)	0.0474 (9.88)	−0.1035 (−2.50)	0.0152 (1.97)	0.0242 (6.17)	0.0530 (10.67)	0.2729
Alcbev	−0.0032 (−1.25)	0.0063 (3.33)	0.1680 (6.11)	0.1914 (4.26)	−0.0010 (−0.63)	0.1056 (8.05)	0.0035 (1.42)	−0.0019 (−1.49)	0.0068 (4.29)	0.1281
Housl	0.0961 (3.29)	0.1426 (6.70)	3.7887 (12.32)	2.7500 (5.43)	0.2202 (12.86)	−0.6234 (−4.20)	0.0246 (0.89)	0.0708 (5.03)	0.2888 (16.40)	0.3777
Telephone	0.0089 (3.71)	−0.0030 (−1.73)	0.1162 (4.57)	−0.0490 (−1.18)	0.0020 (1.42)	0.0507 (4.16)	−0.0005 (−0.21)	0.0015 (1.26)	0.0096 (6.53)	0.2413
Water	0.0154 (9.29)	0.0009 (0.75)	0.0993 (5.64)	0.0745 (2.59)	−0.0023 (−2.36)	−0.0093 (−1.20)	−0.0008 (−0.50)	−0.0006 (−0.78)	0.0024 (2.35)	0.1511
Apparel	−0.0053 (−0.81)	0.0128 (2.70)	0.7979 (11.67)	0.3947 (3.50)	0.0033 (0.85)	−0.0373 (−1.13)	0.0002 (0.02)	0.0097 (3.08)	0.0176 (4.42)	0.1635
Transl	0.0665 (1.27)	0.0682 (1.79)	1.3320 (2.40)	2.5012 (2.77)	0.0549 (1.78)	−0.1838 (−0.69)	0.0056 (0.11)	0.0085 (0.34)	0.0497 (1.55)	0.0490
Gasoil	0.0100 (1.73)	0.0123 (2.90)	0.0476 (0.77)	−0.4125 (−4.09)	0.0012 (0.34)	0.1399 (4.73)	0.0070 (1.27)	−0.0003 (−0.09)	0.0305 (8.56)	0.2417
Health	−1	0.0304 (3.42)	0.8712 (6.74)	1.7848 (8.47)	0.0024 (0.33)	−0.0678 (−1.09)	0.0331 (2.86)	0.0514 (8.79)	0.0189 (2.53)	0.1296
Entertn	0.0572 (3.42)	−1	0.7245 (4.07)	1.1981 (4.13)	0.0085 (0.86)	0.0847 (0.99)	−0.0044 (−0.28)	0.0081 (1.00)	0.0422 (4.11)	0.1107
Percare	0.0077 (6.74)	0.0034 (4.07)	−1	0.1566 (7.89)	−0.0001 (−0.20)	−0.0128 (−2.20)	0.0013 (1.19)	0.0015 (2.71)	0.0053 (7.57)	0.2516
Reading	0.0059 (8.47)	0.0021 (4.13)	0.0588 (7.89)	−1	0.0012 (2.80)	−0.0008 (−0.23)	−0.0005 (−0.77)	0.0014 (3.99)	0.0016 (3.63)	0.1110

(continued)

Table 2.2B (continued)

Coefficients

Category	Health	Entertn	Percare	Reading	Educ	Tobacco	Misc	Contrib	Perins	R^2
Educ	0.0067	0.0129	0.0447	1.0004	−1	−0.0719	0.0186	0.0044	−0.0053	0.0896
	(0.33)	(0.86)	(−0.20)	(2.80)		(−0.69)	(0.95)	(0.45)	(−0.42)	
Tobacco	−0.0026	0.0017	−0.0561	−0.0095	−0.0010	−1	0.0004	−0.0006	−0.0031	0.0252
	(−1.09)	(0.99)	(−2.20)	(−0.23)	(−0.69)		(0.15)	(−0.49)	(−2.10)	
Misc	0.0367	−0.0026	0.1621	−0.1709	0.0072	0.0099	−1	0.0241	−0.0031	0.0169
	(2.86)	(−0.28)	(1.19)	(−0.77)	(0.95)	(0.15)		(3.90)	(−0.39)	
Contrib	0.0221	0.0185	0.7299	1.7528	0.0067	−0.063	0.0940	−1	0.0160	0.0644
	(8.79)	(1.00)	(2.71)	(3.99)	(0.45)	(−0.49)	(3.90)		(1.03)	
Perins	0.2503	0.0595	1.5935	1.2506	−0.0049	−0.212	−0.0074	0.0099	−1	0.2782
	(2.53)	(4.11)	(7.57)	(3.63)	(−0.42)	(−2.10)	(−0.39)	(1.03)		

*Hous1 refers to Housing less Telephone and Water
**Trans1 refers to Transportation less Gasoline & Motor Oil
Source Author's computation

Derivation of Own- and Cross-Price Elasticities

Absent direct information on prices, it is not possible (short of heroic assumptions about the structure of preferences) to estimate price effects from household surveys of consumer expenditures in a framework of conventional demand analysis. Surprisingly, however, this is possible in the present framework, for both own- and cross-price elasticities can be obtained through simulation. The key for doing so is to note that changes in prices can be incorporated into the analysis through the "intercept" vector z in Eq. (2.5). Equation (2.5) is solved for two different values of z, the first to establish a baseline set of expenditures and a second to reflect the assumed change (or changes) in price. The resulting price effects are then represented by the differences between the two solution vectors y.

The procedure involved, using the entries in Table 2.1 as parameters, will be illustrated for an assumed 10% increase in the price of food.[8] The results are tabulated in Table 2.3. To begin with, baseline expenditures are obtained from Eq. (2.5) using the "intercepts" from the last row in Table 2.1 for z, that is, a value of $732 for food, $-$4.40 for alcoholic beverages, etc. These are given in the first column of Table 2.3. The total expenditure implied by this value for z is $13,593.[9] Note that the base expenditure for food is $1705. An increase in the price of food of 10% is represented as a 10% increase in this base expenditure, that is, an increase of $171. The intercept for food in z ($732) is replaced by $903 ($732 + $171). All other elements of z remain the same. Call this new value of z, z^*. Equation (2.5) is then solved for z^*, with total expenditure held results appear in columns 4 and 5 of Table 2.3. Expenditures are seen to increase $83.06 for food, decrease $54.61 for housing, and so on and so forth. Table 2.3 also includes results for a 5% increase in the price of food. Results for 5 and 10% decreases are given in Table 2.4.

The elasticities in columns 3 and 5 of Tables 2.3 and 2.4 are obtained as follows. To begin with, the total-expenditure constraint, in an obvious notation, will be given by:

$$\sum p_i q_{i=1} = y \qquad (2.6)$$

---

[8] All calculations are undertaken in SAS.
[9] This represents a quarterly total; the implied total annual expenditure is about $54,000.

Table 2.3 Own- & cross-price elasticities: for a 5 and 10% increase in price of food, 14 categories of expenditure (total expenditure held constant)

Change in price of food

Base		+5%				+10%		
			Elasticity				Elasticity	
Category	Expenditures	Change	Allocation	Quantity		Change	Allocation	Quantity
Food	$1705.72	$43.42	0.51	−0.49		$82.88	0.49	−0.51
Alcoholic beverages	115.43	0.98	0.01	0.17		1.87	0.01	0.16
Housing	4111.91	−28.55	−0.33	−0.14		−54.49	−0.32	−0.13
Apparel	532.53	4.80	0.06	0.18		9.16	0.05	0.17
Transportation	2847.18	−18.74	−0.22	−0.13		−35.77	−0.21	−0.13
Health	695.93	−13.91	−0.16	−0.40		−26.54	−0.16	−0.38
Entertainment	877.88	4.51	0.05	0.10		8.61	0.05	0.10
Personal care	129.58	−0.86	−0.01	−0.13		−1.65	−0.01	−0.13
Reading	69.55	−0.35	−0.004	−0.10		−0.67	−0.004	−0.10
Education	285.67	6.20	0.07	0.43		11.83	0.07	0.41
Tobacco	78.04	−2.55	−0.03	−0.65		−4.87	−0.03	−0.62
Miscellaneous	151.01	−0.02	−0.0003	−0.003		−0.05	0.0003	−0.003
Contributions	341.13	3.60	0.04	0.21		6.87	0.04	0.20
Personal insurance	1651.63	1.47	0.02	0.02		2.81	0.02	0.02

Total expenditure: $13,593
Source Author's computation

Table 2.4 Own- & cross-price elasticities: for a 5 and 10% decrease in price of food, 14 categories of expenditure (total expenditure held constant)

Category	Base Expenditures	Change	−5% Elasticity Allocation	Quantity	Change	−10% Elasticity Allocation	Quantity
Food	$1705.72	−$48.01	0.56	−0.44	−$101.39	0.59	−0.41
Alcoholic beverages	115.43	−1.08	0.01	0.19	−2.29	0.01	0.20
Housing	4111.91	31.57	−0.37	−0.15	66.66	−0.39	−0.16
Apparel	532.53	−5.31	0.06	0.20	−11.21	0.07	0.21
Transportation	2847.18	20.72	−0.24	−0.15	43.76	−0.26	−0.15
Health	695.93	15.38	−0.18	−0.44	32.48	−0.19	−0.47
Entertainment	877.88	−4.99	0.06	0.11	−10.53	0.06	0.12
Personal care	129.58	0.95	−0.01	−0.15	2.02	−0.01	−0.16
Reading	69.55	0.39	−0.005	−0.11	0.83	−0.005	−0.12
Education	285.67	−6.85	0.08	0.48	−14.47	0.08	0.51
Tobacco	78.04	2.82	−0.03	−0.72	5.96	−0.03	−0.76
Miscellaneous	151.01	0.03	−0.0003	−0.003	0.06	−0.0003	−0.004
Contributions	341.13	−3.98	0.05	0.23	−8.41	0.05	0.25
Personal insurance	1651.63	−1.63	0.02	0.02	−3.44	0.02	0.02

Total expenditure: $13,593
Source Author's computation

Hence, for a change of Δp_i in the price of q_i, with total expenditure y held constant, we will have:

$$\Delta(p_i q_i) + \sum_{j \neq i} p_j \Delta q_j = 0, \tag{2.7}$$

Or

$$\frac{\Delta(p_i q_i)}{q_i \Delta p_i} + \sum_{j \neq i} \frac{p_j \Delta q_j}{q_i \Delta p_i} = 0, \tag{2.8}$$

where $q_i \Delta p_i$ denotes the increase in expenditures for good i (assuming a zero price elasticity) occasioned by Δp_i,[10] and also

$$\frac{\Delta(p_i q_i)}{q_i \Delta p_i} + \sum_{j \neq i} \left(\frac{\Delta q_j}{\Delta p_i} \frac{p_i}{q_j}\right) \left(\frac{p_j q_j}{p_i q_i}\right) = 0. \tag{2.9}$$

Next, let[11]:

$$R_{ii} = \frac{\Delta(p_i q_i)}{q_i \Delta p_i} \tag{2.10}$$

$$R_{ji} = \frac{p_j \Delta q_j}{q_i \Delta p_i},$$

so that Expression (2.9) can be written as

$$R_{ii} + \sum_{j \neq i} R_{ji} = 0. \tag{2.11}$$

Finally, if we let

$$\kappa_{ji} = \frac{p_j q_j}{p_i q_i}, \tag{2.12}$$

Expression (2.9) can now be rewritten in elasticity form as

$$\eta_{ii} + 1 + \sum_{j \neq i} \eta_{ji} \kappa_{ji} = 0, \tag{2.13}$$

---

[10] $q_i \Delta p_i$ also represents the decrease/increase in real income dictated by Δp_i.
[11] For reasons that will become clear shortly, these quantities will be referred to as allocation (or reallocation) elasticities.

where

$$\eta_{ji} = \frac{\Delta q_j}{q_j} \bigg/ \frac{\Delta p_i}{p_i}$$
$$= \frac{p_j \Delta q_j}{p_j q_j} \bigg/ \frac{q_i \Delta p_i}{p_i q_i} \qquad (2.14)$$

denotes the conventional (quantity) elasticity of good j with respect to a change in the price of good i.

Thus, for a 10% increase for the price of food, the own-price elasticity for food [from Expression (2.15)] will be given by

$$\eta_{11} = \frac{82.88 - 170.57}{170.57}$$
$$= -0.51, \qquad (2.15)$$

while the cross-elasticity for housing (say) will be given by

$$\eta_{31} = -0.32 \left(\frac{1705.72}{4111.91} \right)$$
$$= -0.13. \qquad (2.16)$$

both as in column 5 of Table 2.3.[12] Two conclusions are evident in these tables: Elasticities (in absolute value) depend upon the magnitude of price changes and, though similar, are smaller for price decreases than for price increases of like size. Taken together, these results imply that elasticities vary with levels of prices.[13]

---

[12] Note that the own-price elasticity is measured from expenditure after the price change (rather than from the base expenditure), since, with no change in quantity demanded, this is what expenditure (as given by $q_i \Delta p_i$) would be in the absence of a non-zero elasticity. However, a non-zero (negative) elasticity causes some of the revenue [specifically, $q_i \Delta p_i - \Delta(p_i q_i)$] in effect to "melt" away because of the higher price. For cross-elasticities, on the other hand, since prices of other goods are not changed, calculations are made from base expenditures.

[13] At this point, it is useful to step back and contemplate—indeed, even marvel!—at what the procedures behind the formulae in Expressions (2.11) and (2.12) imply and accomplish. *For they allow for price elasticities (both own and cross) to be adduced in circumstances in which all price information is subsumed in expenditures.* While the estimation of own-price elasticities is straightforward in conventional demand analysis (in situations in which explicit price data are available), estimation of cross-elasticities typically flounders because of large inter-correlations amongst prices (i.e., by multicollinearity). Importantly, the

Own- and Cross-Price Elasticities for Telecommunications and Gasoline & Motor Oil

In this section, we turn to application of the procedure just described to the calculation of own- and cross-price elasticities for telecommunications and gasoline & motor oil. This will be done using parameters for the matrix B in Expression (2.2) whose elements are defined as mean values from estimation of the regression equations represented in Expression (2.1) with data from the 28 quarterly BLS Consumer Expenditure Surveys from 2006 Q1 through 2012 Q4. The B matrices in Expression (2.2) used in the calculations are given in the Appendix.

However, before looking at the results, it will be useful to review the standard notions of substitution and complementarity. In traditional demand theory, effects arising from a change in the price of a good are seen as consisting of two components, a substitution effect, in which (on a given indifference surface) there is substitution away from (what are now) relatively more expensive goods, and an income effect, in which (with money income held constant) the change in price causes movement to a lower or higher indifference surface depending upon whether the price change is positive or negative. In a two-good world, the two goods (because of the budget constraint) must necessarily be substitutes, which is to say that their cross-price elasticities must be negative. However, in a multi-good world, while substitution must prevail overall, some goods can also be complements, which is to say that they can have positive cross-price elasticities (ham and eggs being a prime example). While in traditional theory own-price elasticities are expected to be negative, they can in fact be positive if the good in question is an inferior good (i.e., has a negative income elasticity) and also weighs significantly in total expenditure.[14]

In traditional theory, the fact that other goods as a group must be substitutes against the good whose price has changed is imposed by

procedure is almost entirely empirical. No assumptions are made regarding behavior, other than that households spend in the way that data are recorded in the BLS surveys, and that their expenditures are bound by a total-expenditure budget constraint. In short, the analysis is almost entirely statistical and mathematical.

[14] Potatoes in nineteenth-century Ireland, as studied by Giffen, are the standard example of such a good, hence the term "Giffen" good.

the budget constraint, while substitution or complementarity between individual pairs of goods is a consequence of the shape of indifference surfaces. In the present framework, the intra-budget coefficients in A (as has been noted) represent the conjunctive effects on household expenditures of household tastes and preferences together with the constraints imposed by prices and income. Intra-budget coefficients thus in effect represent reduced-form parameters, whose validity in exercises such as the present depend only on stability of the underlying determining structure. The form of this structure does not really matter, in that tastes and preferences can be as postulated in traditional theory or they can be lexicographical, hierarchical, or whatever. The only requirement is that underlying structures—whatever they might be—be *distributionally stable* across households.[15] In such circumstances, the present framework allows for a much more flexible interpretation of substitution and complementation, particularly in environments in which consumption activity based involves a variety of goods as inputs, as in, for example, a dinner party. To the extent that such activities are important in real-world consumption, they will accordingly be reflected in the intra-budget coefficients in A, and (importantly) should not be expected to be the same at all levels of income. Goods that are substitutes at low levels of expenditure can be complements at higher levels of expenditure and vice versa.[16]

Similar reasoning can account for own-price elasticities that are positive rather than negative. In traditional theory, as has been noted, a positive own-elasticity of demand can only arise in the case of an inferior good (i.e., one with a negative income elasticity) that accounts for a significant proportion of households' budgets. Within the present framework, however, a good that, in isolation, would respond inversely to movements in its own price can in fact move positively because of feedbacks from other goods. Such can be the case in a three-good world if the first good has a strong *negative* relationship with the second good which in turn has a strong *positive* relationship with the third good which then feedbacks strongly *positively* on the first good.[17]

---

[15] For a more detailed discussion, see Taylor (2017a, b).

[16] Several instances of this are encountered in Taylor (2017a).

[17] Such can also arise if the relationships amongst the goods are *positive, negative, negative* instead of *negative, positive, positive*. Note that relative importance of budget shares will matter as well.

Table 2.5 Effects of A±10% change in price of telecommunications: 18 categories of expenditure (total expenditure held constant)

Category	Base	10% decrease In price	Change	10% increase In price	Change
Food home	$1205.99	$1232.21	$26.22	$1182.49	−$23.50
Food away	576.54	576.31	−0.23	576.75	0.21
Alcoholic beverages	90.31	91.46	1.15	89.27	−1.03
Housing (less telephone & water)	3700.75	3735.64	34.89	3699.47	−31.28
Telephone	294.60	271.43	−23.16	315.36	20.77
Water	118.82	120.23	1.41	117.56	−1.26
Apparel	313.81	307.64	−6.18	319.36	5.54
Transportation (less gasoline & oil)	1483.61	1465.62	−17.99	1499.73	16.12
Gasoline & oil	603.04	603.03	−0.03	603.10	0.03
Health	752.37	768.48	16.11	737.93	−14.44
Entertainment	602.95	598.87	−4.08	606.60	3.65
Personal care	77.36	76.77	−0.58	77.88	0.52
Reading	29.29	29.52	0.23	29.08	−0.21
Education	265.32	248.45	−16.87	280.44	15.12
Tobacco	82.04	84.48	2.44	79.86	−2.18
Miscellaneous	159.58	158.10	−1.48	160.91	1.32
Contributions	453.88	450.32	−3.56	457.06	3.19
Personal insurance	1410.01	1401.72	−8.29	1417.44	7.43
Total expenditure	$12,220	$12,220	0.00	$12,220	0.00
Change in real income		$29.46		−$29.46	

Source Author's computation

Telecommunications

Own- and cross-price elasticities for telecommunications have been calculated from simulations of Expression (2.5) for changes in the price of telecommunications ±10%. Quantity effects from the simulations are presented in Table 2.5 and elasticities [as calculated from Expression (15)] in Table 2.6.

In Table 2.6, it is seen that, although similar, elasticities differ between a down and up price changes. The own-elasticity for telecommunications is −0.21 for a 10% decrease in price and −0.30 for a 10% increase, while cross-elasticities range from +0.57 for education to −1.18 for housing for the 10% decrease and from 0.55 for transportation to −1.06 for housing for the 10% increase. Strong substitutes with

Table 2.6 Estimated own- and cross-price elasticities: for 10% decrease/increase in price of telecommunications (from.Table 2.5)

Category	10% decrease In price	10% increase In price
Food at home	−0.89	−0.80
Food away	0.01	0.01
Alcoholic beverages	−0.04	−0.04
Housing (less telephone & water)	−1.18	−1.06
Telephone	−0.21	−0.30
Water	−0.05	−0.04
Apparel	0.21	0.19
Transportation (less gasoline & oil)	0.61	0.55
Gasoline & motor oil	0.001	0.001
Health	−0.55	−0.49
Entertainment	0.14	0.12
Personal care	0.02	0.02
Reading	−0.01	−0.01
Education	0.57	0.51
Tobacco	−0.08	−0.07
Miscellaneous	0.05	0.05
Contributions	0.12	0.11
Personal insurance	0.28	0.25

Source Author's computation

telecommunications include transportation and education and strong complements with food, housing, and health.

Gasoline & Motor Oil

In late April 2014, the average price per gallon of regular gasoline in the U. S. was $3.70, but by the end of the year, the average price had fallen to nearly $2.00 a gallon, a drop of close to 50%. In this section an analysis of price changes, using the same procedure as for telecommunications in the preceding section, is undertaken for expenditures for gasoline and motor oil, first, for a price changes of ± of 50%, then, for a price decrease of 50% followed by a "recovery" in price of 75%. Finally, in order to provide a comparison with "small" price changes, simulations are also pursued for changes of ±10%. The results for ±50% are tabulated in Tables 2.7 and 2.8, −50%/+75% in Tables 2.9 and 2.10, and for ±10% in Table 2.11.

Table 2.7 Effects of ±50% change in price of gasoline & motor oil with total expenditure held constant: 18 categories of expenditure

Category	50% decrease			50% increase	
	Base	In price	Change	In price	Change
Food home	$1205.99	$1347.30	$141.31	$1122.72	−$83.27
Food away	576.54	566.81	−9.73	582.28	5.73
Alcoholic beverages	90.31	93.38	3.07	88.50	−1.81
Housing (less telephone & water)	3700.75	4109.66	408.92	3459.79	−240.96
Telephone	294.60	313.84	19.24	283.26	−11.34
Water	118.82	132.39	13.57	110.83	−7.99
Apparel	313.81	293.68	−20.14	325.69	11.87
Transportation (less gasoline & oil)	1483.61	1302.52	−181.08	1590.32	106.71
Gasoline & oil	603.04	274.17	−328.90	796.88	193.81
Health	752.37	865.55	113.19	685.67	−66.70
Entertainment	602.95	559.33	−43.62	628.65	25.70
Personal care	77.36	79.54	2.18	76.07	−1.28
Reading	29.29	32.92	3.63	27.14	−2.14
Education	265.32	201.66	−63.66	302.84	37.51
Tobacco	82.04	91.91	9.86	76.23	−5.81
Miscellaneous	159.58	157.97	−1.61	160.53	0.95
Contributions	453.88	464.89	11.01	447.39	−6.49
Personal insurance	1410.01	1332.76	−77.24	1455.52	45.52
Total expenditure	$12,220	$12,220	0.00	$12,220	0.00
Change in real income		$301.53		−$301.53	

Source Author's computation

Key features of the simulations for ±50% changes in price are as follows:

1. The first thing to note is that real-income effects for the price changes considered are clearly going to be much greater for gasoline & oil than for telecommunications, for the share of gasoline & oil in total expenditure is about double that for telecommunications and a 50% price change is obviously much larger than 10%.
2. For telecommunications, expenditure changes in Table 2.5 are seen to be more or less equal in absolute value. For gasoline & oil, we see that this is not the case in Table 2.7. More will be said about this below.

Table 2.8 Estimated own- and cross-price elasticities for ±50% change in price of gasoline & oil (from Table 2.7)

Category	10% decrease In price	10% increase In price
Food at home	−0.23	−0.14
Food away	0.03	0.02
Alcoholic beverages	−0.07	−0.04
Housing (less telephone & water)	−0.22	−0.13
Telephone	−0.13	−0.08
Water	−0.23	−0.13
Apparel	0.13	0.08
Transportation (less gasoline & Oil)	0.24	0.14
Gasoline & motor oil	0.09	−0.36
Health	−0.30	−0.18
Entertainment	0.14	0.09
Personal care	−0.06	−0.03
Reading	−0.25	−0.15
Education	0.48	0.28
Tobacco	−0.24	−0.14
Miscellaneous	0.02	0.01
Contribution	−0.05	−0.03
Personal insurance	0.11	0.06

Source Author's computation

3. In Table 2.7, the simulated own-price elasticities are seen to be negative for the 50% price increase, but *positive* for the 50% price decrease. Footnote 11

For the ±50% simulations in Table 2.7, the expenditure changes are measured from the same base expenditures (i.e., as listed in the first columns of these tables).[18] In contrast, for the simulations in Table 2.9 involving a 50% decrease in price followed by a 75% increase, expenditure changes are calculated from the equilibrium expenditures associated with the 50% decrease in price.[19] For this scenario, the

---

[18] These simulations, to remind ourselves, are calculated from Expression (2.5) with A and z (for the base expenditures) from Tables 2.13A and 2.13B in the Appendix.

[19] Thus, base expenditures for the 75% increase in price are not the same as the expenditures appearing in column 2 of Table 2.7. The latter are calculated using the procedure described in Footnote 20, while the former are calculated from Expression (2.5) as

Table 2.9 Effects of 50% decrease/+75% increase in price of gasoline & motor oil: 18 categories of expenditure (total expenditure held constant)

Category	50% decrease base change		75% increase base change	
Food home	$1205.99	$141.31	$999.11	−$60.75
Food away	576.54	−9.73	420.33	4.18
Alcoholic beverages	90.31	3.07	69.24	−1.32
Housing (less telephone & water)	3700.75	408.92	3047.57	−175.80
Telephone	294.60	19.24	232.73	−8.27
Water	118.82	13.57	98.17	−5.83
Apparel	313.81	−20.14	217.78	8.66
Transportation (less gasoline & oil)	1483.61	−181.08	965.90	77.85
Gasoline & oil	603.04	−328.90	203.31	141.40
Health	752.37	113.19	641.86	−48.66
Entertainment	602.95	−43.62	414.77	18.75
Personal care	77.36	2.18	58.98	−0.94
Reading	29.29	3.63	24.41	−1.56
Education	265.32	−63.66	149.54	27.37
Tobacco	82.04	9.86	68.15	−4.34
Miscellaneous	159.58	−1.61	117.15	0.69
Contributions	453.88	11.01	344.74	−4.73
Personal insurance	1410.01	−77.24	988.33	33.21
Total expenditure	$12,220	0.00	$9062.10	0.00
Change in real income	$301.53		−$152.48	

Source Author's computation

simulated own-price elasticities for the 75% increase following the 50% decrease are similar, and with the own-elasticity once again positive.

While a positive own-price elasticity for gasoline is not something that we would ordinarily expect, its occurrence in a three-good world in the present framework is, as noted above, relatively straightforward to rationalize. Things are obviously more complicated with 18 categories of consumption, nevertheless the question of whether a positive own-price elasticity in this context is a consequence of feedbacks from other categories can be examined using an iterative scheme based upon the structural representation of the framework in Expression (2.3), specifically, in the form:

$$y^1 = z + Ay^0 \qquad (2.17)$$

$$y^k = z + w^k A y^{k-1}, \qquad (2.18)$$

Table 2.10 Effects of 50% decrease/+75% increase in price of gasoline & motor oil: 18 categories of expenditure (total expenditure held constant)

Category	50% decrease In price	75% increase In price
Food at home	−0.23	−0.08
Food away	0.03	0.01
Alcoholic beverages	−0.07	−0.03
Housing (less telephone & water)	−0.22	−0.08
Telephone	−0.13	−0.05
Water	−0.23	−0.08
Apparel	0.13	0.05
Transportation (less gasoline & oil)	0.24	0.11
Gasoline & motor oil	0.09	−0.07
Health	−0.3	−0.1
Entertainment	0.14	0.06
Personal care	−0.06	−0.02
Reading	−0.25	−0.09
Education	0.48	0.24
Tobacco	−0.24	−0.08
Miscellaneous	0.02	0.01
Contribution	−0.05	−0.02
Personal insurance	0.11	0.04

Source Author's computation

Table 2.11 Simulated own- and cross-price elasticities with respect to 50% decrease/75% increase in price of gasoline & oil (18 categories of expenditure)

Iteration	Change in expenditures	Elasticity
1	−293.91	−0.0253
2	−279.11	−0.0744
3	−307.58	0.0200
4	−309.75	0.0273
5	−315.19	0.0453
6	−317.45	0.0528
7	−319.69	0.0602
8	−321.32	0.0656
9	−322.71	0.0702
10	−323.83	0.0739
.	.	.
.	.	.
∞	−328.90	0.0908

Source Author's computation

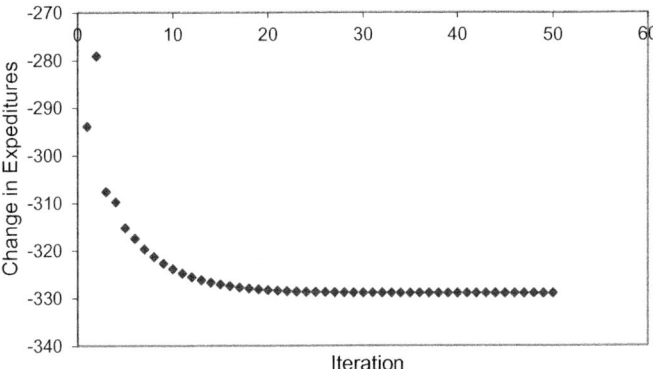

Fig. 2.1 Iterative paths for gasoline & motor oil expenditures for 50% increase in price of gasoline & oil (*Source* Authors' computation)

where y^0 denotes the vector of base expenditures, k denotes iterations, and w^k denotes the factor applied to Ay^{k-1} to impose the budget constraint.[20] Results for a 50% increase in the price of gasoline & motor oil are given for the first 10 iterations in Table 2.9 and for 30 iterations in Fig. 2.1. Note that convergence effectively occurs by the 26th iteration. Note, too, that the own-elasticity is initially negative, but switches to positive after two iterations because of feedbacks from other categories.

Since elasticities adduced in this framework are not independent of either levels of expenditure or direction or of size of price changes, the final exercise in this section is to compare simulations for price changes of ±10% in the price of gas and oil with those from price changes of ±50%. The results, for the simulated elasticities only, are listed in Table 2.11. The values are seen to be generally similar except for the gasoline & motor oil own-elasticity for a 50% price decrease (Table 2.12).

---

[20]Specifically, w^k is calculated as Ay^0/Ay^{k-1}. Discussion and details of this as well as other methods of taking the budget constraint into account can be found in Chapter 4 of *Internal Structure*.

Table 2.12 Estimated own- and cross-price elasticities with respect to ±10% and ±50% changes in price of gasoline & motor oil: 18 categories of expenditure

Category	Price decrease		Price increase	
	10%	50%	10%	50%
Food at home	−0.18	−0.23	−0.17	−0.14
Food away	0.03	0.03	0.02	0.02
Alcoholic beverages	−0.05	−0.07	−0.05	−0.04
Housing (less telephone)	−0.17	−0.22	−0.16	−0.13
Telephone	−0.10	−0.13	−0.09	−0.08
Water	−0.18	−0.23	−0.16	−0.13
Apparel	0.10	0.13	0.09	0.08
Transportation (less gas & oil)	0.19	0.24	0.17	0.14
Gasoline & motor oil	−0.15	0.09	−0.23	−0.36
Health	−0.24	−0.30	−0.21	−0.18
Entertainment	0.11	0.14	0.10	0.09
Personal care	−0.04	−0.06	−0.04	−0.03
Reading	−0.19	−0.25	−0.17	−0.15
Education	0.38	0.48	0.34	0.28
Tobacco	−0.19	−0.24	−0.17	−0.14
Miscellaneous	0.02	0.02	0.01	0.01
Contributions	−0.04	−0.05	−0.03	−0.03
Personal insurance	0.09	0.11	0.08	0.06
Changes in real income	$20.33	$301.53	−$20.33	−$301.53

Source Author's computation

SUMMARY AND CONCLUSIONS

The primary purpose of the present communication has been to demonstrate how a framework developed from data collected in surveys of households' consumer expenditures can be used to calculate complete arrays of own- and cross-price elasticities.[21] The procedure is illustrated

---

[21] Since the focus in the exercises of this paper is primarily on methodology and demonstration, a comparison of the elasticities that have been obtained with those found in the literature is not undertaken. However, for those who might be interested in such, see for telecommunications, amongst others, Crandall and Waverman (1995) and the many studies cited and reviewed in my 1994 book, *Telecommunications Demand in Theory and Practice*. For gasoline & oil, see Dahl (1982), Epsey (1996), Houthakker et al. (1974), Hughes et al. (2008), Schmalensee and Stoker (1999), Puller and Greening (1999), and Taylor and Houthakker (2010).

for simulated changes of ±10% in the price of telecommunications and for ±50% and −50%/+75% changes in the price of gasoline & motor oil. The engine for the analysis, whose construction is described in detail in Taylor (2013), is a matrix of "intra-budget" coefficients (A) that represent the direct relationships amongst the different categories of expenditure in households' budgets. In the present exercise, the elements of the matrix A, for 18 categories of expenditure, are constructed from the information in 28 quarters of data (2006 through 2012) from the ongoing BLS Survey of Consumer Expenditure, and are defined as the mean values over the 28 quarters of data.

The parameters in A, which can be seen as reflecting the conjunctive effects on household expenditures of household tastes and preferences together with the constraints imposed by prices and income, are defined as mean values over the 28 quarters of coefficients estimated from sequences of regression equations in which expenditures for each category of expenditure are related to the expenditures for each of the other 17 categories. Price elasticities can then be calculated via simulations, with total expenditure held constant, in which "exogenous" expenditures (in the present context) for telecommunications and gasoline & motor oil are altered by amounts that correspond to the postulated price changes. Since off-diagonal elements in the A matrix are all pretty much non-zero, the procedure accordingly allows not only for calculation of own-price elasticities but for cross-price elasticities as well. Moreover, since calculations proceed from the total-expenditure constraint, the elasticities thus calculated satisfy a modified form of the Hicks-Allen-Schultz condition that own- and cross-price elasticities sum to -1 with *money* total expenditure held constant and to 0 with *real* total expenditure held constant.

A strength of calculations in this framework is that price changes can be translated immediately into real-income effects, which in turn allows for straightforward separation of income and substitution effects. Indeed, the only conceptual difference between income and substitution effects in this framework and in the traditional Hicks-Allen setting is that compensation for the real-income effect is with respect to the budget constraint rather than with respect to the original indifference surface. Also, an interesting empirical result of the present exercise is the appearance of apparent "Giffen" goods in unexpected places, that is, of *positive* own-price elasticities (as in Table 2.8 for gasoline & motor oil). In the conventional theory of consumer behavior, Giffen goods can arise only in the context of inferior goods (i.e., goods that have a negative income elasticity) which account for a significant share of total expenditure. By this criteria, one would accordingly not expect expenditures for gasoline & oil to qualify.[22] However, when feedbacks from the effects on other categories of expenditure (as represented in the off-diagonal terms of the A matrix) are taken into account, own-price elasticities that, *ceteris paribus*, are negative (cf. Tables 2.12) can switch to positive.

Acknowledgments Dedicated to Gary Madden, telecom and IT researcher extraordinaire. I am grateful to Dennis Cory and Timothy Tardiff for comments and criticisms. Problems that might remain are of course my responsibility.

APPENDICES

Appendix 1

Definitions of Expenditures in BLS Survey of Consumer Expenditure: 14 Categories of Expenditure.

Food includes expenditures on food at home referring to expenses for grocery stores and food by the consumer on trips; food away from home accounting for all meals at fast food, take-out, delivery, concession stands, buffet and cafeteria, full-service restaurants, vending machines, and mobile vendors; and other miscellaneous venues.

Alcoholic beverages include beer and ale, wine, whiskey, gin, vodka, rum, and other alcoholic beverages.

Housing includes expenditures on owned dwellings like interest on mortgages, property taxes, and repairs and maintenance; rented dwellings like rent and maintenance, utilities like natural gas and electricity; fuels like fuel oil and coal; public services like water, garbage, and telephone; housekeeping supplies; household textiles; furniture; floor coverings; major and small appliance; and other miscellaneous equipment purchases.

---

[22] However, in a 2005 article in *The Nation*, Sasha Abramsky conjectures that gasoline, in certain circumstances, may indeed be a Giffen good. Similarly, Bopp (1983) suggest the same for kerosene (a low-quality fuel used for heating) and *shochu* (a Japanese distilled beverage). See also Jensen and Miller (2008) and Kagel et al. (1995).

Apparel includes expenditures on men's and boy's apparel; women's and girl's apparel; apparel for children under 2; footwear; and other apparel products and services.

Transportation includes expenditures for vehicle purchases; vehicle finance charges; gasoline and motor oil; maintenance and repairs; vehicle insurance; public transportation; and vehicle rental, leases, licenses, and other charges.

Health Care includes expenditures on health insurance; medical services like hospital services and physicians' services; eye and dental care; lab tests and X-rays, convalescent and nursing home care; medical appliances and equipment; and both non-prescription and prescription drugs.

Entertainment includes expenditures on fees and admissions; television, radio, and sound equipment; pets, toys, hobbies, and playground equipment; and other miscellaneous entertainment equipment and services like bicycles, hunting and fishing equipment, boats, photographic equipment and supplies, fireworks, electronic video games, etc.

Personal care products and services includes products for the hair, oral hygiene products, shaving needs, cosmetics and bath products, electric personal care appliances, other personal care products, and personal care services for males and females.

Reading includes subscriptions for newspapers and magazines; books through book clubs; and the purchase of single-copy newspapers, magazines, newsletters, books, and encyclopedias and other reference books.

Education includes tuition; fees; and textbooks, supplies, and equipment for public and private nursery schools, elementary and high schools, colleges and universities, and other schools.

Tobacco products and smoking supplies includes cigarettes, cigars, snuff, loose smoking tobacco, chewing tobacco, and smoking accessories (such as cigarette or cigar holder pipis flint and pipe cleaner).

Miscellaneous includes safety deposit box rental, checking account fees and other bank service charges, credit card memberships, legal fees, accounting fees, funerals, cemetery lots, union dues, occupational expenses, expenses for other properties, and finance charges other than those for mortgages and vehicles.

Cash contributions includes cash contributed to persons or organizations outside the consumer unit, including alimony and child support payments; care of students away from home; and contributions to religious, educational, charitable, or political organizations.

Personal insurance includes expenditures on life insurance; endowments; mortgage insurance; and other premiums for personal liability, accident, and disability, and other non-health insurance other than for homes and vehicles. (Source: U.S. Bureau of Labor Statistics, available at http://www.bls.gov/cex/csxgloss.htm.)

Appendix 2

When the large discrepancy (in absolute value) in the change in expenditures for gasoline & motor oil for all households for the ±50% price changes was first encountered (−$329 vs. $194) in Table 2.7, my reaction was that a coding error almost certainly had to be involved. However, this was not the case, for the difference in fact arises from the manner in which the budget constraint is imposed, specifically, in simulations using Expression (2.3), whether the budget constraint is imposed as in Expression (2.18), or as in:

$$y^k = w^k z + Ay^{k-1}, \qquad (2.19)$$

[where (as before) $w^k = \frac{y^0}{y^{k-1}}$], or as in:

$$y* = w * \hat{y}, \qquad (2.20)$$

where

$$\hat{y} = \lim(z + Ay^{k-1}), k \to \infty \qquad (2.21)$$

and where

$$w^* = \frac{y^0}{\hat{y}}. \qquad (2.22)$$

In the event, it turns out that y^* in Expression (2.20) is equal to the solution vector y from Expression (2.5)—as in Table 2.7 for the ±50% price changes. That this is the case can be seen numerically in Tables 2.13A and 2.13B. The numbers appearing in columns 1 and 3 of this show changes in gasoline & motor oil expenditures from the first 10 iterations of Expression (2.18) for a 50% decrease in the gasoil price

Table 2.13A Means of intra-budget expenditures regression coefficients: 18 categories of U.S. Consumption expenditure (2006 Q1–2012 Q4)

Coefficients

Category	Intercept	Fdhome	Fdaway	Alcbev	Housl	Telephone	Water	Apparel	Transl	Gasoil
Fdhome	594.35	−1.00000	0.04292	−0.00036	0.47908	0.03539	0.01738	0.03879	0.00609	0.12742
Fdaway	46.24	0.04263	−1.00000	0.05315	0.39899	0.01181	0.00143	0.06814	0.22669	0.05607
Alcbev	8.24	−0.00419	0.67182	−1.00000	0.97922	−0.03545	−0.00793	0.26937	0.41628	−0.01273
Housl	956.77	0.02719	0.02270	0.00426	−1.00000	0.00536	0.00484	0.02295	0.06642	0.00005
Telephone	119.61	0.42095	0.14074	−0.03250	1.12980	−1.00000	0.05424	0.10980	0.72730	0.51438
Water	33.92	0.48900	0.04478	−0.01904	2.33983	0.12815	−1.00000	−0.02772	0.43709	0.18261
Apparel	−77.80	0.06201	0.10832	0.03010	0.57842	0.01367	−0.00221	−1.00000	0.31280	0.01647
Transl	−29.99	0.00002	0.00658	0.00095	0.03347	0.00173	0.00043	0.00639	−1.00000	0.00816
Gasoil	147.99	0.28188	0.11712	−0.00146	0.01177	0.09365	0.01430	0.02739	0.63247	−1.00000
Health	288.83	0.03041	0.01053	−0.00486	0.09935	0.00493	0.01028	−0.00398	0.05087	0.01232
Entertn	−84.98	0.01730	0.05857	0.00974	0.21858	0.00216	0.00330	0.04053	0.11321	0.01882
Percare	−0.20	0.20141	0.67904	0.11380	2.69887	0.11970	0.05927	1.02715	0.60225	0.05145
Reading	2.20	0.28771	0.33711	0.19286	2.33316	0.00555	0.05342	0.51252	0.60519	−0.26636
Educ	193.10	0.00740	0.03511	−0.00048	0.09869	0.00527	−0.00098	0.01179	0.02552	0.00456
Tobacco	53.94	0.15538	−0.09249	0.09715	−0.67354	0.04068	−0.00859	−0.01312	0.12407	0.14164
Misc	−17.29	−0.00380	0.00794	0.00378	0.18688	0.00379	0.00243	0.00609	0.04837	0.00570
Contrib	−49.48	−0.00328	0.01809	−0.00137	0.05754	0.00034	0.00159	0.01667	0.02998	0.00008
Perins	−79.49	0.03213	0.03962	0.00953	0.34616	0.00736	0.00379	0.01044	0.09734	0.03471

Source Author's computation

Table 2.13B Means of intra-budget expenditures regression coefficients: 18 categories of U.S. Consumption expenditure (2006 Q1–2012 Q4)

Coefficients

Category	Health	Entertn	Percare	Reading	Educ	Tobacco	Misc	Contrib	Perins
Fdhome	0.06553	0.04461	0.00460	0.00234	0.03847	0.01682	−0.00339	−0.01367	0.16726
Fdaway	0.02513	0.15975	0.01580	0.00288	0.19241	−0.01021	0.01416	0.13856	0.21169
Alcbev	−0.13324	0.36471	0.03317	0.01878	−0.02885	0.12861	0.06338	−0.11166	0.63564
Hous1	0.01251	0.03731	0.00356	0.00103	0.02968	−0.00409	0.01436	0.02567	0.10432
Telephone	0.12772	0.05131	0.03221	0.00057	0.28236	0.05284	0.05834	0.00385	0.45739
Water	0.62456	0.23034	0.03723	0.01151	−0.15363	−0.02318	0.05879	0.29300	0.58569
Apparel	−0.00751	0.17031	0.03470	0.00581	0.06159	−0.00272	0.01702	0.17546	0.11985
Transl	0.00339	0.01036	0.00038	0.00014	0.00384	0.00046	0.00248	0.00443	0.01518
Gasoil	0.05801	0.12123	0.00292	−0.00463	0.03791	0.03310	0.01972	−0.03727	0.39750
Health	−1.00000	0.04055	0.00884	0.00758	0.00636	−0.00069	0.03168	0.11222	0.02929
Entertn	0.03003	−1.00000	0.00406	0.00372	0.01471	0.00162	0.01076	0.05794	0.08247
Percare	0.84631	0.46673	−1.00000	0.04994	−0.05726	−0.09051	0.05331	0.52643	1.12741
Reading	2.12189	1.22056	0.14271	−1.00000	0.68314	−0.03611	0.19916	1.55769	1.59247
Educ	0.00132	0.00920	−0.00001	0.00117	−1.00000	−0.00297	0.00152	0.06642	0.03833
Tobacco	−0.01771	0.02884	−0.01919	−0.00277	−0.12260	−1.00000	0.06639	−0.10604	−0.25428
Misc	0.05015	0.02284	0.00086	0.00122	−0.00165	0.00594	−1.00000	0.17877	0.02812
Contrib	0.03773	0.01061	0.00216	0.00191	0.03491	−0.00164	0.02957	−1.00000	0.04114
Perins	0.01172	0.05033	0.00481	0.00240	0.03283	−0.00512	0.00659	0.02607	−1.00000

Source Author's computation

(column 1) and a 50% increase (column 3). Columns 2 and 4 show the same for simulations using

$$y^k = z + Ay^{k-1}. \qquad (2.23)$$

Convergence values (i.e., −$297.10, −$399.76, etc.) are shown as well.[23] At the bottom of the table, it is seen that the total expenditure that is associated with values of −$399.76 and $399.76 corresponding to Expression (2.18)—in which the budget constraint of $12, 220 is not imposed—are $9062 and $15,378, respectively, yielding values for w^* in Expression (2.22) of 1.3485 (12,220/9062) and 0.7846 (12,220/15,378). When these two factors are applied to the $\hat{y}$s [from Expression (2.21)] in Expression (2.17) to take the budget constraint of $12,220 into account, the result is budget-constrained expenditures for gasoline & motor oil of $274.17 for the 50% price decrease and $786.88 for the 50% price increase.[24] Subtraction of base expenditures of $603.04 from these quantities then yields the −$328.90 and $193.81 of Expression (2.3) (Table 2.14).

Amongst other things, what this exercise illustrates is that (as discussed in detail in Chapter 4 of *Internal Structure*) the way in which a budget constraint is imposed in this framework is a matter of choice. This is both a strength and a weakness, a weakness because different methods lead to different results, but a strength in that the method used can be tailored to the question at hand. The method to be used can turn on whether the question of interest involves dynamics or a steady state. Dynamics can then be identified with iterative solutions involving the structural form per Expression (2.3) and a steady state with reduced-form solutions per Expression (2.5). Dynamics, accordingly, can usefully be associated with the sequential effects on expenditures from a price change under the assumption that total expenditure is constant. This allows for the initial effect on the good whose price has changed

---

[23] Simulations using Expression (2.15) are seen to be near convergence by iteration 10. Expression (2.20), in contrast, requires about 70 iterations for y^k to converge.

[24] From line 9 in Table 2.11.

Table 2.14 Changes in gasoline & motor oil expenditures from ±50% changes in gasoline & motor oil price: first 10 iterations of Expressions (2.15) and (2.18) (all households)

Iteration	Changes in expenditures			
	50% price decrease		50% price increase	
	$y^k = z + w^k y^{k-1}$	$y^k = z + Ay^{k-1}$	$y^k = z + w^k y^{k-1}$	$y^k = z + Ay^{k-1}$
1	−$293.91	−$301.53	$279.75	$301.53
2	−274.02	−301.53	250.08	301.53
3	−295.59	−338.34	269.41	338.34
4	−294.46	−348.79	267.53	348.79
5	−296.9	−360.32	269.18	360.32
6	−296.78	−367.73	269.03	367.73
7	−297.07	−374.00	269.18	374.00
8	−297.06	−378.88	269.17	378.88
9	−297.09	−382.86	269.18	382.86
10	−297.09	−386.07	269.18	386.07
.	.	.	.	.
.	.	.	.	.
.	.	.	.	.
∞	−297.1	−399.76	269.18	399.76

Total expenditure (unconstrained): $9062 $15,376
Total expenditures(constrained): $12,220 $12,220
Change in expenditures (unconstrained): −$328.90 $193.81
Change in expenditures [from Expression (2.3)]: −$328.90 $193.81
Change in real income: $301.53 − $301.53
Source Author's computation

to be isolated from the effects that arise through feedbacks from other goods. Goods that might appear to be Giffen in the long run can thus be seen to behave normally in the short run. The simulations exercises for both telecommunications and gasoline & motor oil, using Expression (2.18), represented in Table 2.12 and Fig. 2.1 provide illustration.

References

Abramsky, S. 2005. Running on Fumes. *The Nation*, October 17, pp. 15–19.
Bopp, A. 1983. The Demand for Kerosene: A Modern Giffen Good. *Applied Economics* 15 (4): 459–467.
Crandall, R.W., and L. Waverman. 1995. *Talk Is Cheap: The Promise of Regulatory Reform in North American Telecommunications*. Washington, DC: Brookings Institution Press.

Dahl, C. 1982. Do Gasoline Demand Elasticities Vary? *Land Economics* 58: 374–382.

Epsey, M. 1996. Explaining the Variation in Elasticity Estimates of Gasoline Demand in the United States: A Meta-Analysis. *The Energy Journal* 17: 49–60.

Houthakker, H.S., P.K. Verleger, and D. Sheehan. 1974. Dynamic Demand Analysis for Gasoline and Electricity. *American Journal of Agricultural Economics* 56: 412–418.

Hughes, J.E., C.R. Knittel, and D. Sperling. 2008. Evidence of a Shift in the Short-Run Price Elasticity of Gasoline Demand. *The Energy Journal* 29 (1): 113–134.

Jensen, R., and N. Miller. 2008. Giffen Behavior and Subsistence Consumption. *American Economic Review* 97 (1): 188–192.

Kagel, J.H., R.C. Battalio, and L. Green. 1995. *Economic Choice Theory: An Experimental Analysis of Animal Behavior*, 25–28. Cambridge: Cambridge University Press.

Puller, S., and L. Greening. 1999. Household Adjustment to Gasoline Price Change: An Analysis Using 9 Years of US Survey Data. *Energy Economics* 21: 37–52.

Schmalensee, R., and T.M. Stoker. 1999. Household Gasoline Demand in the United States. *Econometrica* 67 (3): 645–662.

Taylor, L.D. 1994. *Telecommunications Demand in Theory and Practice*. Dordrecht: Kluwer Academic Publishers.

Taylor, L.D. 2013. *The Internal Structure of U.S. Consumption Expenditures*. New York: Springer-Verlag.

Taylor, L.D. 2017a. Notes on Measurement of Consumer Expenditures (With Special Reference to the On-Going BLS Consumer Expenditure Survey. Department of Economics, University of Arizona, Tucson, AZ, myros@att.net.

Taylor, L.D. 2017b. The Internal Structure of Consumption II. Department of Economics, University of Arizona, Tucson, AZ, myros@att.net.

Taylor, L.D., and H.S. Houthakker. 2010. *Consumer Demand in the United States: Prices, Income, and Consumer Behavior*, 3rd ed. New York: Springer-Verlag.

CHAPTER 3

A Cross-Country Analysis of ICT Diffusion, Economic Growth, and Global Competitiveness

Aniruddha Banerjee, Paul N. Rappoport, and James Alleman

Introduction

Forty years ago, virtually the entire telecommunications sector was state owned, managed, and controlled since its inception.[1] In the mid-1980s, a movement toward privatization, liberalization, and deregulation took hold. Now, the sector has been privatized in most countries and subjected

---

[1] There were some exceptions, namely, the United States and Canada (where the inventor of the telephone started companies) and some Scandinavian systems.

This paper is dedicated to the memory of Professor Gary Madden (1952–2017) who inspired all of us with his academic rigor and energy. Not only was he a prolific researcher, he freely helped the rest of us in our research. Blessed with a wry sense of humor, Gary was a great colleague, mentor, and friend, not to mention a very decent human being. We feel the pain of his untimely passing.

A. Banerjee (✉)
IHS Markit, Inc., Boston, USA

P. N. Rappoport
Temple University, Philadelphia, USA

© The Author(s) 2020
J. Alleman et al. (eds.), *Applied Economics in the Digital Era*,
https://doi.org/10.1007/978-3-030-40601-1_3

to regulatory reform, with major such reform occurring in the late 1990s (Estache et al. 2006). Since then the internet and cellular-mobile industries have advanced significantly. Mobile service has exploded, particularly in the developing world. This has changed the dynamics of the sector dramatically.

This paper uses cross-country analysis of selected countries to update and expand prior research on the contribution of Information and Telecommunications Technology (ICT) to the growth of Gross Domestic Product (GDP). To do so, we follow the framework of Czernich et al. (2011) in our use of instrumental variables and of Farhadi et al. (2012) in our use of the ICT Development Index (IDI) available from the International Telecommunication Union (ITU). Moreover, we enhance the analysis by using more recent data and, in particular, the Global Competitiveness Index (GCI) from the World Economic Forum (WEF). Finally, we rely extensively on the Conference Board's Total Economic Database.

Using data from recent years, we evaluate empirically the impact of ICT on economic growth and vice versa in selected countries. For the 2008–2017 period, we fail to find any evidence of "bidirectional causality," i.e., causality in both directions. This finding contrasts with results obtained in prior research conducted for earlier periods. A central question, therefore, is that if the latest findings in this paper continue to hold in the future, then what happened to change earlier trends?

The paper is organized as follows. The section on "Literature Review" reviews the economic literature on the impact of ICT on economic growth and development. The section on "Data" describes the data used in this study and their sources. The research plan and the analytical methodology used are discussed in the section on "Research Plan and Methodology." All findings of this research are presented in the section on "Estimation/Results." Finally, the section on "Discussion of Findings and Conclusions" discusses the findings, examines remaining questions, and considers directions for future research.

Literature Review

Prior research on ICT's contribution to growth and development is rich and varied. At various times, ICT has been defined to include internet, broadband, fixed, and mobile services.

J. Alleman
University of Colorado, Boulder, USA

ICT and Growth

The determinants of economic growth have been a focus of economic researchers since the beginning of the discipline. Only relatively recently, beginning in the early 1980s, has the contribution of the ICT sector to economic growth become a matter of significant interest.[2] Surveys of ICT and growth that focus on the impact(s) of ICT on a firm's productivity include Brynjolfsson and Yang (1996), Brynjolfsson and Hitt (1995, 2000), Cardona et al. (2013), and Greenstein and McDevitt (2009, 2012). Cardona et al. (2013) noted that "[a] broad consensus attributes a large share of growth to investment in ICT. However, not all countries share the same benefits from ICT." That consensus about ICT contributing to productivity—and hence to growth—is also found in studies of specific industries and in cross-country studies. The generally positive impact of ICT on economic growth is found in both simple and dynamic regression models and in studies of causality between ICT investment and economic growth. Earlier work by Alleman et al. (1991) addressed the research on telephony's impact on economic growth and development and provided a comprehensive review of the work conducted up to 1991. More recently, research on ICT and productivity has been conducted by Madden and Savage (2000), Dutta (2001), Röller and Waverman (2001), Dewan and Kraemer (2000), OECD (2003), Sridhar and Sridhar (2007), Venturini (2009), Koutroumpis (2009), Vu (2011), Farhadi et al. (2012), Cardona et al. (2013), Spiezia (2012), Jung (2014), Bertschek and Niebel (2016), Niebel (2018), and Toader et al. (2018). Recent studies have expanded the analysis from a general view of ICT to that of the relationship between broadband or wireless investment and economic growth. Studies that have focused on broadband include Minges (2016), Crandall et al. (2007), Katz et al. (2009), and Czernich et al. (2009).

Magnitude

Early work on ICT's contribution to growth was relatively modest in its approach, using simple regression models of GDP growth against telephone penetration (in logarithmic transformations) or similar variables. The more recent work has attempted to account for dynamics by

---

[2] Robert M. Solow famously stated that "You can see the computer age everywhere but in the productivity statistics;" this became known as Solow's Paradox (Brynjolfsson 1993). The ICT literature addressed this "Paradox."

specifying lags when testing for magnitude and direction of Granger-causality between ICT and economic growth. Econometric approaches have included the use of instrumental variables (Czernich et al. 2009) and of structural models of the sector (Röller and Waverman 2001). Röller and Waverman's seminal work provided a strong critique of earlier models and constructed instead a micro supply and demand model before addressing macroeconomic impacts. Spiezia (2012) used panels to examine the contribution of different types of ICT investment to economic growth, focusing on 26 industries across 18 OECD countries over the 1995–2007 period. He found that ICT investment contributes between 0.4 and 1.0% to value-added growth per year, a range that includes the estimate of 0.45% in Röller and Waverman (2001). Koutroumpis (2009) found the impact to be 0.25% for the 2002–2007 period. Generally, these studies found a significant positive causal impact of ICT investment on productivity and economic growth. After the advent and diffusion of mobile telephony, Waverman et al. (2005) estimated the impact of mobile telephone service on economic growth, finding that it contributes significantly in low income countries (probably twice as much in developing countries as in developed countries).

Broadband and Mobile

Katz and his coauthors (2009a, b, c; Katz et al. 2009; and Katz and Suter 2009) have conducted several studies on the economic impact of broadband in Latin America and elsewhere. The global, national, and regional studies of the economic impact of broadband are reviewed in Katz (2009a, b, c). More recently, Vu (2011) showed that the marginal effect on economic growth of the penetration of internet users was larger than that of cellular phones, which was, in turn, larger than that of personal computers for the average country.

Jung (2014) conducted a comprehensive review of the literature on the impact of ICT infrastructures (focusing on broadband) on economic growth and, in particular, the ways in which ICT can enhance economic activities as suggested by the literature.

Toader et al. (2018) focused on a more recent period (2000–2017) to determine the impact of ICT on economic growth across 28 European Union countries. A positive and strong effect was found between ICT infrastructure investment and economic growth; however, that effect was dependent on the type of technology examined. Three proxies for ICT were examined: household broadband penetration, penetration of internet access, and mobile subscriptions per 100 people. Mobile penetration was found to have the highest impact on economic growth.

A survey by Bertschek and Niebel (2016) addressed previous research on how broadband availability and adoption impact economic growth, centering around two questions: (1) Does broadband infrastructure investment have a positive effect on economic indicators? and (2) Do the economic impacts associated with wireline and wireless infrastructure differ?

Causality

Several studies have addressed the simultaneity between ICT investment and economic growth. Is ICT both a cause and a consequence of economic growth? The pioneering work of Cronin et al. (1991) found bidirectional causality between telecommunication investment and GDP in the US, using data for the 1958–1988 period.

Similarly, using data from 27 Central and Eastern European countries for the 1990–1995 period, Madden and Savage (1998) found bidirectional causality between telecommunications investment and GDP. In Madden and Savage (2000), analysis of the effects of telecommunication investment on GDP was expanded to 43 countries for the 1975–1990 period.

Dutta (2001) analyzed the causal link between telecommunication infrastructure and economic growth, using data from 1960–1993 for 15 developing and 15 developed countries. He found that a unidirectional effect of telecommunications on growth was more probable than bidirectional causality.

Using data for 92 countries over the 1996–2003 period, Waverman et al. (2005) found that higher mobile penetration led to significant increases in economic growth compared to fixed-line penetration, but bidirectional causality was not present.

Chakraborty and Nandi (2011) examined data from 93 developing countries over the 1985–2007 period and found unidirectional causality between access lines and economic growth in the short run but bidirectional causality in the long run.

Looking forward, the impact of ICT on economic growth appears to be changing as technologies become more diffuse. Mobile communications and mobile broadband now play a more dominant role in our communications environment. A GSMA report, The Mobile Economy (2019), describes the scale of contributions of mobile telephony to economic growth. Supporting research on the positive effects of mobile penetration on economic growth appears in Waverman and Koutroumpis (2011), Ward and Zheng (2016), and Bahrini and Qaffas (2019). From a historical perspective, economic historian Joel Mokyr (2005) put the relationship between economic growth and technology (alternately, ICT) in perspective. "The experience of the past two centuries in the western world supports the view that useful knowledge and its application to production went through a phase transition in which it entered a critical region in which equilibrium concepts may no longer apply. This means that as far as future technological progress and economic growth are concerned, not even the sky is the limit."

Contribution of This Paper

In most cases, the studies cited above do not cover the entire ICT sector. One of the contributions of this paper is our use of IDI as a proxy for investment in the ICT sector. Farhadi et al. (2012) is the only study to use this variable and is closest to this paper. It applies Generalized Method of Moments (GMM) estimation to a dynamic panel of 159 countries from 2000–2009. Our paper takes a similar approach for a more recent period (2008–2017) and enhances the model by including various macroeconomic and market-level control variables that fall within the aegis of the GCI. In addition, we test for Granger-causality.

The paper also explores how the impact of ICT is likely changing over time. The traditional view of ICT concerned mainly fixed-line telecommunications investment and, only recently, began including broadband and mobile technologies. IDI is a good example of a widely used ICT index.[3] It gives equal weight to fixed telephone line subscriptions, mobile subscriptions, and internet penetration (subscriptions). This paper considers alternative definitions of an ICT index, one that focuses more heavily on emerging technologies such as mobile broadband. Does a mobile-based ICT measure exhibit simultaneity or bidirectional causality with economic growth? This is an important issue given the emergence of 5G mobile technology.

---

[3] See International Telecommunication Union (2018).

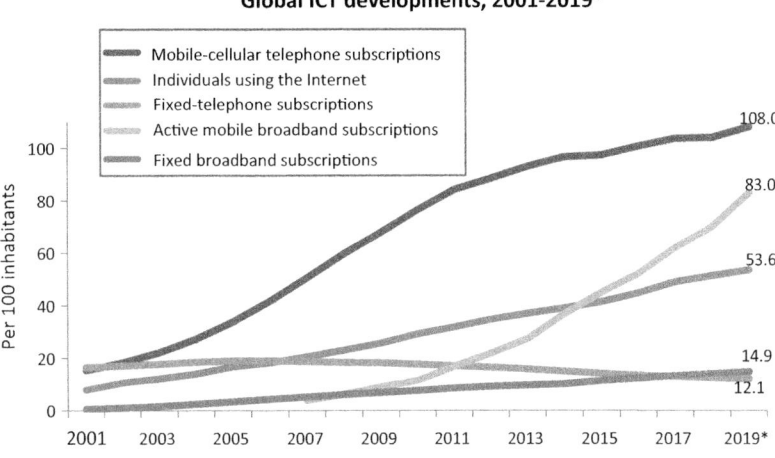

Fig. 3.1 Information and Communications Technology (ICT) per 100 inhabitants (*Source* ITU World Telecommunications/ICT Indicators Database, http://www.itu.int/ict/statistics)

ICT Environment

A summary of the ICT environments is shown in Fig. 3.1. Traditional fixed-line telephone service has declined while mobile service has increased substantially. Indeed, mobile service appears to have emerged as a substitute for fixed-line service, although the latter does offer the opportunity to provide internet, even broadband, services. Even in developing countries, individuals are "cutting the cord," i.e., giving up their fixed-line service in favor of mobile service (Banerjee et al. 2014). Figure 3.1 shows that the growth of mobile telephone service has been spectacular. Also, while the pattern for internet penetration is not as dramatic as that for mobile service, it has made considerable progress as well over the last dozen years or so.

The pattern is different for fixed-line telephone service. Penetration rates have decreased over time in many countries. Undoubtedly, this is due to the substitution of mobile for fixed-line services.

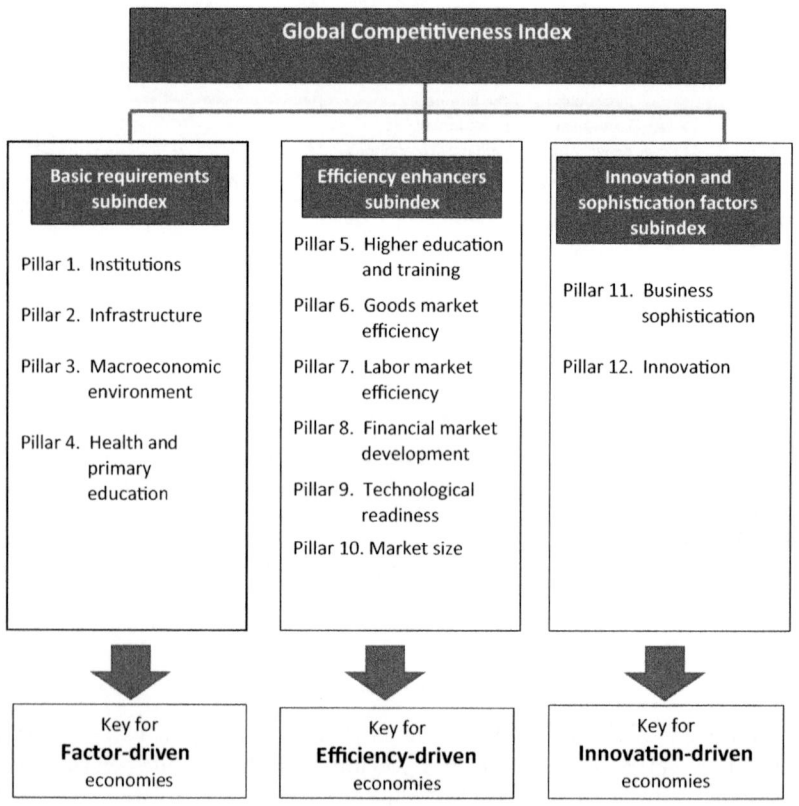

Fig. 3.2 Global Competitiveness Index (*Source* World Economic Forum Global Competitive Report)

Data

For this study, data on GCI, IDI, and GDP per capita growth are assembled from multiple sources. GCI (available from the WEF) is a composite of "The Twelve Pillars of Competitiveness," each of which is constructed from underlying indicators. These indicators, as shown in Fig. 3.2, reflect various facets of the overall economy that can be placed into three groups: (1) institutional and infrastructural indicators, (2) efficiency enhancers, and (3) innovation indicators. Subindices formed from

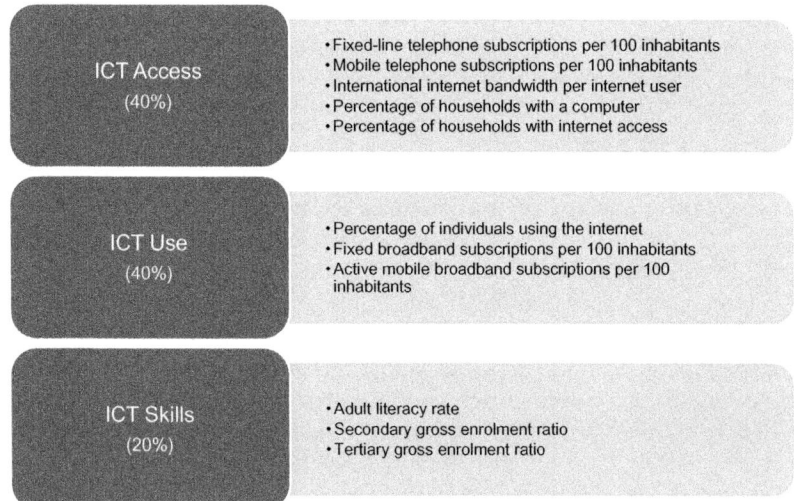

Fig. 3.3 ICT Development Index (IDI) (*Source* ITU, https://www.itu.int/en/ITU-D/Statistics/Pages/publications/mis2017/methodology.aspx)

these three groups of indicators are aggregated to produce the composite GCI. In the panel data analysis conducted for this study, GCI is used as a control measure.

IDI (available from the ITU) is a weighted composite of ICT readiness (infrastructure, access), ICT use (intensity), and ICT impact/skills (outcomes), as shown in Fig. 3.3. Subindices formed for the three sets of ICT indicators are aggregated using weights of 40, 40, and 20%, respectively. IDI is not, however, a measure of ICT investment per se, which many other studies have used to investigate the ICT–economic growth relationship. ICT investment is not easy to measure and, in many cases, has to be estimated from underlying data (see Vu 2011). It is also simply not available for developing or less affluent countries. By using IDI in this study, the focus is more on ICT development than ICT investment per se. But, as will become evident later, we also investigate how some specific components of IDI (mobile internet penetration, fixed broadband penetration, and percentage of households with broadband access) cause, or are caused by, economic growth.

We use GDP per capita at purchasing power parity (available from the Output Conference Board) to construct our measure of economic growth. This serves as a better metric than GDP per se because of obvious scale effects in the latter. The percent change in GDP per capita reflects annual economic growth.

Our final panel data sample consists of 107 countries for which annual observations on all indices and variables are available for a recent 10-year period (2008–2017). These countries represent all four income groups used by the World Bank to classify countries by income status, namely, high income, upper middle income, lower middle income, and low income (see World Development Indicators 2012). In this sample, 46 are high income countries, 27 are upper middle income countries, 25 are lower middle income countries, and nine are low income countries.

There is one potential issue with trying to isolate a relationship between GDP per capita growth and IDI: the early years in our sample span the period just after the Great Recession (2007–2009) and growth rates show up as persistently negative for several years in many countries. Thus, a part of what happened to economic growth may have less to do with ICT and more with other shocks. This is clearly an empirical issue.

Research Plan and Methodology

We examine the impact of ICT on economic growth in selected countries following the approach in Farhadi et al. (2012). For this study, we use data from 107 selected countries over the 2008–2017 period. As noted in the section on "Literature Review," previous research has shown that ICT (most recently, in the form of mobile phone and broadband services) has had an impact on economic growth, as did fixed-line service in an earlier period.

Research Questions

In this study, we examine two specific questions.

- Whatever the pattern of causality between ICT and economic growth in the past, has that pattern changed over time, especially after the Great Recession which ended in 2009?

- Has the rapid leapfrogging of fixed-line networks by mobile telephony and internet access changed the nature of ICT development itself and, therefore, any causal relationship with economic growth?

As indicated earlier, both broadband internet and mobile telephone services have grown spectacularly in the last decade. Privatizations have had time to settle and regulation has had time to mature. Thus, it is appropriate to examine the combined impact of broadband and mobile services as the primary components of ICT.

This is only the second study to use IDI to represent the state of ICT,[4] rather than the more traditional "ICT investment" variable which is constructed in various ways by different researchers. IDI is broader than any of the traditional investment measures. Apart from indicators of infrastructure access and usage (which the traditional measures also try to capture), IDI incorporates, as well, internet use skills which, arguably, capture the potential of a population to use ICT for GDP-affecting economic activities.

Given the increasing ease in leapfrogging older technology and the reduced costs of entering a market to provide ICT services, that market potential for ICT use is the key measure here. If more people use broadband services (fixed or mobile) then that implies that new value-added services may be in play. That is, the supply of ICT itself creates demand for those downstream end-use services and affects GDP in the process. Of course, this can also be disruptive and have negative impacts on GDP, such as when online shopping/e-commerce services enabled by broadband and mobile telephony end up substituting for (or even replacing) brick-and-mortar businesses. However, a well-functioning mobile network provides signals to—and can stimulate—those who are thinking about other investments.

The primary hypothesis tested in our study is that any causality between GDP growth and ICT development is likely a two-way street, i.e., bidirectional, rather than solely from one to the other. The literature is replete with such tests, but there appears to be little consistency in the pattern of findings. Much depends on the types and numbers of countries used in sample data, the time periods studied (because leapfrogging next-generation technologies are emerging with increasing frequency),

---

[4] Farhadi et al. (2012) was the first to do so, to the best of our knowledge.

and whether controls are used in models to account for other factors that affect the environment in which the GDP–ICT relationship arises. In this study, rather than relying on unobserved panel fixed effects as those controls, the GCI is explicitly used as a control to isolate the GDP–IDI relationship.

If GDP is a proxy for income (and is the source of demand), then we would expect GDP to influence ICT usage. However, as has been shown in past research, the high cost to build out telecommunications or ICT networks has had to be overcome by sufficient GDP growth in order for ICT development to occur. This potential two-way relationship prompts our investigation in this study, using both more recent and more comprehensive data.

Model Specification

Given the need to test for Granger-causality between our measure of ICT development and economic growth, we choose a dynamic panel data model (with fixed effects) specification. As explained below, the two main variables of interest (IDI and GDP per capita) are used in two variants of the model, once with IDI as the dependent variable and GDP per capita as the independent variable, and once with the order of the two variables reversed. In both variants of the model, GCI enters as an independent variable, intended to be a control for all other factors that can otherwise limit our ability to isolate the relationship between IDI and GDP per capita.

The actual variables used in our study are:

- Ln(GDPPC), natural log of real GDP per capita
- Ln(IDI), natural log of IDI
- Ln(GCI), natural log of GCI.

We also consider and test models in which IDI is replaced by alternative measures of the ICT variable, particularly those representing mobile technologies and fixed broadband, These are:

- PMI, penetration rate of mobile internet service
- PFB, penetration rate of fixed broadband service
- PctBA, percentage of households with any broadband access

Granger-Causality Test in Dynamic Panel Data Fixed Effects Model

Consider the dynamic panel data model with fixed effects

$$y_{it} = \alpha_0 + \sum_{k=1}^{m} \alpha_k y_{i,t-k} + \sum_{k=0}^{m} \delta_k x_{i,t-k} + \mu_i + \varepsilon_{it} \quad i = 1, 2, \ldots, N, t = m+1, \ldots, T$$

(time-specific terms omitted)

In this model, y and x are the dependent variable and independent variable, respectively. μ_i are cross-section-specific (here, country-specific) fixed effects that account for unobserved independent influences on y. ε_{it} is a random error term that is independently and identically distributed (IID) with zero mean and finite variance. i and t are indexes for cross-sectional units (from 1 to N) and time (from m to T), where m is the maximum lag (indexed by k) on both the lagged dependent variable and independent variable. α_k and δ_k are coefficients of the lag terms for y and x, respectively.

A Granger-causality test is set up by the null hypothesis.

H_0: x does not Granger-cause y if all δ_k are jointly zero.

A parallel test of the null hypothesis H_0: y does not Granger-cause x is conducted in a similar manner by reversing x and y in the above equation and testing whether the coefficients of all lagged y terms are jointly zero. Following such tests,

- Unidirectional Granger-causality is established if one of the null hypotheses is rejected but the other is not.
- Bidirectional Granger-causality is established if both null hypotheses are rejected.
- Contemporaneous Granger-causality is established if the coefficient at the zero lag is statistically significant.

Consistent Estimation of the Dynamic Panel Data Fixed Effects Model

Inclusion of lagged y and x terms makes the specified regression model dynamic. In testing for Granger-causality, it is customary to specify k (number of lags) generously subject to data availability (at the risk of over-specifying) and then trimming the lag structure as needed.

For estimation of the model, we choose the Arellano and Bond (1991) procedure based on a GMM estimator. This estimator was proposed specifically for dynamic panel data models with lagged dependent variables.

In this model, by construction, the fixed effects μ_i are correlated with lagged dependent terms, so they are first "swept out" by first-differencing the equation. As a consequence, this introduces serial correlation in the differenced errors $\varepsilon_{it} - \varepsilon_{i,t-1}$. More seriously, this introduces correlation between the differenced lagged dependent variable and the differenced error term, thus causing a simultaneous equations bias.

Arellano and Bond (1991) tackled this problem by using a GMM estimator that is akin to an instrumental variable estimator. To consistently estimate the dynamic panel data model with lagged dependent variables, Arellano and Bond suggested instruments that are both levels and differences of various lags of the dependent variable.[5]

While in traditional panel data estimation, taking deeper lags means the loss of some observations, the Arellano–Bond GMM estimator compensates for this by increasing the number of moment restrictions on the lags of the dependent variable as the number of time periods increases.

The Arellano–Bond GMM estimator of dynamic linear panel data models to test for Granger-causality is well-suited to circumstances in which the number of units (here, countries) is large but there are relatively few time periods (here, 2008–2017). Usually, it is assumed that the error term is IID, so first-differencing to eliminate the fixed effects causes the differenced errors to become serially correlated. Arellano and Bond (1991) proposed a test for autocorrelation of orders 1 and 2. If the test finds autocorrelation of order 1 but none of order 2, then the model has been properly specified and estimated. Sometimes, differencing may induce the error terms to behave like a low-order moving average process, in which case Arellano–Bond suggested an alternative set of instruments to use at the estimation step.[6]

Finally, the initial hypothesis is of homoscedastic errors, but clustering by countries (e.g., by different income groups) can cause heteroskedasticity to arise and render the estimators inefficient. For this reason, robust heteroskedastic-consistent standard errors should be used.

---

[5] Anderson and Hsiao (1981) were the first to propose an instrumental variable estimator in this circumstance, but their estimator proved to be asymptotically inefficient.

[6] Holtz-Eakin et al. (1988) formulated an early prototype of the Arellano–Bond estimator, and that estimator was used specifically for Granger-causality testing in panel data by Nair-Reichert and Weinhold (2000).

ESTIMATION/RESULTS

We conduct our Granger-causality tests using an unbalanced panel of 107 countries observed over 10 years (2008–2017). In this sample, the panels (countries) contain varying numbers of observations over time (6.8 per country on average). Also, 73 of those countries (roughly two-thirds of the full sample) are classified as "more affluent countries," i.e., high income and upper middle income countries. The remaining 34 countries (roughly a third of the full sample) are classified as "less affluent countries," i.e., lower middle income and low income countries.

Relationship Between Economic Growth and IDI

Table 3.1 shows several models in which Ln(GDPPC) is regressed on Ln(IDI) and Ln(GCI) (the control). Different lags are used in accordance with the GMM procedure. Results of the Arellano–Bond autocorrelation test are also shown. Separate models are estimated for all countries, more affluent countries, and less affluent countries. The purpose of this regression is to test for Granger-causality from IDI to GDP per capita.

Similarly, Table 3.2 shows several models with the reverse regression of Ln(IDI) on Ln(GDPPC) and Ln(GCI). These too are estimated separately for all countries, more affluent countries, and less affluent countries. The purpose of this regression is to test for Granger-causality in the opposite direction, i.e., from GDP per capita to IDI.

The regression results in Tables 3.1 and 3.2 can be summarized in the form of specific (Yes or No) Granger-causality findings in both directions between IDI and GDP per capita.

Table 3.3 shows that with IDI representing ICT development, there is no statistical evidence of Granger-causality from IDI to GDP per capita in the full sample (all countries) or in the sample with more affluent countries. This finding for the 2008–2017 period (i.e., mainly after the Great Recession) is in sharp contrast to those from studies from earlier periods, which consistently found at least unidirectional causality from ICT investment to GDP per capita. This raises two possibilities. First, IDI, which is a composite index that measures and combines different communications technologies and their use, may be "doing too much." A more direct measure of ICT investment or development may yield better insights. Second, IDI may be exaggerating the importance

Table 3.1 Test of Granger-causality from IDI to GDP per capita, 2008–2017

Dep Var: Ln(GDPPC)	All countries			More affluent countries			Less affluent countries		
	Coeff	Std Err	Sig	Coeff	Std Err	Sig	Coeff	Std Err	Sig
Ln(GDPPC) Lag 1	0.9921	0.1708	***	0.9575	0.2078	***	0.8800	0.1203	***
Ln(GDPPC) Lag 2	−0.1374	0.0983		−0.1362	0.1070		−0.0670	0.1087	
Ln(IDI)	0.0022	0.0030		0.0010	0.0042		−0.0078	0.0032	**
Ln(IDI) Lag 1	−0.0035	0.0029		0.0004	0.0031		−0.0127	0.0037	***
Ln(IDI) Lag 2	−0.0019	0.0035		0.0028	0.0040		−0.0144	0.0039	***
Ln(IDI) Lag 3	−0.0007	0.0036		0.0061	0.0039		−0.0177	0.0043	***
Ln(GCI)	0.0454	0.0542		0.1241	0.0872		0.0198	0.0628	
Ln(GCI) Lag 1	−0.0395	0.0128	***	−0.0368	0.0155	**	−0.0238	0.0166	
Ln(GCI) Lag 2	0.0191	0.0129		0.0117	0.0145		0.0211	0.0265	
Year	0.0047	0.0027	*	0.0048	0.0033		0.0101	0.0027	***
Wald chi-squared	4581.19		***	4359.10		***	3369.27		***
Observations	618			428			190		
Groups	107			73			34		
Instruments	41			41			41		
AR(1)—z statistic	−2.697		***	−2.138		**	−2.310		**
AR(2)—z statistic	−0.394			0.144			−0.668		

Source Authors' computation

Note ***significant at 1% level; **significant at 5% level; *significant at 10% level

Table 3.2 Test of Granger-causality from GDP per capita to IDI, 2008–2017

Dep Var: Ln(IDI)	All countries			More affluent countries			Less affluent countries		
	Coeff	Std Err	Sig	Coeff	Std Err	Sig	Coeff	Std Err	Sig
Ln(IDI) Lag 1	−0.0084	0.0305		−0.0363	0.0448		0.0940	0.0343	***
Ln(IDI) Lag 2	−0.6548	0.0292	***	−0.5877	0.0419	***	−0.6926	0.0452	***
Ln(GDPPC)	0.0942	0.5414		−0.1730	0.5362		−1.1317	0.8481	*
Ln(GDPPC) Lag 1	0.4899	0.4502		0.3769	0.4944		1.5314	0.9228	
Ln(GDPPC) Lag 2	−0.3149	0.3992		0.1821	0.4644		−0.7144	0.7184	
Ln(GCI)	0.2437	0.1633		0.7426	0.2155	***	−0.5545	0.3329	*
Ln(GCI) Lag 1	0.1011	0.1028		−0.1375	0.0980		0.7916	0.2472	***
Ln(GCI) Lag 2	0.0925	0.1494		−0.2208	0.1889		0.3050	0.2059	
Year	0.0478	0.0121	***	0.0548	0.0147	***	0.0616	0.0171	***
Wald chi-squared	1051.46		***	1209.53		***	832.18		***
Observations	723			500			223		
Groups	107			73			34		
Instruments	42			42			42		
AR(1)—z statistic	−5.715		***	−4.610		***	−3.572		***
AR(2)—z statistic	0.302			−0.681			0.951		

Source Authors' computation
Note ***significant at 1% level; **significant at 5% level; *significant at 10% level

Table 3.3 Summary of Granger-causality findings (IDI to GDP per capita), 2008–2017

Lag	All countries	More affluent countries	Less affluent countries
0	No	No	Yes
1	No	No	Yes
2	No	No	Yes
3	No	No	Yes
Reject null hypothesis that all lag coefficients are jointly zero? (Wald test)			
	No	No	Yes

Source Authors' computation

of "legacy" trends (such as in the subscriptions to fixed-line telephony) at a time when mobile/wireless communications had become far more prominent.[7] Perhaps for some of these reasons, the IDI was revamped substantially in 2018.[8] However, because our sample period ends in 2017, the revision to the IDI does not affect our study.

For less affluent countries, however, a Granger-causal impact is found from IDI to GDP per capita, with coefficients of current and lagged IDI that are highly statistically significant. At the same time, those coefficients imply a generally negative impact of IDI on GDP per capita. This finding is hard to interpret, particularly if the belief is that ICT development boosts economic growth.[9] It compels us (see below) to consider alternatives to IDI for representing the ICT variable.

Results of Granger-causality testing in the opposite direction, i.e., from GDP per capita to IDI, are summarized in Table 3.4. Once again, there is no clear evidence of any such causality for more affluent countries or, indeed, for all countries taken together. However, for less affluent countries, a causal effect at lag 1 of GDP per capita does appear to exist (at the 10% level of significance). That effect is positive, suggesting

---

[7] As Fig. 3.1 shows, the post-2008 period is marked by rapid acceleration of mobile telephony and broadband, a more moderate acceleration of fixed broadband, and a steady decline in fixed-line telephony. In many ways, this is a very different period for communications technologies and their penetration than the earlier periods for which most causality tests have been performed in the past.

[8] See International Telecommunication Union (2018).

[9] See Farhadi et al. (2012) for another instance of such a finding.

that less affluent countries experience a boost in ICT development because of GDP per capita growth, but with a year's lag.

To summarize, for the period studied, we are unable to establish statistical evidence of Granger-causality in either direction between GDP per capita and ICT development in more affluent countries. This finding is also true for the full sample of 107 countries, but that result is likely driven by the fact that more than two-thirds of those countries fall into the "more affluent" category. However, we do find weak bidirectional causality in less affluent countries (both contemporaneous and lagged from IDI to GDP per capita, and lagged from GDP per capita to IDI).

Relationship Between GDP Per Capita and Alternative Measures of ICT

We noted earlier that IDI is a composite measure of ICT access, use, and skills. There is some question of how clearly any causal relationship can be established between such a composite index and economic

Table 3.4 Summary of Granger-causality findings (GDP per capita to IDI), 2008–2017

Lag	All countries	More affluent countries	Less affluent countries
0	No	No	No
1	No	No	Yes
2	No	No	No
Reject null hypothesis that all lag coefficients are jointly zero? (Wald test)			
	No	No	Yes

Source Authors' computation

Table 3.5 Pairwise Pearson correlation among mobile internet penetration, fixed broadband penetration, and percentage of households with broadband access

	More affluent countries			Less affluent countries		
	$Ln(PMI)$	$Ln(PFB)$	$Ln(PctBA)$	$Ln(PMI)$	$Ln(PFB)$	$Ln(PctBA)$
$Ln(PMI)$	1.0000			1.0000		
$Ln(PFB)$	0.4251	1.0000		0.3719	1.0000	
$Ln(PctBA)$	0.7082	0.8212	1.0000	0.6667	0.7222	1.0000

Source Authors' computation

Table 3.6 Test of Granger-causality from mobile internet penetration to GDP per capita, 2008–2017

Dep Var: Ln(GDPPC)	All countries			More affluent countries			Less affluent countries		
	Coeff	Std Err	Sig	Coeff	Std Err	Sig	Coeff	Std Err	Sig
Ln(GDPPC) Lag 1	0.8583	0.0336	***	0.8850	0.0299	***	0.8044	0.0712	***
Ln(PMI)	0.0070	0.0033	**	0.0086	0.0054		0.0091	0.0047	*
Ln(PMI) Lag 1	−0.0015	0.0028		−0.0002	0.0049		−0.0022	0.0028	
Ln(PMI) Lag 2	0.0050	0.0024	**	0.0038	0.0039		0.0060	0.0026	**
Ln(GCI)	0.0674	0.0265	**	0.0466	0.0355		0.0955	0.0234	***
Ln(GCI) Lag 1	−0.0325	0.0130	**	−0.0413	0.0161	***	−0.0321	0.0212	
Constant	1.3736	0.3280	***	1.2141	0.3172	***	1.6255	0.5991	***
Wald chi-squared	2609.81		***	2432.02		***	1281.81		
Observations	590			418			172		
Groups	106			73			33		
Instruments	41			41			41		
AR(1)—z statistic	−2.447		**	−1.978		**	−1.857		*
AR(2)—z statistic	−1.187			−0.327			−1.596		

Source Authors' computation
Note ***significant at 1% level; **significant at 5% level; *significant at 10% level

Table 3.7 Summary of Granger-causality findings (mobile internet penetration to GDP per capita), 2008–2017

Lag	All countries	More affluent countries	Less affluent countries
0	Yes	No	Yes
1	No	No	No
2	Yes	No	Yes
Reject null hypothesis that all lag coefficients are jointly zero? (Wald test)			
	Yes	No	Yes

Source Authors' computation

growth, particularly if there are differences between more and less affluent countries. Rather, a case can be made for exploring causality between economic growth and some of the components of IDI, instead of IDI itself. The post-2008 period has seen a rapid rise in availability and adoption of mobile technologies (for both telephony and broadband internet access) and, to a lesser extent, of fixed broadband access. These more directly represent the advances made in ICT in recent years, and we surmise that using them as alternatives to the composite IDI is more likely to reveal a Granger-causal relationship with economic growth.

To investigate this possibility, we replace IDI with the three measures mentioned earlier, namely, mobile internet penetration (PMI), fixed broadband penetration (PFB), and percentage of households with broadband access (PctBA).[10] However, even though PMI and PFB address different technologies/services, they can still be correlated over time (see Fig. 3.1). Table 3.5 shows moderate correlation between the two measures in both more and less affluent countries. In contrast, the correlations between PctBA and PMI and between PctBA and PFB are considerably higher.

We conduct Arellano–Bond regressions in three variants:

- Regression of Ln(GDPPC) on Ln(PMI) and a reverse regression of Ln(PMI) on Ln(GDPPC).
- Regression of Ln(GDPPC) on both Ln(PMI) and Ln(PFB) but no reverse regression.

---

[10] The last measure includes both mobile and fixed access to broadband. The measures considered here are similar to those in the Toader et al. (2018) study of 28 European Union countries.

Table 3.8 Test of Granger-causality from fixed broadband penetration to GDP per capita, 2008–2017

Dep Var: Ln(GDPPC)	All countries			More affluent countries			Less affluent countries		
	Coeff	Std Err	Sig	Coeff	Std Err	Sig	Coeff	Std Err	Sig
Ln(GDPPC) Lag 1	0.7733	0.0803	**	0.7724	0.0892	***	0.7373	0.0914	***
Ln(PFB)	0.0003	0.0032		0.0059	0.0159		−00033	0.0038	
Ln(PFB) Lag 1	−0.0001	0.0041		−0.0059	0.0214		0.0010	0.0038	
Ln(PFB) Lag 2	0.0083	0.0052		0.0092	0.0117		0.0013	0.0059	
Ln(GCI)	0.0495	0.0151	***	0.0481	0.0190	**	0.0578	0.0198	***
Ln(GCI) Lag 1	−0.0230	0.0101	**	−0.0272	0.0125	**	−0.0106	0.0186	
Ln(GCI) Lag 2	0.0130	0.0127		0.0099	0.0141		0.0211	0.0284	
Year	0.0055	0.0027	**	0.0055	0.0031	*	0.0092	0.0039	**
Constant	−8.8215	4.7447	*	−8.6869	5.3009	*	−16.408	7.2587	**
Wald chi-squared	3563.44		***	2458.34		***	2076.62		***
Observations	680			498			182		
Groups	104			73			31		
Instruments	43			43			43		
AR(1)—z statistic	−2.605		***	−2.088		**	−1.868		*
AR(2)—z statistic	−0.990			−0.803			−0.557		

Source Authors' computation
Note ***significant at 1% level; **significant at 5% level; *significant at 10% level

Table 3.9 Summary of Granger-causality findings (fixed broadband penetration to GDP per capita), 2008–2017

Lag	All countries	More affluent countries	Less affluent countries
0	No	No	No
1	No	No	No
2	No	No	No
Reject null hypothesis that all lag coefficients are jointly zero? (Wald test)			
	No	No	No

Source Authors' computation

- Regression of Ln(GDPPC) on Ln(PctBA) and reverse regression of Ln(PctBA) on Ln(GDPPC).

Causal Impacts on Economic Growth

To assess whether there are any causal impacts on economic growth from the three measures of ICT, we first conduct the regressions with Ln(GDPPC) as the dependent variable. Detailed results from these four regressions are reported in Tables 3.6, 3.8, 3.10, and 3.12. As before, these results are summarized in the form of "Yes/No" Granger-causality findings in Tables 3.7, 3.9, 3.11, and 3.13, respectively.

Tables 3.6 and 3.7 provide statistical evidence on the question of Granger-causality from mobile internet penetration to economic growth during the 2008–2017 period. Mobile internet penetration appears to boost GDP per capita, both contemporaneously and at a two-year lag, in less affluent countries. This finding is plausible for two reasons. First, it is well documented that less affluent countries with historically poorly developed or nonexistent fixed-line networks have relied on mobile telephony and internet access to build out their communications infrastructure.[11] Second, in these countries, mobile telephony and internet services have been instrumental in developing commerce and organizing economic and market activities that ultimately contributed to economic growth.[12] The same does not, however, appear to be true of

---

[11] Because all variables are expressed in logarithms, the coefficients have the usual interpretation as elasticities. For example, in Table 3.6, the coefficient for Ln(PMI) in less affluent countries is 0.0091. This implies that a 1% increase in mobile internet penetration boosts GDP per capita contemporaneously by 0.91%.

[12] See Gruber et al. (2011), Edquist et al. (2018), and James (2012).

Table 3.10 Test of Granger-causality from mobile internet penetration and fixed broadband access (jointly) to GDP per capita, 2008–2017

Dep Var: Ln(GDPPC)	All countries			More affluent countries			Less affluent countries		
	Coeff	Std Err	Sig	Coeff	Std Err	Sig	Coeff	Std Err	Sig
Ln(GDPPC) Lag 1	0.8531	0.0390	***	0.8687	0.0333	***	0.7893	0.0935	***
Ln(PMI)	0.0048	0.0034		0.0092	0.0063		0.0070	0.0053	
Ln(PMI) Lag 1	−0.0002	0.0028		−0.0015	0.0046		−0.0005	0.0030	
Ln(PMI) Lag 2	0.0036	0.0022		0.0034	0.0036		0.0035	0.0022	*
Ln(PFB)	0.0003	0.0038	*	0.0413	0.0209	**	−0.0011	0.0045	
Ln(PFB) Lag 1	−0.0022	0.0033		−0.0366	0.0236		0.0001	0.0036	
Ln(PFB) Lag 2	0.0048	0.0053		0.0068	0.0113		0.0112	0.0066	*
Ln(GCI)	0.0670	0.0274	**	0.0437	0.0365		0.0713	0.0208	***
Ln(GCI) Lag 1	−0.0279	0.0137	**	−0.0379	0.0170	**	−0.0122	0.0238	
Ln(GCI) Lag 2	0.0176	0.0144		0.0107	0.0175		0.0169	0.0280	
Constant	1.3969	0.3726	***	1.3299	0.3252	***	1.7455	0.7989	**
Wald chi-squared	3112.90		***	2781.78		***	2768.03		***
Observations	559			416			143		
Groups	103			73			30		
Instruments	45			45			45		
AR(1)—z statistic	−2.392		**	−1.979		**	−1.793		*
AR(2)—z statistic	−0.924			−0.220			−1.282		

Source Authors' computation
Note ***significant at 1% level; **significant at 5% level; *significant at 10% level

Table 3.11 Summary of Granger-causality findings (mobile internet penetration and fixed broadband penetration jointly to GDP per capita), 2008–2017

Lag	All countries	More affluent countries	Less affluent countries
Mobile internet penetration			
0	No	No	No
1	No	No	No
2	No	No	Yes
Fixed broadband penetration			
0	Yes	Yes	No
1	No	No	No
2	No	No	Yes
Reject null hypothesis that all lag coefficients are jointly zero? (Wald test)			
	Yes	Yes	Yes

Source Authors' computation

more affluent countries in the 2008–2017 period. The contribution of mobile internet service to overall economic growth in these countries is not statistically evident.[13]

Tables 3.8 and 3.9 provide statistical evidence on the question of Granger-causality from fixed broadband penetration to economic growth during the 2008–2017 period. During this period, fixed broadband penetration appears to have no statistically discernible causal impact on GDP per capita. A likely explanation is that, in both more affluent and less affluent countries, fixed broadband grew slowly over this period (see Fig. 3.1) and made only marginal contributions to economic growth. Even though fixed broadband had not yet become a mature or saturated service in many less affluent countries, it did not provide the same impetus to economic growth that mobile telephony did.

Tables 3.10 and 3.11 provide statistical evidence on the question of Granger-causality from mobile internet penetration and fixed broadband penetration considered jointly to economic growth during the 2008–2017 period. In more affluent countries, fixed broadband penetration appears to have a contemporaneous causal impact on GDP per capita.

---

[13] See GSMA The Mobile Economy (2019), P. Rouvinen (2006), and Saibal (2016).

Table 3.12 Test of Granger-causality from percentage of households with broadband access to GDP per capita, 2008–2017

Dep Var: Ln(GDPPC)	All countries			More affluent countries			Less affluent countries		
	Coeff	Std Err	Sig	Coeff	Std Err	Sig	Coeff	Std Err	Sig
Ln(GDPPC) Lag 1	0.7435	0.0881	***	0.7776	0.0948	***	0.7528	0.0941	***
Ln(PctBA)	0.0060	0.0045		0.0496	0.0139	***	−0.0012	0.0020	
Ln(PctBA) Lag 1	−0.0017	0.0018		−0.0131	0.0128		.0007	0.0022	
Ln(PctBA) Lag 2	0.0057	0.0038		−0.0186	0.0148		−0.0002	0.0024	
Ln(GCI)	0.0487	0.0137	***	0.0462	0.0211	**	0.0590	0.0166	***
Ln(GCI) Lag 1	−0.0233	0.0104	**	−0.0319	0.0132	**	−0.0163	0.0184	
Ln(GCI) Lag 2	0.0177	0.0131		0.0116	0.0148		0.0299	0.0281	
Year	0.0057	0.0028	**	0.0056	0.0031	*	0.0079	0.0037	**
Constant	−9.0905	4.8595	*	−9.0502	5.3171	*	−13.843	6.7773	**
Wald chi-squared	3461.77		***	2234.15		***	2393.44		***
Observations	725			498			227		
Groups	107			73			34		
Instruments	43			43			43		
AR(1)—z statistic	−2.301		**	−2.117		**	−1.846		*
AR(2)—z statistic	−0.954			−0.558			−0.901		

Source Authors' computation
Note ***significant at 1% level; **significant at 5% level; *significant at 10% level

Table 3.13 Summary of Granger-causality findings (percentage of households with broadband access to GDP per capita), 2008–2017

Lag	All countries	More affluent countries	Less affluent countries
0	No	Yes	No
1	No	No	No
2	No	No	No
Reject null hypothesis that all lag coefficients are jointly zero? (Wald test)			
	No	Yes	No

Source Author's computation

In less affluent countries, there is evidence of a causal impact on GDP per capita, albeit with a two-year lag, from the penetration of both types of services.

Tables 3.12 and 3.13 provide statistical evidence on the question of Granger-causality from the percentage of households with broadband access to economic growth during the 2008–2017 period. The measure of ICT used here is described by the ITU as a form of "ICT access," rather than of "ICT use" (which is how the other two measures are characterized). Thus, as the percentage of households with access to broadband increases, the potential clearly exists for greater broadband use creating favorable prospects for online commerce and, eventually, economic growth. Table 3.13 shows, however, that a contemporaneous causal impact on economic growth only occurs in more affluent countries. While this finding is limited to the 2008–2017 period, it serves as a caution that sheer access to broadband, whether fixed or mobile, is not a guarantee of economic growth benefits, especially in less affluent countries.

Causal Impacts from Economic Growth

Reverse regressions to test for causal impacts of economic growth on PMI, PFB, and PctBA are reported in Tables 3.14, 3.16, and 3.18, respectively, and the causality findings are summarized in Tables 3.15, 3.17, and 3.19, respectively. The hypothesis is that economic growth may itself feed back into ICT development and growth by creating additional demand for ICT services.

Table 3.14 Test of Granger-causality from GDP per capita to mobile internet penetration, 2008–2017

Dep Var: Ln(PMI)	All countries			More affluent countries			Less affluent countries		
	Coeff	Std Err	Sig	Coeff	Std Err	Sig	Coeff	Std Err	Sig
Ln(PMI) Lag 1	0.7425	0.0502	***	0.6425	0.0511	***	0.7438	0.0798	***
Ln(GDPPC)	1.7769	0.6999	**	1.7225	0.6613	***	2.9865	1.5749	*
Ln(GDPPC) Lag 1	−0.7824	0.7375		−0.3783	0.5814		−1.4725	2.3530	
Ln(GDPPC) Lag 2	−0.7833	0.4288	*	−1.0160	0.4720	**	−0.6412	1.1755	
Ln(GCI)	−0.6927	0.4758		−0.6496	0.4339		−0.2732	0.8359	
Ln(GCI) Lag 1	0.0874	0.2338		0.6039	0.2303	***	−0.4759	0.5052	
Constant	−1.2590	4.1551		−3.4120	3.4923		−6.6048	8.2880	
Wald chi-squared	865.69		***	835.62		***	484.55		***
Observations	664			465			199		
Groups	106			73			33		
Instruments	41			41			41		
AR(1)—z statistic	−3.741		***	−3.175		***	−2.797		***
AR(2)—z statistic	1.017			0.864			0.826		

Source Authors' computation

Note ***significant at 1% level; **significant at 5% level; *significant at 10% level

Tables 3.14 and 3.15 provide statistical evidence on the question of Granger-causality from economic growth to mobile internet penetration during the 2008–2017 period. Such causality is indicated in both more affluent and less affluent countries, more strongly in the former. A contemporaneous positive impact exists in both types of countries, whereas an attenuating negative impact with a two-year lag also occurs in more affluent countries. However, the positive causal impact more than offsets the negative lagged impact.

Tables 3.16 and 3.17 provide statistical evidence on the question of Granger-causality from economic growth to fixed broadband penetration during the 2008–2017 period. There is no evidence of a causal impact from economic growth to fixed broadband penetration in less affluent countries, although there is evidence of a two-year lagged positive impact in more affluent countries. This delayed impact, albeit positive, suggests that even in more affluent countries with already well-established fixed broadband networks, additional demand creation is slow to materialize.

Tables 3.18 and 3.19 provide statistical evidence on the question of Granger-causality from economic growth to the percentage of households with any broadband access during the 2008–2017 period. Once again, the only causal impact that occurs is in more affluent countries, although that impact is contemporaneous. This suggests that, even though the fixed broadband use in more affluent countries may have reached saturation points, economic growth can still create additional demand for some form of broadband access. The findings in Tables 3.14 and 3.15 suggest that greater access to mobile broadband

Table 3.15 Summary of Granger-causality findings (GDP per capita to mobile internet penetration), 2008–2017

Lag	All countries	More affluent countries	Less affluent countries
0	Yes	Yes	Yes
1	No	No	No
2	Yes	Yes	No
Reject null hypothesis that all lag coefficients are jointly zero? (Wald test)			
	Yes	Yes	Yes

Source Author's computation

Table 3.16 Test of Granger-causality from GDP per capita to fixed broadband penetration, 2008–2017

Dep Var: Ln(PFB)	All countries			More affluent countries			Less affluent countries		
	Coeff	Std Err	Sig	Coeff	Std Err	Sig	Coeff	Std Err	Sig
Ln(PFB) Lag 1	0.5585	0.2278	**	0.4759	0.1346	***	0.4361	0.1718	**
Ln(GDPPC)	0.0799	0.3623		0.2069	0.2453		−0.6900	1.4325	
Ln(GDPPC) Lag 1	0.2046	0.4953		−0.1961	0.3312		1.2574	1.7880	
Ln(GDPPC) Lag 2	0.4616	0.3440		0.3470	0.1492	**	1.1562	1.0697	
Ln(GCI)	0.1902	0.2041		0.2287	0.1182	*	0.1009	0.2633	
Ln(GCI) Lag 1	−0.3092	0.2574		0.0428	0.0572		−1.0411	1.0510	
Constant	−6.1030	3.9435		−2.7696	0.9810		−12.967	6.7004	*
Wald chi-squared	359.26		***	654.44		***	196.25		***
Observations	695			501			193		
Groups	105			73			32		
Instruments	41			41			41		
AR(1)—z statistic	−1.680		*	−1.613			−1.615		*
AR(2)—z statistic	0.613			−0.675			1.184		

Source Authors' computation
Note ***significant at 1% level; **significant at 5% level; *significant at 10% level

Table 3.17 Summary of Granger-causality findings (GDP per capita to fixed broadband penetration), 2008–2017

Lag	All countries	More affluent countries	Less affluent countries
0	No	No	No
1	No	No	No
2	No	Yes	No
Reject null hypothesis that all lag coefficients are jointly zero? (Wald test)			
	No	Yes	No

Source Author's computation

may have occurred during the study period, which also spans the period of transition from 3G to faster and higher-capacity 4G mobile networks.

Discussion of Findings and Conclusions

This study is designed to answer two questions about the relationship between ICT development (including access and use) and economic growth. These concern (1) the direction of causality, if any, between them and (2) changes in the nature of ICT itself which, in turn, have affected its relationship with economic growth. The study spans the post-Great Recession period of 2008–2017, which is both a period of economic tumult and slow recovery and significant pathbreaking developments in ICT technologies.

In answering these questions, the study distinguishes between more affluent (high and upper middle income) countries and less affluent (lower middle and low income countries). That is because, although significant research has already occurred on the central question of the existence of the ICT–economic growth relationship, most past studies have failed to uncover sharp differences between more affluent and less affluent countries. Econometric models constructed instead by pooling all those countries can yield larger samples but can also miss the consequences of the different trajectories on which ICT development has occurred in more affluent and less affluent countries. This is particularly true in a period (such as the one we study) in which mobile technologies have grown at a disproportionate pace, at first complementing and then substituting for more mature fixed-line technologies that were once preeminent in more affluent countries.

Table 3.18 Test of Granger-causality from GDP per capita to percentage of households with broadband access, 2008–2017

Dep Var: Ln(PctBA)	All countries			More affluent countries			Less affluent countries		
	Coeff	Std Err	Sig	Coeff	Std Err	Sig	Coeff	Std Err	Sig
Ln(PctBA) Lag 1	0.6669	0.0984	***	0.7590	0.0423	***	0.3886	0.1014	***
Ln(GDPPC)	0.4721	0.2687	*	0.5621	0.3314	*	0.0316	1.2743	
Ln(GDPPC) Lag 1	−0.0313	0.5663		−0.5687	0.4032		3.1550	2.7287	
Ln(GDPPC) Lag 2	0.4016	0.2301	*	0.0090	0.1292		0.2977	0.5825	
Ln(GCI)	0.0575	0.0939		0.1018	0.0739		−0.1204	0.3624	
Ln(GCI) Lag 1	−0.0510	0.2109		0.1272	0.0839		−0.6383	0.8998	
Constant	−6.8792	5.8136		0.6470	0.7934		−26.5056	9.7975	***
Wald chi-squared	2691.10		***	4504.41		***	1066.85		***
Observations	731			502			229		
Groups	107			73			34		
Instruments	41			41			41		
AR(1)—z statistic	−1.635		*	−3.300		***	−1.612		*
AR(2)—z statistic	−1.698		*	−0.879			−1.365		

Source Authors' computation
Note ***significant at 1% level; **significant at 5% level; *significant at 10% level

Table 3.19 Summary of Granger-causality findings (GDP per capita to percentage of households with broadband access), 2008–2017

Lag	All countries	More affluent countries	Less affluent countries
0	Yes	Yes	No
1	No	No	No
2	Yes	No	No
Reject null hypothesis that all lag coefficients are jointly zero? (Wald test)			
	Yes	Yes	No

Source Authors' computation

The less frequently studied question of causality in the ICT–economic growth relationship features prominently in this study. We adopt the commonly used metric of Granger-causality and look for causal relationships, whether unidirectional or bidirectional, contemporaneous or lagged. To this end, the study employs a dynamic panel data with fixed effects econometric model, to which it applies a GMM estimator that utilizes instrumental variables to produce consistent estimates in the presence of lagged dependent variables. The Arellano–Bond version of this estimator is widely used for such models and has good diagnostic tests; moreover, it creates all the requisite instruments by taking lags of levels and first differences of the variables already specified for the model. This method of estimation is well-suited for samples that have a large number of panels measured over a relatively short time period and may be unbalanced. Our sample, with 107 countries and 10 years, satisfies that requirement.

We begin the study by exploring causal relationships between ICT (measured by the ITU's composite IDU index) and economic growth (measured by GDP per capita at purchasing power parity). We also use the WEF's GCI index to control for a large number of macroeconomic and market-related factors that can also influence ICT and economic growth. As the study progresses and we evaluate the relationships discovered between IDI and GDP per capita, we conclude that, as a composite of many ICT-related features, IDI may not be the most suitable measure for sharply isolating causal relationships between ICT and economic growth. We replace IDI with some of its important components, namely, mobile internet penetration, fixed broadband penetration, and the percentage of households with any broadband access. When run through the same econometric modeling framework, we find more distinct causal relationships that vary by which measure of ICT is used and whether countries analyzed are more affluent or less affluent. This is important because these variations answer the questions about competing ICT technologies, particularly after 2008, and the different technology adoption histories in more affluent and less affluent countries.

Given our study objectives, we conduct Granger-causality tests in both directions, both from ICT to economic growth and from economic growth to ICT. The first direction, which has been studied most often in past research, concerns the ways in which ICT adoption create opportunities for electronic commerce and market development and, hence, for economic growth. The second direction, although studied relatively infrequently, is also important because it concerns how economic growth can generate the resources to invest in ICT technologies and to serve additional ICT-related demand created by rising income. Intuitively, a feedback loop (or bidirectional causality) may be expected, especially in economies of less affluent countries which rely on the advent and acceptance of new and disruptive technologies for economic growth. In more mature and relatively more affluent countries which already have well-established technologies in place, the demand creation feedback loop from economic growth to ICT may be less likely to materialize.

The core findings in our paper are presented in Tables 3.6–3.19. To summarize:

- Stronger and statistically significant causal impacts exist in less affluent countries from mobile internet penetration to GDP per capita. Any comparable effect for more affluent countries (where mobile internet penetration is already quite high during the study period) is not found.
- No comparable causal impacts are found in both more affluent and less affluent countries from fixed broadband penetration to GDP per capita. During the study period, we hypothesize that fixed broadband growth had reached saturation levels in more affluent countries, and little of it occurred in less affluent countries which had already opted for mobile broadband as a leapfrog technology. Consequently, any positive causal impact of fixed broadband on economic growth in 2008–2017 is likely to have been marginal and statistically undetectable.

- A contemporaneous causal impact on economic growth from the percentage of households with broadband access only occurs in more affluent countries. Although this finding only pertains to the 2008–2017 period, it suggests that broadband access (whether fixed or mobile) alone—and not its use—need not boost economic growth, especially in less affluent countries.
- Causal impacts are statistically detected from GDP per capita growth to mobile internet penetration in both more affluent and less affluent countries. During the study period, significant advancements occurred in mobile communications technology, with the advent of 4G and LTE services and the beginnings of 5G. These had profound demand-side effects on the use of mobile broadband in both types of countries. More importantly, this confirms that Granger-causality between mobile internet penetration and GDP per capita is, indeed, bidirectional, at least in less affluent countries.
- Only a weak and delayed causal impact occurs from GDP per capita to fixed broadband penetration in more affluent countries, while there is none at all in less affluent countries. Once again, if the question is about economic growth creating additional demand for fixed and mobile broadband, any evidence of that happening is loaded disproportionately on the mobile side. While even a weak causal impact from economic growth to fixed broadband in more affluent countries qualifies as evidence of some feedback, the evidence is not strong enough to accept it without reservation.

Despite the findings of this paper, there remain several researchable questions. First, does this study conclusively prove the existence of bidirectional causality between mobile internet penetration and economic growth? The evidence appears to be stronger on that question in less affluent countries which have leapfrogged fixed-line technologies with rapidly evolving mobile technologies. But, with the advent of 5G, we expect that bidirectionality to remain in place and even include more affluent countries.

Second, how much does the time period studied matter? Past research that focused mainly on the years before 2008 almost universally found causal impacts from ICT to economic growth, although ICT investment (not the all-embracing measure of ICT development) was most widely studied. Have we uncovered a break from that tradition by failing to find meaningful causal impacts from composite measures such as IDI and having to resort to more component-level measures of ICT instead? More importantly, in future research, should the omnibus concept of ICT itself yield to specific communication technologies?

Third, in conducting research of this nature, is it meaningful to work with large samples of countries without differentiating among them in substantive ways? For example, many studies pool countries that fall all along the spectrum of economic development, per capita income, and technological advancement. While such pooling may be convenient for the purposes of econometric modeling, unless sufficient controls are built into the models to account for the diversity of the included countries, causal findings can be difficult to verify statistically or even interpret. We are aware of a few studies that have distinguished among countries classified by the World Bank according to per capita income, and we have tried to continue that tradition. However, diversity among countries and, particularly, their policy and technology choices go beyond just per capita income. We have used the multifaceted GCI index to account for that diversity, but more research may be needed to understand the full value of doing so.

Finally, what are the policy implications of our findings? The causality findings in our study are more reflective (of the consequences of past policy and technology choices) than prescriptive. As our study clearly demonstrates, new technology directions can be disruptive in a number of ways. Not least among them is the econometric modeler's dilemma about the selection of variables. Beginning with that age-old question about whether GDP per capita truly captures a country's state of development—a country with a large population may nevertheless have relatively low per capita income despite ranking high on the overall GDP scale—questions may arise elsewhere as well. To what extent should governments or policy-makers feel obliged to make technology choices in order to take advantage of findings about the feedback loop between economic growth and ICT technologies? Or, should they refrain from tipping the scales in favor of one technology or another? Can emerging ICT technologies aggravate the Digital Divide problem or actually attenuate them? Should governments and policy-makers intervene in the matter? These policy questions are likely to persist in the future, and new questions may arise as technologies evolve and generate new end-use services. That will create opportunities to address many of the more normative policy-related issues that are adjacent to findings about causal impacts.

REFERENCES

Alleman, J., C. Hunt, D. Michaels, M. Mueller, P.N. Rappoport, and L. Taylor. 1991. Telecommunications and Economic Development: Empirical Evidence from Southern Africa. Available at www.colorado.edu/engineering/alleman/print_files/soafrica_paper.pdf (21.03.2014).

Anderson, T.W., and C. Hsiao. 1981. Estimation of Dynamic Models with Error Components. *Journal of the American Statistical Association* 76 (375): 598–606.

Arellano, M., and S. Bond. 1991. Some Tests of Specification for Panel Data: Monte Carlo Evidence and an Application to Employment Equations. *The Review of Economic Studies* 58 (2): 277–297.

Bahrini, R., and A. Qaffas. 2019. The Impact of Information and Communication Technology on Economic Growth: Evidence from Developing Countries. *Economies*, Open Access Journal, MDPI 7 (1): 1–13.

Banerjee, A., P.N. Rappoport, and J. Alleman. 2014. Forecasting Video Cord-Cutting: The Bypass of Traditional Pay Television. In *Demand for Communications Services—Insights and Perspectives: Essays in Honor of Lester D. Taylor*, ed. J. Alleman, Á. Ní-Shúilleabháin, and P.N. Rappoport. Berlin: Springer Verlag.

Bertschek, I., and T. Niebel. 2016. Mobile and More Productive? Firm-Level Evidence on the Productivity Effects of Mobile Internet Use. *Telecommunications Policy* 40 (9): 888–898.

Brynjolfsson, E. 1993. The Productivity Paradox of Information Technology: Review and Assessment. *Communications of ACM* 36 (12): 66–77.

Brynjolfsson, E., and L.M. Hitt. 1995. Information Technology as a Factor of Production: The Role of Differences Among Firms. *Economics of Innovation and New Technology* 3 (3): 183–200.

Brynjolfsson, E., and L.M. Hitt. 2000. Beyond Computation: Information Technology, Organizational Transformation and Business Performance. *Journal of Economic Perspectives* 14 (4): 23–48.

Brynjolfsson, E., and S. Yang. 1996. Information Technology and Productivity: A Review of the Literature. *Advances in Computers* 43: 179–214.

Cardona, M., T. Kretschmer, and T. Strobel. 2013. ICT and Productivity: Conclusions from the Empirical Literature. *Information Economics and Policy* 25: 109–125.

Chakraborty, C., and B. Nandi. 2011. 'Mainline' Telecommunications Infrastructure, Levels of Development and Economic Growth: Evidence from a Panel of Developing Countries. *Telecommunications Policy* 35 (5): 441–449.

Crandall, R.W., W. Lehr, and R. Litan. 2007. The Effects of Broadband Deployment on Output and Employment: A Cross-Country Analysis of US

Data. *Issues in Economic Policy*, No. 6, The Brookings Institution, Washington, DC. Retrieved from http://www.brookings.edu/papers/2007/06labor_crandall.aspx.

Cronin, J.F., E.B. Parker, E.K. Colleran, and M.A. Gold. 1991. Telecommunications Infrastructure and Economic Growth: An Analysis of Causality. *Telecommunications Policy* 15 (6): 529–535.

Czernich, N., O. Falck, T. Kretschmer, and L. Wöessmann. 2009. Broadband Infrastructure and Economic Growth. CESifo Working Paper Series No. 2861. http://papers.ssrn.com/sol3/Delivery.cfm/SSRN_ID1516232_code 459177.pdf?abstractid=1516232&mirid=1 (13.04.2014).

Czernich, N., O. Falck, T. Kretschmer, and L. Wöessmann. 2011. Broadband Infrastructure and Economic Growth. *The Economic Journal* 121 (552): 505–532.

Dewan, S., and K.I. Kraemer. 2000. Information Technology and Productivity: Evidence from Country-Level Data. *Management Science* 46 (4): 548–562.

Dutta, A. 2001. Telecommunications and Economic Activity: An Analysis of Granger Causality. *Journal of Management Information Systems* 17 (4): 71–95.

Edquist, H., P. Goodridge, J. Haskel, X. Li, and E. Lindquist. 2018. How Important Are Mobile Broadband Networks for the Global Economic Development? *Information Economics and Policy* 45: 16–29.

Estache, A., A. Goicoechea, and M. Manacorda. 2006. Telecommunications Performance, Reforms, and Governance. World Bank Policy Research Working Paper No. 3822.

Farhadi, M., R. Ismail, and M. Fooladi 2012. Information and Communication Technology Use and Economic Growth, November 12. Retrieved from https://journals.plos.org/plosone/article?id=10.1371/journal.pone.0048903.

Greenstein, S., and R. McDevitt. 2009. The Broadband Bonus: Accounting for Broadband Internet's Impact on US GDP. National Bureau of Economic Research Working Paper No. 14758. http://www.nber.org/papers/w14758 (24.03.2014).

Greenstein, S., and R. McDevitt. 2012. Measuring the Broadband Bonus in Thirty OECD Countries. OECD Digital Economy Papers 197, OECD Publishing. http://dx.doi.org/10.1787/5k9bcwkg3hwf-en (29.03.2014).

Gruber, H., P. Koutroumpis, T. Mayer, and V. Nocke. 2011. Mobile Telecommunications and the Impact on Economic Development. *Economic Policy* 26 (67): 389–426.

GSMA. 2019. *The Mobile Economy 2019*. Retrieved from https://www.gsmaintelligence.com/research/?file=b9a6e6202ee1d5f787cfebb95d-3639c5&download.

Holtz-Eakin, D., W. Newey, and H.S. Rosen. 1988. Estimating Vector Autoregressions with Panel Data. *Econometrica* 56 (6): 1371–1395.

International Telecommunication Union. 2018. The ICT Development Index. Retrieved from https://www.itu.int/en/ITU-D/Statistics/Documents/statistics/ITU_ICT%20Development%20Index.pdf.

James, J. 2012. The Distributional Effects of Leapfrogging in Mobile Phones. *Telematics and Informatics* 29 (3): 294–301.

Jung, J. 2014. Regional Inequalities in the Impact of Broadband on Productivity: Evidence from Brazil. IBEI Working Paper.

Katz, R.L. 2009a. Estimating Broadband Demand and Its Economic Impact in Latin America. Paper presented at the 3rd ACORN-REDECOM Conference, Sept 4–5, Mexico City, Mexico.

Katz, R.L. 2009b. *The Contribution of Information and Communications Technologies to Economic Development: Latin American Proposals to Current Economic Challenges.* Madrid, Spain: Ariel.

Katz, R.L. 2009c. The Economic and Social Impact of Telecommunications Output: A Theoretical Framework and Empirical Evidence for Spain. *Intereconomics* 44 (1 January/February): 41–48.

Katz, R.L., and S. Suter. 2009. Estimating the Economic Impact of the Broadband Stimulus Plan. Columbia Institute for Tele-Information Working Paper, February.

Katz, R.L, S. Vaterlaus, P. Zenhäusern, and S. Suter. 2009. The Impact of Broadband on Jobs and the German Economy. Columbia Institute for Tele-Information Working Paper.

Koutroumpis, P. 2009. The Economic Impact of Broadband on Growth: A Simultaneous Approach. *Telecommunications Policy* 33 (9): 471–485.

Madden, G., and S.J. Savage. 1998. CEE Telecommunications Investment and Economic Growth. *Information Economics and Policy* 10 (2): 173–195.

Madden, G., and S.J. Savage. 2000. Telecommunications and Economic Growth. *International Journal of Social Economics* 27: 893–906.

Minges, M. 2016. Exploring the Relationship Between Broadband and Economic Growth. World Development Report: Digital Dividends, 1–21.

Mokyr, J. 2005. Long Term Economic Growth and the History of Technology, Chapter 17. In *Handbook of Economic Growth*, vol. 1b, ed. P. Aghion and S. Durlauf. North-Holland: Elsevier.

Nair-Reichert, U., and D. Weinhold. 2000. Causality Tests for Cross-Country Panels: New Look at FDI and Economic Growth in Developing Countries. Retrieved from https://www.scheller.gatech.edu/centers-initiatives/ciber/projects/workingpaper/1999/99_00–12.pdf (16.06.2018).

Niebel, T. 2018. ICT and Economic Growth—Comparing Developing, Emerging and Developed Countries. *World Development* 104: 197–211.

OECD. 2003. ICT and Economic Growth: Evidence from OECD Countries, Industries and Firms. Retrieved from https://www.oecd-ilibrary.org/science-and-technology/ict-and-economic-growth_9789264101296-en (05.06.2018).

Röller, H., and L. Waverman. 2001. Telecommunications Infrastructure and Economic Development: A Simultaneous Approach. *American Economic Review* 91 (4): 909–923.

Rouvinen, P. 2006. Diffusion of Digital Mobile Telephony: Are Developing Countries Different? *Telecommunications Policy* 30 (1): 46–63.

Saibal, G. 2016. How Important Is Mobile Telephony for Economic Growth? Evidence from MENA Countries. *Info* 18 (3): 58–79.

Spiezia, V. 2012. ICT Investments and Productivity: Measuring the Contribution of ICTS to Growth. *OECD Journal: Economic Studies* 2012 (1): 199–211.

Sridhar, K.S., and V. Sridhar. 2007. Telecommunications Infrastructure and Economic Growth: Evidence from Developing Countries. *Applied Econometrics and International Development* 7 (2): 37–61.

Toader, E., B. Firtescu, A. Roman, and S.G. Anton. 2018. Impact of Information and Communication Technology Infrastructure on Economic Growth: An Empirical Assessment for the EU Countries. *Sustainability* 10 (10): 3750.

Venturini, F. 2009. The Long-Run Impact of ICT. *Empirical Economics* 37 (3): 497–515.

Vu, K.M. 2011. ICT as a Source of Economic Growth in the Information Age: Empirical Evidence from the 1996–2005 Period. *Telecommunications Policy* 35 (4): 357–372.

Ward, M., and S. Zheng. 2016. Mobile Telecommunications Service and Economic Growth: Evidence from China. *Telecommunications Policy* 40 (2): 89–101.

Waverman, L., and P. Koutroumpis. 2011. Benchmarking Telecoms Regulation—The Telecommunications Regulatory Governance Index (TRGI). *Telecommunications Policy* 35 (5): 450–468.

Waverman, L., M. Meschi, and M. Fuss. 2005. The Impact of Telecoms on Economic Growth in Developing Countries. Vodafone Policy Paper Series, 2, 10–24.

World Development Indicators. 2012. data.worldbank.org/data-catalog/world-development-indicators/wdi-2012. (18.03.2014).

CHAPTER 4

Adoption of E-Commerce by Individuals and Digital-Divide

Ángel Valarezo, Rafael López, and Teodosio Pérez Amaral

Introduction

A standard definition of e-commerce (Eurostat 2018a) and INE (2018) is the placing of orders of goods or services via the internet, excluding orders via manually typed e-mails or text messages. Electronic payment or delivery are not required for an e-commerce transaction. Only purchases made for personal reasons are considered.

According to the European Commission's Digital Economy and Society Index (DESI) of 2018 (European Commission 2018a), Spain ranks tenth among the 28 EU Member States, belonging to the group of medium-performing countries. This index tracks the evolution of

This research is funded by the Autonomous Community of Madrid—Spain. Project: Finance, Innovation and Strategies: Economic Analysis of Business Productivity and its Determinants (PRODECON-CM HM S2015/HUM-3491), 2016–2018.

Á. Valarezo (✉) · R. López · T. Pérez Amaral
Instituto Complutense de Análisis Económico (ICAE), Madrid, Spain
e-mail: angel.valarezo.unda@ucm.es

© The Author(s) 2020
J. Alleman et al. (eds.), *Applied Economics in the Digital Era*,
https://doi.org/10.1007/978-3-030-40601-1_4

EU countries in digital competitiveness, assessing their digitization around five dimensions or policy areas: Connectivity,[1] Human Capital[2] (or digital skills), Use of internet Services by citizens,[3] Integration of Digital Technology,[4] and Digital Public Services.[5] Spain has improved in almost all domains, except for Human Capital. Although Spanish citizens perform a wide variety of internet activities, Spain fell back from rank 16 to rank 17 in the internet use section. Online shopping is below the EU-28[6] average and several indicators suggest it is due to a weak demand side, especially on the private user side.

The modeling of e-commerce adoption based on disaggregate measures of socioeconomic factors on an individual level has received scarce attention. Much of the research about e-commerce adoption has been done from business and technological standpoints, focusing on the process of how the customer decides to shift from offline to online shopping. The aim is usually to identify the main characteristics and determinants, which can be targeted by marketing strategies or by technical implementations. Surveys and theoretical models have been specially designed or adapted to capture behavioral determinants and moderator effects on the decision to buy online.

The effects of socioeconomic dimensions on the adoption of e-commerce is closely related to the concept of "digital-divide," the unequal access of individuals to digital technology (Cerno and Pérez-Amaral 2006b). As Srinuan and Bohlin (2011) and Demoussis and Giannakopoulos (2006), among others, pointed out, digital-divide is a multifaceted phenomenon that does not depend only on technological determinism. Socioeconomic, institutional, and psy`chological factors have been included in these studies.

Much of the purely online supply is composed of high value-added products and services, so not using e-commerce can end up being a source of social exclusion. Helsper and van Deursen (2015) found that the internet remains more beneficial for those with a higher social status,

---

[1] Fixed broadband, mobile broadband, and prices.
[2] Internet use, basic and advanced digital skills.
[3] Citizens' use of content, communication, and online transactions.
[4] Business digitization and e-commerce.
[5] eGovernment and eHealth.
[6] EU-28: Member States of the European Union.

meaning that in terms of social implications existing offline inequalities could potentially be exacerbated.

The aim of this paper is to identify the main socioeconomic and demographic factors that influence an individual's adoption of online shopping. This will be the basis for discussing public policies and private strategies to promote the use of e-commerce and bridging the digital-divide by specifically targeting those groups with the lowest penetration rates of e-commerce and other ICT services. In order to account for the determinants on the individual decision of becoming an online buyer for private use, descriptive analysis and quantitative models are formulated.

The pool dataset used in this paper has a total of 174,776 observations and was constructed using the annual representative survey on ICT usage by households and individuals (years 2008–2017) carried out by the National Institute of Statistics of Spain (INE).

The rest of the paper is organized as follows. Section "Literature Review" presents a literature review. Section "The ICT-H Survey: Pooled Data" describes the evolution and current situation of adoption of e-commerce in Spain. Section "Theoretical and Empirical Models" presents the data to be used and its main characteristics. In section "Linear Probability Model" the estimations of the linear probability, logistic regression and Heckman selection models are presented, showing the effects of the selected explanatory variables on the probability of adopting e-commerce. Section "Conclusions" contains conclusions, and policy recommendations.

LITERATURE REVIEW

Models of ICT Adoption

Adoption of technology, online consumer behavior, and the specific decision of adopting e-commerce have been studied based on several frameworks. By far, the most used approaches are applications, adaptations, and/or unifications of models and theories of individual acceptance and intention. Among the most sounded theoretical backgrounds are: Innovation Diffusion Theory (IDT), Rogers (2003), Theory of Reasoned Action (TRA), Ajzen and Fishbein (1977, 1980), Theory of Planned Behavior (TPB), Ajzen (1991), Technology Acceptance Model, TAM (Davis et al. 1989; Davis and Warshaw 1989; Davis 1989, 1993), Theoretical Extension of the Technology Acceptance Model (TAM2), Venkatesh and Davis (2000), Decomposed Theory of Planned Behavior

(DTPB), Taylor and Todd (1995), Expectation Confirmation Theory (ECT), Oliver (1980), and the Unified Theory of Acceptance and Use of Technology (UTAUT), Venkatesh et al. (2003).

Most of the rich body of theory and applied research about ICT's diffusion, are based on IDT, TRA, TAM, and TPB, which includes variables that affect an individual's motivation to accept a new technology and helps to explain the decision-making process of doing so. If the goal is to go further than the initial acceptance, meaning the study of post-purchase behavior, ECT is widely used to analyze the individual to continue using a technology or performing transactions. The contributions of the above-mentioned literature are considered, as precedents of this study.

Digital-Divide

Not long ago the main indicators of digital-divide were related with the "haves" and "have nots," however National Telecommunications and Information Administration (NTIA) (1995) stated the necessity to go beyond the traditional focus on telephone penetration and infrastructure access. At that time, the importance of studying computer and modem penetration according to socioeconomic, demographic, and geographic variables was clear. A few years later several works (DiMaggio and Hargittai 2001; Hargittai 2001; Norris 2001; OECD 2001; among others), pointed out that the research community should further expand their focus to the full range of digital inequality, not only accounting for equipment and internet access, but also for digital skills, technological evolution, and the widening scope of internet usage, without avoiding the complexity of characterizing individuals at different levels of analysis.

The shift from studying the inequalities of access to the differences in the extent of use brought new conceptual dimensions. Hargittai (2001), among others, suggested that there is a "second-level of digital-divide" where the people's online skills to achieve internet related tasks (as finding information online) play a key role, instead of just considering whether someone is or is not an internet user. A "third-level of digital-divide" has been proposed, as Helsper and van Deursen (2015) accounted for the link between inequality and the achievement of tangible outcomes of internet use.

Internet Use and E-Commerce Adoption

The determinants of internet use and online shopping in Spain have been empirically analyzed before, based on the INE's survey on ICT equipment and usage by households and individuals. The studies have mainly used cross-sectional data and pools of cross sections. One of the early studies in Spain about the demand of internet access and use was carried out by Cerno and Pérez-Amaral (2006); two models were estimated using data from 2003, one for broadband access at home and the other for internet use intensity. Lera-López et al. (2011) explored the impact of socioeconomic, demographic, and regional factors to explain not only internet use but also the frequency of use by individuals. Cerno and Pérez-Amaral (2009) analyzed the survey for 2003, characterizing the e-consumer profile, for both the number of purchases and the expenditure. Garín-Muñoz and Pérez-Amaral (2011) used data from 2007 to model the adoption and use of e-commerce identifying the effects of key sociodemographic variables, attitudes, and beliefs. Covering the period 2004–2009, Pérez-Hernández and Sánchez-Mangas (2011) focused on having internet at home and its implications for e-commerce adoption. A working paper of Alonso and Arellano (2015) explored the heterogeneity and diffusion of the digital economy in Spain from 2003 to 2014, estimating the effects of innovation and adoption of internet, e-commerce, and e-banking. During the 2008–2014 period, Correa et al. (2015) also constructed a multiple-stage model to identify the importance of the individual determinants on the online purchasing decision. The dataset of 2016 has been studied by Garín-Muñoz et al. (2019), to model the individual adoption of e-commerce, e-banking, and e-government; and Valarezo et al. (2018), to explore the determinants of the individual decision to perform cross-border e-commerce for private ends.

The above-cited empirical studies have mostly specified binary response models. Demographic and socioeconomic characteristics of households and individuals have been used and interpreted. Although most of the works that have dealt with INE's survey on ICT account for the explanatory factors, they don't analyze specific subcategories. Besides controlling for aggregate time effects, the current paper analyzes an individual's economic incentives and determinants of e-commerce

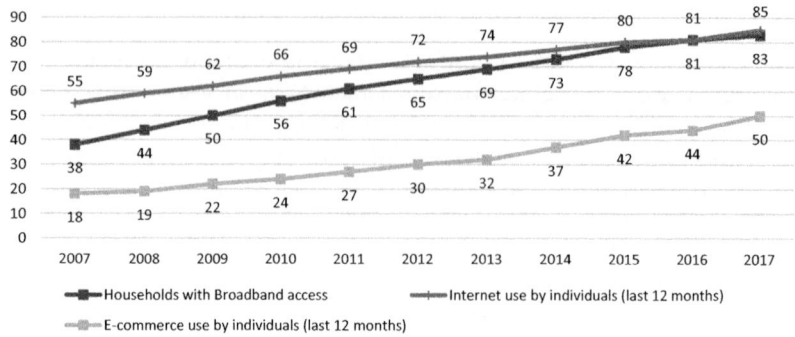

Fig. 4.1 Households with private broadband access at home, penetration rates of internet use, and e-commerce. As percentage of people aged 16–74 (2007–2017) (*Source* Authors' creation from *INE* (2018). Survey on Equipment and Use of Information and Communication Technologies in Households)

adoption, categorizing the explanatory variables as follows: sociodemographic (which includes, gender, age, population size, number of household members, nationality[7]), skills (attained levels of education, levels of digital competence), economic (employment situation, level of household income), and geography and time (region of residence and year).

Adoption of E-Commerce in Spain

In 2017, 82.7% of Spanish households had broadband connection and 85% of the Spanish population (between 16 and 74 years[8]) used internet at least once in the last twelve months, although only 69.0% did so on a daily basis. While e-commerce penetration rate reached 40%, namely four out of 10 people bought online for private use at least once in the last 3 months. It goes up to 49.9% if the considered period is extended to 12 months (Fig. 4.1). Spain still lags behind the EU average of e-commerce adoption, not to mention when the comparison is made with respect to the best performers which include: Finland, Germany, Iceland,

---

[7] Having Spanish nationality or not.

[8] For comparability with the data published by EUROSTAT, INES's ICT survey results are referred to households with at least one person between 16 and 74 years old.

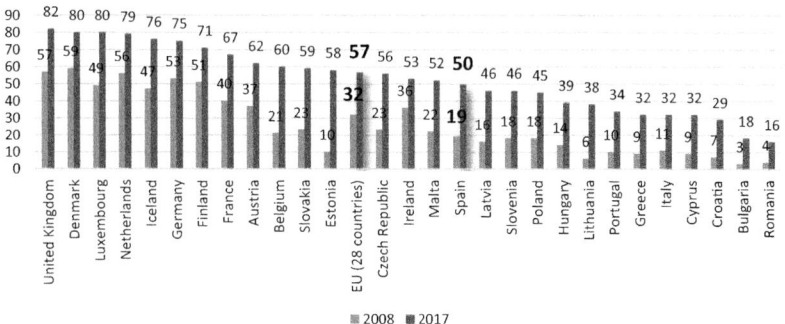

Fig. 4.2 Penetration rates of e-commerce in the EU-28 as percentage of people aged 16–74 (2008 and 2017) (*Source* Authors' creation from *Eurostat* (2018b). Survey on Equipment and Use of Information and Communication Technologies in Households)

Netherlands, Luxembourg, Denmark, and United Kingdom (UK), all above 70%, considering those who bought online in the last 12 months.

Relevant indicators on Europe's digital performance are summarized by DESI, developed within the Digital Single Market strategy of the European Commission (2018b).

According to the country-specific profile report (European Commission, 2018a) Spain belongs to the medium-performance group of countries[9] and ranks tenth among the 28 EU Member States in DESI 2018, improving two positions with respect to 2017.[10] However, this position represents an average that hides important differences when comparing the most and least advanced EU countries. This is the case of the Spanish penetration rate of online shopping by individuals for private use, ranking 17th, 4 places behind the EU average (Fig. 4.2).

---

[9] DESI's medium-performing countries are Spain, Austria, Lithuania, Germany, Slovenia, Portugal, Czech Republic, France, and Latvia.

[10] Small increases of penetration rates are observed in 2018. Households with broadband connection reached up to 86.1%, matching internet use. E-commerce adoption has also increased up to 43.5%.

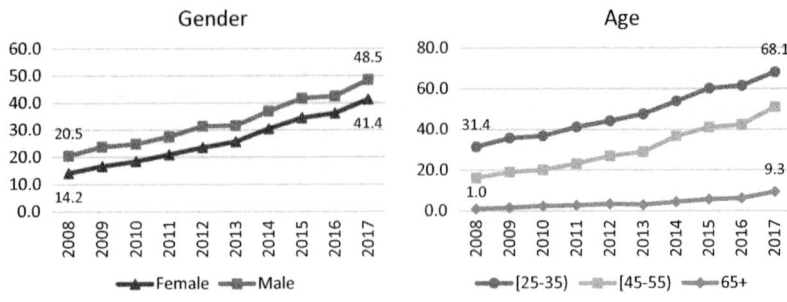

Fig. 4.3 Evolution of individual penetration rates of e-commerce in Spain (2008–2017): Gender and Age (*Source* Authors' creation from *INE* [2018]. Survey on Equipment and Use of Information and Communication Technologies in Households)

Aggregate data seems to suggest that the penetration rates of e-commerce, as other internet services, in Spain have had a healthy evolution throughout the last few years, which might imply that the digital-divide is narrowing. However, when we consider specifically the evolution of penetration rates by different variables and their categories, several divides are widening when the differences between groups with the lowest and largest penetration rates in 2008 and 2017 are measured in absolute values.[11]

The distance between male and female has increased slightly in 2017 with respect to 2008. It is more evident when we compare the age groups between the highest and the lower levels of adoption of online shopping. The group aged 25–35 years old goes from 31.4% in 2008 to 68.1% in 2017, while those aged over 65 with 1% of penetration rate in 2008 reached up to 9.3% in 2017 (Fig. 4.3).

Figure 4.4 shows the evolution of individual penetration rates of e-commerce for the main categories of Education and Digital Skills.

---

[11] All comparisons assessed in absolute values show greater distances. On the other hand, relative distances (the difference between the lowest and the highest penetration rate, as a percentage of the highest penetration rate) show a rapid growth in e-commerce adoption, but slow narrowing of the digital-divide.

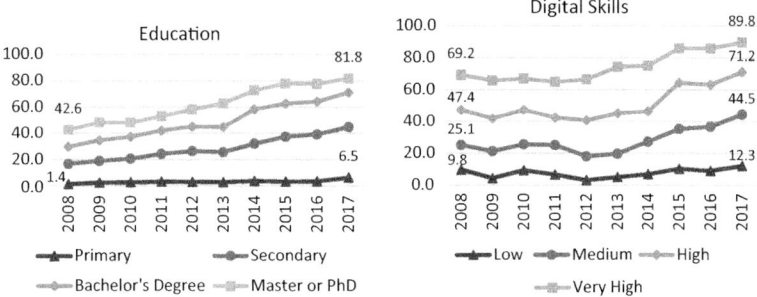

Fig. 4.4 Evolution of individual penetration rates of e-commerce in Spain (2008–2017): by Education and Digital Skills (*Source* Authors' creation from *INE* [2018]. Survey on Equipment and Use of Information and Communication Technologies in Households)

The trend for both dimensions is similar, the distances between categories are wider in 2017 than in 2008. The group with primary or less education goes from 1.4% in 2008 to 6.5% in 2017, while those with a master or Ph.D. went from 20.5 to 83.5%. Individuals with low digitals skills went from 9.8% in 2008 to 12.3% in 2017, while those with a high level went from 69.2% in 2008 to 89.8% in 2017.

Income and Employment situation also reflects a positive evolution in terms of growth of penetration rates, but not in terms of bridging the digital-divide. 3.6% of individuals with low income bought online in 2008, reaching 41% in 2017; and the group of people with high income went from 23.6% in 2008 to 79.7% in 2017. Unemployed and retired had a penetration rate of 4% in 2008 and 25.1% in 2017, while employed went from 13.1% in 2008 to 63.9% in 2017 (Fig. 4.5).

An aggregate perspective suggests a positive evolution of e-commerce adoption in Spain. But, contrasting among groups of categories for different factors reveals that growth in adoption is not yet reducing the distance in absolute values between individuals with lowest penetration rates compared to those with the highest, on the contrary it is widening. And, as expected, most of the distances are slowly narrowing when relative values are compared (Table 4.1).

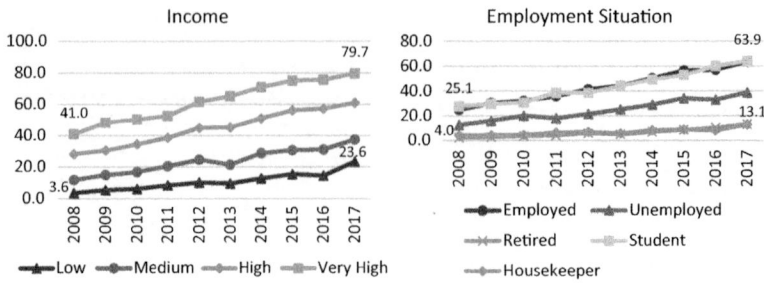

Fig. 4.5 Evolution of individual penetration rates of e-commerce in Spain (2008–2017): Income and Employment Situation (*Source* Authors' creation from *INE* [2018]. Survey on Equipment and Use of Information and Communication Technologies in Households)

Table 4.1 Individual penetration rates of e-commerce in Spain (2008 and 2017)

Characteristics	Category	Relative distance 2008 (%)	Penetration rates		Relative distance 2017 (%)
			2008	2017	
Gender	Female	30.70	14.2	41.4	14.60
	Male		20.5	48.5	
Age	<25	96.80	25.1	64.5	↓ 86.3
	(25, 35)		31.4	68.1	
	(35, 45)		21.3	62.3	
	(45, 55)		16.2	51	
	(55, 65)		7.2	31.8	
	65+		1	9.3	
Education	Primary	96.70	1.4	6.5	↓ 92.1
	Secondary		16.6	44.7	
	Bachelor's degree		29.7	71.1	
	Master or Ph.D.		42.6	81.8	
Digital skills	Low	85.80	9.8	12.3	↑ 86.3
	Medium		25.1	44.5	
	High		47.4	71.2	
	Very high		69.2	89.8	
Habitat	<20,000	41.40	13	38.7	↓ 24.4
	20,000–100,000		16.3	45.4	
	100,000–500,000		18.2	42.2	
	>500,000		22.2	51.2	

(continued)

Table 4.1 (continued)

Characteristics	Category	Relative distance 2008 (%)	Penetration rates 2008	Penetration rates 2017	Relative distance 2017 (%)
Household members	1	42.30	12.4	31.1	↑ 47.0
	2		14.6	32.5	
	3		18.1	51.2	
	4		21.5	58.7	
	5+		14.9	43	
Nationality	Foreigner	18.10	14.5	41	↓ 9.4
	Spanish		17.7	45.3	
Employment situation	Employed	85.60	25.1	63.9	↓79.6
	Unemployed		12.7	38.8	
	Retired		2.2	13.3	
	Student		27.8	64.2	
	Housekeeper		4	13.1	
	Other		11.2	27.6	
Income	Low	91.20	3.6	23.6	↓ 70.4
	Medium		11.8	37.5	
	High		28.3	61	
	Very high		41	79.7	
Autonomous community	Andalucía	49.80	13.1	41	↓33.2
	Aragón		18.7	49.1	
	Asturias		16.1	44.3	
	Baleares		24.3	52.9	
	Canarias		12.2	37.6	
	Cantabria		18.6	49.8	
	Castilla y León		14.6	39.4	
	Castilla La Mancha		12.6	41.7	
	Cataluña		21.2	49.1	
	Valencia		15.3	43.7	
	Extremadura		14.4	37.1	
	Galicia		13.9	36	
	Madrid		23.8	53.9	
	Murcia		14.2	39.2	
	Navarra		20.1	53.2	
	País Vasco		21.1	46.7	
	La Rioja		18.3	44	
Autonomous cities	Ceuta	29.60	12.6	48.6	↓0.8
	Melilla		17.9	47.8	
Total			17.3	44.9	

Source Authors' creation from Instituto Nacional de Estadística (INE) (2018). Survey on Equipment and Use of Information and Communication Technologies in Households

Note Weighted data: The sampling weight corresponds to the number of subjects in the population represented by each observation. Penetration rates as a percentage of individuals aged 16 and up. An individual is considered as an online buyer in any of the categories if he or she has carried out an online purchase for private use within the last 12 months. Relative distances are the differences between the lowest and the highest penetration rates, as a percentage of the highest penetration rate

The ICT-H Survey: Pooled Data

The dataset of 174,776 observations comes from pooling the annual cross sections from 2008 to 2017 of the survey on Equipment and Use of Information Technologies in Households (ICT-H), carried out annually by the Spanish National Statistical Institute (INE 2018). The survey is a panel-based study, that collects information about ICT's equipment and usage in Spanish households, following the methodological recommendations of Eurostat.

As information and communication technologies change continuously, the questionnaire[12] undergoes slight changes every year based on Eurostat's annual model questionnaires. The survey gathers information about the following subjects: access to and use of ICT by individuals and/or in households, use of the internet by individuals and/or in households, ICT security and trust, ICT competence and skills, barriers to use of ICT and the internet, perceived effects to ICT usage on individual and/or on households, use of e-government, and ubiquitous connectivity. The data allows to classify households by region of residence, geographical location, population density, type of household, and by household monthly net income. Individuals can also be classified by gender, country of birth, country of citizenship, educational level, occupation, employment situation, age, and legal or de facto marital status (Eurostat 2018b).

Since this paper examines the socioeconomic and demographic factors that influence the odds of becoming an online shopper, only some of the questions from the questionnaire are used for the statistical description and for the specification and estimation of the models. Table 4.1 shows the penetration rates as a percentage of individuals aged 16 and above.

Overall penetration rates of e-commerce in Spain has more than doubled in a decade, going from 17.3% in 2008, to 44.9%. As expected, the rate is higher when calculated as a percentage of individuals aged 16–74, who used the internet within the last three months, or even within the last year.

Differences of adoption are observable within specific demographic factors (Gender, Age, Habitat and number of Household Members).

---

[12] From 2008 to 2017 used questionnaires have included between 50 and 70 questions, many of them with several possible answers, meaning more than 200 variables, depending on the composition of the questionnaires for each year.

Attributes as being male, being between 16 and 45 years old, living in urban areas of more than 100,000 inhabitants and belonging to households with three or more members, are associated with higher penetration rates of e-commerce.

The higher the level of education, digital skills, and income, the higher the penetration rates of online shopping. These are the variables where the gaps between groups of each category are more evident. In 2017 only 6.5% of people with primary or less education have bought online, while 81.8% did so in the case of those who hold a Master's or a Ph.D. degree. In the case of Digital Skills, the penetration is 89.8% for high level, compared with 12.3% for low level. For low, medium, high, and very high income, the penetrations rates were 23.6, 37.5, 61.0, and 79.7%, respectively

Comparing regions, Madrid, Navarra, Baleares, Cantabria, and Cataluña are the top five autonomous communities with the highest penetration rates. País Vasco, that was 4th in 2008, dropped to the 9th position in 2017; and the 4th place now is occupied by Cantabria, which was 7th in 2008. It is remarkable how the autonomous city of Ceuta (in northern Africa) has moved from 17th to 7th, going from 12.6% (in 2008) up to 48.6% (in 2017) of e-commerce adoption, above the national average.

Differences within categories suggest that there are significant divides in e-commerce adoption. Penetration rates diverge across educational, digital skills, and income levels. The same happens with gender, age, nationality, household members, habitat, and regions.

THEORETICAL AND EMPIRICAL MODELS

Adopting e-commerce is an individual economic decision that will depend mostly on economic conditions as well as other sociodemographic and geographic variables. Like most economic decisions related to individual consumption it depends on income, cost, and cost of substitutes/complements, and size of the network (supply and customer base). A discrete choice framework is adopted, derived from utility theory, following an approach similar to Varian (2002) for modeling the demand for bandwidth, and Demoussis and Giannakopoulos (2006) used to model the individual decision to adopt the internet.

The adoption of this simple behavioral model assumes that the individual has an economically predictable behavior, allowing the analysis of choices made among a finite set of alternatives, which in this case are: buying online or not.

The e-commerce adoption decision y_i is determined by the surplus that the decision-maker (individual i) obtains from choosing whether to buy online for private use or not. This is, when costs of buying online are exceeded by the benefits the individual derives utility, U_i, from using the internet to buy goods and services.

$$\text{E-commerce adoption} \begin{cases} y_i = 1 & U_i(buyonline) > U_i(notbuyonline) \\ y_i = 0 & otherwise \end{cases} \quad (4.1)$$

The model of individual adoption based on Utility Maximization Framework:

$$U_i = f(\beta z_i) + \varepsilon_i \quad (4.2)$$

$$z_i = z(x_i, s_i) \quad (4.3)$$

- β is the vector of coefficients of the observed explanatory variables.
- z_i depicts the vector of observed variables related to person i (where the attributes of the alternatives interact with the specific characteristics of the decision-maker).
- ε_i is the influence of all unobserved factors.
- x_i attributes of the alternatives.
- s_i specific characteristics of the person i.

The decision of buying online for private use or not is represented the binary dependent variable defined below in Table 4.2.

Explanatory variables are grouped as follows in Table 4.3: Sociodemographic, Individual Skills, Economic and Geographic, and Time variables:

Since Gender, Nationality, Yearly Dummies, and Regional Dummies are binary variables, no categorization was needed. The rest of the variables were categorized as shown in Table 4.1.

Sociodemographic variables can be interpreted as follows: Gender allows for a gap between males and females; Age lowers or increases the cost, depending on the considered age ranges; Habitat increases or reduces the benefits, conditional on the size of the populations to which the individual belongs; Household Members is related with the benefits and also with the costs (that may be due to a learning effect); and, Nationality might

Table 4.2 E-commerce

E-commerce	= 1, if the individual bought online for private use in the last 12 months
	= 0, otherwise

Source Author's creation

Table 4.3 Explanatory variables

Sociodemographic	Gender:	2 groups: 1 if male, 0 if female
	Age:	6 groups
	Habitat:	4 groups
	Household Members:	5 groups
	Nationality:	2 groups: 1 if Spanish, 0 if Foreign
Individual skills	Education:	4 levels of study
	Digital Skills:	4 levels
Economic	Employment Situation:	6 groups
	Income:	4 groups, monthly net income
Geographic and time	Yearly Dummies:	1 for each year
	Regional Dummies:	17 Autonomous Communities

Source Author's creation

signal effects dependent on costs (e.g., different language) and benefits (e.g., access to a wider range of products and services). Variables such as Education and Digital Skills[13] are expected to diminish the costs of using internet services; Economic variables, Income and Employment Situation, are supposed to increase the benefits. And, Yearly and Regional Dummies point out benefits and costs across regions, and also those associated with changes (e.g., entry of new suppliers, legislation and institutional measures to foster e-commerce) that can occur over the years associated to belonging to specific regions.

A Linear Probability Model (LPM), depicted in Table 4.2. (1), has been estimated in the first place, allowing to measure the change in the probability of an individual adopting e-commerce when the observed explanatory variables change, holding other factors fixed.

---

[13] Digital Skills is a self-elaborated index, based on the answers where the respondent declares whether he or she used specific internet services and/or performed specific computer and internet-related tasks.

Table 4.4 Models of adoption of e-commerce by individual internet users. Linear Probability Model (LPM), Logistic Regression Model (LRM), and Heckman Selection Model (HSM). Pool data (2008–2017)

		(1) E-commerce adoption. Linear Probability Model		(2) E-commerce adoption. Logistic Regression Model		(3) E-commerce adoption. Heckman Selection Model	
		Coef.	t	Odds ratios	z	Coef.	z
Gender	Female						
	Male	0.04	10.33	1.29	10.4	0.04	10.2
Age	16–24						
	25–34	−0.03	−1.89	0.82	−0.98	−0.03	−1.5
	35–44	−0.09	−5.05	0.52	−3.49	−0.08	−4.64
	45–54	−0.11	−6.39	0.46	−4.13	−0.11	−5.86
	55–64	−0.16	−9.19	0.31	−6.00	−0.15	−8.35
	> 65	−0.19	−9.39	0.24	−6.38	−0.18	−8.37
Education	Primary or less						
	Secondary	0.04	5.77	1.45	6.36	0.04	5.13
	Bachelor	0.09	11.13	1.9	10.25	0.09	10.15
	Master/Ph.D.	0.13	14.88	2.34	13.45	0.12	13.52
Digital skills	Low						
	Medium	0.11	5.51	2.48	4.78	0.11	5.21
	High	0.26	12.65	4.88	8.71	0.25	12.2
	Very high	0.47	23.89	12.52	13.84	0.47	23.19
Habitat	<20,000						
	20,000–100,000	−0.01	−1.44	0.95	−1.57	−0.01	−1.78
	100,000–500,000	−0.04	−5.07	0.78	−5.24	−0.04	−5.38
	>500,000	0.00	0.86	1.02	0.73	0.00	0.57
Household Members	One						
	Two	−0.01	−2.1	0.93	−1.85	−0.02	−2.35
	Three	−0.04	−6.72	0.78	−6.36	−0.05	−6.86
	Four	−0.05	−7.00	0.76	−6.71	−0.05	−7.26
	Five or more	−0.08	−8.6	0.62	−8.34	−0.08	−8.72
Nationality	Foreigner						
	Spanish	0.03	3.3	1.16	2.82	0.03	3.09

(continued)

Table 4.4 (continued)

		(1) E-commerce adoption. Linear Probability Model		(2) E-commerce adoption. Logistic Regression Model		(3) E-commerce adoption. Heckman Selection Model	
		Coef.	t	Odds ratios	z	Coef.	z
Employment situation	Employed						
	Unemployed	−0.05	−8.19	0.74	−7.8	−0.05	−7.96
	Retired	−0.04	−3.18	0.82	−2.72	−0.03	−2.49
	Student	−0.06	−4.71	0.73	−5.00	−0.06	−5.05
	Housekeeper	−0.02	−2.58	0.84	−2.65	−0.02	−1.91
	Other	−0.02	−1.6	0.9	−1.36	−0.02	−1.16
Income	Low						
	Medium	0.04	6.69	1.32	6.97	0.04	5.8
	Medium-high	0.11	15.78	1.91	15.14	0.1	14.45
	High	0.17	20.28	2.71	19.45	0.16	18.92
Digital skills × Age	High × 55–64	0.14	5.64	2.4	4.15	0.14	5.44
	High × 65 or more	0.12	3.83	2.4	3.61	0.11	3.59
	Very high × 55–64	0.08	3.5	2.04	3.3	0.08	3.26
	Very high × 65 or more	0.14	4.54	3.07	4.27	0.13	4.19
Year	2008						
	2009	−0.01	−1.58	0.87	−2.41	−0.02	−1.82
	2010	0.03	2.64	1.14	2.32	0.02	2.32
	2011	0.02	1.54	1.05	0.78	0.01	1.15
	2012	0.01	1.35	1.06	1.01	0.01	0.74
	2013	0.04	3.81	1.23	3.72	0.03	3.32
	2014	0.09	9.29	1.61	8.43	0.08	8.46
	2015	0.18	19.19	2.89	18.68	0.17	18.14
	2016	0.18	19.21	2.83	18.56	0.17	18.15
	2017	0.24	26.31	4.17	25.81	0.23	24.71
Autonomous community	Andalucía						
	Aragón	0.01	1.43	1.10	1.56	0.01	−1.82
	Asturias	0.03	3.44	1.22	3.65	0.03	2.32
	Baleares	0.09	7.46	1.68	7.49	0.08	1.15
	Canarias	−0.05	−4.46	0.75	−4.31	−0.05	0.74
	Cantabria	0.05	5.26	1.40	5.42	0.05	3.32
	Castilla la Mancha	0.02	1.79	1.11	1.68	0.02	8.46
	Castilla León	0.00	−0.05	1.01	0.19	0	18.14

(continued)

Table 4.4 (continued)

		(1) E-commerce adoption. Linear Probability Model		(2) E-commerce adoption. Logistic Regression Model		(3) E-commerce adoption. Heckman Selection Model	
		Coef.	t	Odds ratios	z	Coef.	z
	Cataluña	0.05	**5.68**	**1.32**	**5.68**	0.05	**18.15**
	Extremadura	0.01	1.07	1.06	0.98	0.01	**24.71**
	Galicia	0.00	0.00	1.00	0.09	0	−1.82
	La Rioja	0.03	**2.59**	**1.21**	**2.72**	0.03	**2.32**
	Madrid	0.03	**3.07**	**1.17**	**3.20**	0.02	1.15
	Navarra	0.05	**5.47**	**1.35**	**5.47**	0.05	0.74
	País Vasco	0.05	**5.82**	**1.39**	**5.86**	0.05	**3.32**
	Murcia	−0.02	**−2.52**	0.87	**−2.33**	−0.02	**8.46**
	Valencia	−0.02	−1.99	0.91	−1.81	−0.02	**18.14**
	Ceuta	−0.04	−1.55	0.78	−1.64	−0.04	**18.15**
	Melilla	−0.02	−0.58	0.88	−0.73	−0.01	**24.71**
Constant		0.01	0.44	0.05	−15.36	0.04	1.65
N. observations		75,960		75,960		77,362	
F		556.10 DF: 70					
Wald x^2				10581.81 DF: 70		29332.2 DF: 70	
R^2		0.328					
Psuedo R^r				0.2759			
Correctly classified				75.95%			
Wald x^2: H_0 independent equations						54.96 DF: 1	

Source Authors' creation from *Eurostat* (2018b). Survey on Equipment and Use of Information and Communication. Technologies in Households

Note Coefficients and statistics (*t* and *z*) significant at the 5% are represented in bold. Linear Probability (1), Logistic Regression (2), and Heckman Selection (3) models are estimated. Weighted estimations. In all three models, estimations refer to those internet users (people who have used the internet, at least once, within the last year) who have purchased online (or not) during the last year. Robust estimates. The R^2 and the pseudo R^2 are measures of the in-sample goodness of fit. The relevant statistics for testing the significance of the estimated coefficients is the F test and the χ^2 test of joint significance of all slope coefficients, which are high and have low p-values: below 0.001 in all three cases. The percentages of correct classifications are above 75%. A polychoric correlation matrix has been computed to explore the possible collinearity among the regressors of the model. As shown in the Appendix, Table 4.4, correlations are moderate

Since the dependent variable is dichotomous, a Logistic Regression Model (LRM), Table 4.2. (2), has been estimated.[14] And to correct a possible sample selection bias, a Heckman Selection Model (HSM), Table 4.2. (3), has also been estimated. For all three models the specification is equivalent in terms of the considered dependent and explanatory variables and the corresponding base categories.[15,16]

LINEAR PROBABILITY MODEL

The linear probability model is a first approach to modeling the relationship between a binary variable and its determinants. It suffers from well-known limitations such as predicted probabilities that may lie outside the [0, 1] interval. Nevertheless, it is recognized as a reliable indication of signs and significance of estimated coefficients. Moreover, it provides a straightforward interpretation of the estimated coefficients in terms of differences in predicted probabilities.

Each significant coefficient of the independent variables in the LPM estimation represents changes in the probability that a person adopts e-commerce for private use (E-commerce = 1), everything else held constant. Being male increases the probability of buying online by 0.04. Higher levels of educations are related with higher probabilities of adoption, going up to 0.09 for Bachelor and 0.13 for Master or Ph.D. A Spanish person has a probability 0.03 higher than a foreigner. Any other employment situation than being employed, decreases the probability of becoming an online buyer. The higher the income, the higher the probability: 0.04, 0.06, and 0.17, for Medium, Medium-high, and High levels, respectively.

The year dummies seem to be especially relevant for the last five years, increasing the probability of an individual buying online from 0.04 in 2013 up to 0.24 in 2017. Something similar happened with the regional

---

[14] All three models are estimated using Stata 15.

[15] Base categories for each variable have been set as follow: Gender, Male; Age, 16–24 years old; Education, Primary or less; Digital Skills, Low; Habitat, lower than 20,000 inhabitants; Household Members, one member; Nationality, Foreigner; Employment Situation, Employed; Income, Low; Year, 2008; and Autonomous Community, Andalucía. The interaction term "Digital Skills × Age" uses the same base categories already fixed for interacted explanatory variables.

[16] Complete estimates for the interactive terms are show in the Appendix, Table 4.5.

dummies, where living in Canarias lowers the probability by 0.05, while living in Baleares increases by 0.09 the probability of adopting e-commerce.

In the LPM estimates, the interaction between Digital Skills and Age is statistically significant. The interaction term allows us to infer how the effect of Age on the e-commerce adoption is moderated by the level of Digital Skills.[17] High Digital Skills seem to have a positive influence, partly counteracting the negative effects on probability of e-commerce adoption of some age groups. The parameters on the interacted original variables, Age and Digital Skills, have no direct interpretation. Coefficients for each Age group represent the effect of Age on e-commerce adoption for an internet user with Low Digital Skills; that is why all Age groups have negative effects. It is similar at analyzing Digital Skills, its coefficients represent the effect of Digital Skills on e-commerce adoption for an internet user aged 16–24 years old.

Logistic Regression Model

The LRM is an alternative to the LPM. One of the objectives is to obtain predicted probabilities in the interval [0, 1]. This is achieved by fitting a nonlinear model, the LRM, which is widely used in the literature.

Odds ratios[18] estimated from the LRM also allow to establish the relationship between observed variables and the odds of an individual to become an online shopper. In line with LPM, the outcomes of LRM confirm that being male, account for higher levels of education, being Spanish, being employed, and having higher income levels are positively related with the adoption of e-commerce.

Odds ratios of the Interaction term in LRM are not straightforward interpretable, as the magnitude of the interaction effect in nonlinear models does not equal the marginal effect of the interaction term (Ai and Norton 2003).

---

[17] If it is needed to be analyzed the other way around: Interaction term allows us to infer how the effect of Digital Skills on the e-commerce adoption depends on the magnitude of Age.

[18] Odds ratios of the explanatory variables are depicted in Figs. 4.7 and 4.8, in the Appendix section. Odds ratios greater than 1 describe a positive relationship, whereas if they are less than 1, they indicate a negative relationship.

Yearly dummies denote mostly positive and significant effects along time. Odds ratios for 2015 and 2016 are almost twice the odds ratio for 2013; and, for 2017 its odds ratio is more than 4 times higher when it is compared with the base year, 2008. Meaning that the odds of buying online have increased substantially over the years and suggesting unobserved factors that have had important effects in the years 2015–2017.

As with LPM estimates, odds ratios of the regions reflect considerable differences among Spanish Autonomous Communities. Positive and significant effects: Baleares, 1.68; Cantabria, 1.40; País Vasco, 1.39; Navarra, 1.35; Cataluña, 1.32; Asturias, 1.22; La Rioja, 1.21; and Madrid, 1.17. Those negatively related with the odds of shopping online are: Canarias and Murcia, with odds ratios of 0.75 and 0.87, respectively.

The magnitude of the coefficient estimates in the LPM and those of the LRM are not directly comparable. However, the signs are comparable and also the magnitudes of t and z statistics, which show a similar pattern in models (1) and (2).

Heckman Selection Model

The use of e-commerce is only asked to those who answered having used the internet at least once. Hence, models of e-commerce adoption are estimated with the resulting subset of the sample obtained by a previous decision individual made about using the internet, Fig. 4.6. As this might imply a selection bias, Heckman's Selection Model (HSM) is estimated,[19] as Cerno and Pérez-Amaral (2006a), Lera-López et al.(2011) and Pérez-Hernández and Sánchez-Mangas (2011), among others, did at modeling similar individual decision processes.

Wald´s test for independent equations, with $\chi^2 = 54.96$ and low p-value, below 0.0001, suggests using Heckman's method to estimate a model that allows for selection (Heckman 1977) with this dataset. However, the magnitudes of the coefficients of the second stage estimation of HSM, Table 4.2 (3), are comparable and remarkably similar to the results of the LPM, Table 4.2 (1). Most of the categories of the explanatory variables coincide in their statistical significance in the three models. The main differences are observed in some of the regional

---

[19] Stata–Heckman selection model: "*Heckman*" command and "*maximum likelihood estimates*" option (which allows to include the sample elevation factor for all observations, as well as compute robust standard errors.).

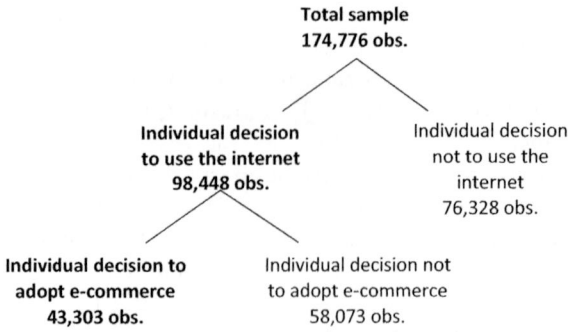

Fig. 4.6 Decision process of the *internet* and e-commerce adoption (2008–2017) (*Source* Authors' creation based on processed data from *INE* [2018]. Survey on Equipment and Use of Information and Communication Technologies in Households)

variables (Baleares, Canarias, Madrid, and Navarra) that are no longer statistically significant, and others that become statistically significant, as Castilla la Mancha, Castilla Leon, Extremadura, Ceuta, and Melilla.

As a selection equation[20] shown in Table 4.3 of the appendix, a probit model of the decision of using the internet has been estimated. All the explanatory variables in the second stage estimation also include as covariates in the first stage equation, except for Digital Skills, to avoid simultaneity, since Digital Skills account for some internet-related activities. Computer availability is used as exclusion restriction, as it is highly correlated with using internet and weakly correlated with the adoption of e-commerce; besides, it might pick up the unobserved individual factors that affect the tendency to use internet services.

Except for Gender and Autonomous Community, all the coefficients of the variables included in the selection model are statistically significant, at the 5%, to explain the individual decision of using the internet. χ^2 test of joint significance of all slope coefficients is 2631.86, which is high and has a p-value below 0.001.

---

[20] Estimation results of first stage of HSM: Appendix, Table 4.7.

Conclusions

General Conclusions

When the proportion of e-buyers in Spain is compared with EU-28 average for 2017, it may seem that the gap is being bridged, but this is an illusion due to data aggregation. If we zoom in specific factors (Age, Education, Digital Skills, Habitat, Household Members, Nationality, Employment Situation, Income, Time, and Regions), it has been improved overall, however, the distances between category groups within those factors still exist and are especially remarkable, as: younger than 25, primary vs. those with higher education, individuals with low digital skills vs. high level, employed vs. any other employment situation, and lower vs. higher income.

Previous works that have used cross-sectional data from 2016, already pointed out how sociodemographic individual characteristics and economic factors explain the adoption of three internet services (e-commerce, e-banking, and e-government) (Garín-Muñoz et al. 2018); and, many of them are statistically significant at explaining the individual decision to perform cross-border e-commerce (Valarezo et al. 2018). The results of the present work reinforce the conclusions and allow to consider aggregate time effects and unobserved economic environment characteristics through the introduction of year dummies.

To the best of our knowledge, this is the first time the period 2008–2017 of this survey has been analyzed using pooled data to model and estimate the determinants of e-commerce adoption in Spain.[21]

Signs and statistical significance of the determinants mostly coincide for Linear Probability, Logistic Regression, and Heckman models. The findings have implications for consumers, policy-makers, and firms.

Some of the determinants that reduce the probability of buying online can hardly be directly affected by policy measures. Among others, these are the factors that have a negative impact: being female, age, not being Spanish, any other employment situation than being employed, and belonging to Autonomous Communities of Canarias, Murcia, and Valencia.

---

[21] Correa et al. (2015) and Pérez-Hernández and Sánchez-Mangas (2011), estimated the determinants of buying online, using pooled data for periods 2008–2014 and 2004–2009, respectively.

On the other hand, factors such as Education, Digital Skills, and living in specific regions, have a positive influence and are more susceptible to be targeted by policy measures for reducing the e-commerce divide. And, as expected, Higher Income is also positively related with e-commerce adoption.

The interaction between Digital Skills and Age points out that high digital skills have a positive influence, partly counteracting the negative effect of some age groups.

Policy Recommendations

There are different types of actors interested in fostering e-commerce. Specific policy recommendations are proposed for demand-side actions, to promote governmental initiatives and to address specific strategies by the private sector.

On the demand side, the main goal for policy-makers should be to narrow the existing digital-divide between groups with low and high penetration rates of e-commerce. Considering that even small increases in Education and Digital Skills substantially improve the probability of buying online, the economic policy could be targeted to specific groups, especially to individuals with lower income, lower levels of digital skills and education, older people, unemployed, retired, housekeepers, and in general for those that belong to groups with lower odds of adopting e-commerce. Training on specific Digital Skills, technical support (online, by phone, and in person) and diffusion through conventional channels may bring about the untapped opportunities of e-commerce.

From the government standpoint relevant initiatives can be carried out, such as: to promote the adoption of complementary services (e.g., e-government, e-health, e-learning, among others), to identify and reedit successful supply-side programs, to reduce transaction costs (e.g., red tape, trade and taxation barriers, geographic and linguistic barriers) and to foster the development of efficient e-commerce platforms. Measures of policy success should be based on adoption gains on those groups that are worst placed, instead of only accounting at an aggregate level.

The private sector can play an active role, adopting easy-to-use platforms, encouraging customer reviews, designing promotions and communication campaigns for groups with low penetration, and streamlining payment and transaction process. Technology already allows e-commerce companies to serve websites and highly customized user interfaces based on artificial intelligence systems trained with user data. It can also adjust and customize front pages and web application user interfaces to match the needs of those groups with lower penetration rates. Another helpful initiative could be the implementation of in-store demonstration of online buying; and for those pure-play e-commerce,[22] physical points of purchase equipped with display devices, where prospective customers can go through the online purchase process (as if they did it from their own devices) with the help of shopping assistants.[23]

Caveats

Some of the limitations found in this study are related with the lack of specific information about barriers that stopped internet users from buying online, since the ICT-H INE's survey was not specifically designed for this purpose. Controlling for individual (unobserved) effects would be useful and not having this represents another important caveat. Supply-side data also may be useful to perform a complementary analysis to get richer context for results and conclusions.

Further Research

The next stage in the further research agenda will be analyzing dynamic models, using the full panel data set. Doing so we will be able to control for individual (unobserved) effects and account for possible simultaneity and dynamic relationships. Specific online transactions and digital activities that are performed or affect individuals also deserve attention such as: buying and selling online different types of goods and services, cloud computing, use of digital certificates, sharing economy services, digital transformation, and the use of ICTs at work, among others.

Acknowledgements An earlier version of this paper was presented at the 29th ITS European Conference, Trento—Italy, 2018.

---

[22] A pure play e-commerce is an e-commerce business that only sell through the internet.

[23] A good example are the physical stores of the third largest Spanish e-commerce firm, Pcomponentes.com, where customers and prospective customers get advice and guide for later purchasing online.

Appendix

See Tables 4.5, 4.6 and 4.7 and Figs. 4.7 and 4.8.

Table 4.5 Models of adoption of e-commerce by individual internet users. Linear Probability Model (LPM) and Logistic Regression. Complete estimates for the interaction term Digital Skills × Age. Pooled data (2008–2017)

		(1) E-commerce adoption. Linear Probability Model		(2) E-commerce adoption. Logistic Regression		(3) E-commerce adoption. Heckman Selection Model	
		Coef.	t	Odds ratios	Z	Coef.	z
Digital skills × Age	Medium × 16–24						
	Medium × 25–34	0.02	0.79	1.09	0.4	0.02	0.65
	Medium × 35–44	0.07	**2.90**	**1.65**	**2.43**	0.07	**2.85**
	Medium × 45–54	**0.06**	**2.57**	**1.57**	**2.16**	**0.06**	**2.57**
	Medium × 55–64	**0.08**	**3.23**	**1.89**	**2.94**	**0.08**	**3.17**
	Medium × 65 or more	0.04	1.63	1.73	**2.23**	0.04	1.57
	High × 16–24						
	High × 25–34	**0.09**	**3.88**	1.46	1.79	**0.09**	**3.57**
	High × 35–44	**0.13**	**6.00**	**2.12**	**3.79**	**0.13**	**5.83**
	High × 45–54	**0.12**	**5.38**	**2.01**	**3.47**	**0.12**	**5.26**
	High × 55–64	**0.14**	**5.64**	**2.40**	**4.15**	**0.14**	**5.44**
	High × 65 or more	**0.12**	**3.83**	**2.40**	**3.61**	**0.11**	**3.59**
	Very high 16 × 24						
	Very high × 25–34	**0.08**	**3.58**	**1.59**	**2.19**	**0.07**	**3.21**
	Very high × 35–44	**0.11**	**5.06**	**2.21**	**3.98**	**0.1**	**4.84**
	Very high × 45–54	**0.11**	**4.8**	**2.16**	**3.76**	**0.1**	**4.61**
	Very high × 55–64	**0.08**	**3.5**	**2.04**	**3.3**	**0.08**	**3.26**
	Very high × 65 or more	**0.14**	**4.54**	**3.07**	**4.27**	**0.13**	**4.19**

Source Authors' creation from *INE* (2018). Survey on Equipment and Use of Information and Communication Technologies in Households
Coefficients, Odd ratios, t and z statistics significant at the 5% are represented in bold

Table 4.6 Polychoric correlations among selected independent variables

	Gender	Age	Education	Digital skills	Household members	Income
Gender	1.00					
Age	0.02	1.00				
Education	−0.07	−0.02	1.00			
Digital skills	0.07	−0.31	0.44	1.00		
Household members	0.02	−0.32	−0.02	0.07	1.00	
Income	0.08	0.05	0.45	0.31	0.24	1.00

Source Authors' creation from *INE* (2018). Survey on Equipment and Use of Information and Communication Technologies in Households

Note For assessing possible multicollinearity. Computed polychoric correlations are moderate. Little sign of multicollinearity is found

Table 4.7 Model of internet use. Heckman first stage estimates. Probit Regression Model. Pooled data (2008–2017)

Internet use. Probit Regression Model–Heckman first stage estimates							
		Coef.	z			Coef.	z
Gender	Female			Computer availability	No		
	Male	0.04	1.1		Yes	1.04	29.34
Age	16–24			Year	2008		
	25–34	−0.18	−2.25		2009	0.17	2.53
	35–44	−0.36	−4.54		2010	0.2	3.18
	45–54	−0.56	−6.97		2011	0.26	3.79
	55–64	−0.7	−8.44		2012	0.45	6.35
	>65	−0.87	−9.02		2013	0.3	4.4
Education	Primary or less				2014	0.55	7.84
	Secondary	0.11	2.51		2015	0.57	7.58
	Bachelor	0.33	5.62		2016	0.59	8.67
	Master/Ph.D.	0.59	9.09		2017	0.94	12.84
Habitat	<20,000			Autonomous Community	Andalucía		
	20,000–100,000	0.14	3.13		Aragón	0.07	0.92
	100,000–500,000	0.2	3.24		Asturias	−0.03	−0.45
	>500,000	0.09	2.55		Baleares	0.13	1.51
Household Members	One				Canarias	0	−0.01
	Two	−0.01	−0.32		Cantabria	0	0.04
	Three	−0.09	−2.09		Castilla la Mancha	0.03	0.42
	Four	−0.07	−1.27		Castilla León	−0.04	−0.66
	Five or more	−0.07	−0.91		Cataluña	0.01	0.12
Nationality	Foreigner				Extremadura	0.08	1.11
	Spanish	0	0.02		Galicia	0.02	0.27

(continued)

Table 4.7 (continued)

Internet use. Probit Regression Model–Heckman first stage estimates

		Coef.	z			Coef.	z
Employment situation	Employed				La Rioja	0.05	0.57
	Unemployed	−0.09	-1.9		Madrid	0.05	0.75
	Retired	**−0.39**	**−7.2**		Navarra	−0.06	−0.92
	Student	**1.16**	**7.5**		País Vasco	0.01	0.22
	Housekeeper	**−0.36**	**−6.21**		Murcia	**−0.15**	**−2.09**
	Other	**−0.34**	**−4.24**		Valencia	0.05	0.8
Income	Low				Ceuta	−0.04	−0.2
	Medium	**0.17**	**4.26**		Melilla	−0.1	−0.62
	Medium-high	**0.33**	**6.68**	Constant		**0.86**	**7.48**
	High	**0.56**	**8.11**				
N. observations				77,362			
Wald χ^2				2631.86 DF: 53			
Pseudo R^2				0.2363			
Correctly classified				97.08%			
Wald χ^2: H_0: independent equations				54.96 DF: 1			

Source Authors creation from *INE* (2018). Survey on Equipment and Use of Information and Communication Technologies in Households

Note Coefficients and statistics (*t* and *z*) significant at the 5% are represented in bold. Heckman Selection first step is estimated. Robust estimates. Pseudo R^2 is a measure of the in-sample goodness of fit. The relevant statistics for testing the significance of the estimated coefficients is the χ^2 test of joint significance of all slope coefficients equal to zero, which is high and has low *p*-value, below 0.001. The percentages of correct classifications are above 97%

4 ADOPTION OF E-COMMERCE BY INDIVIDUALS AND DIGITAL-DIVIDE   131

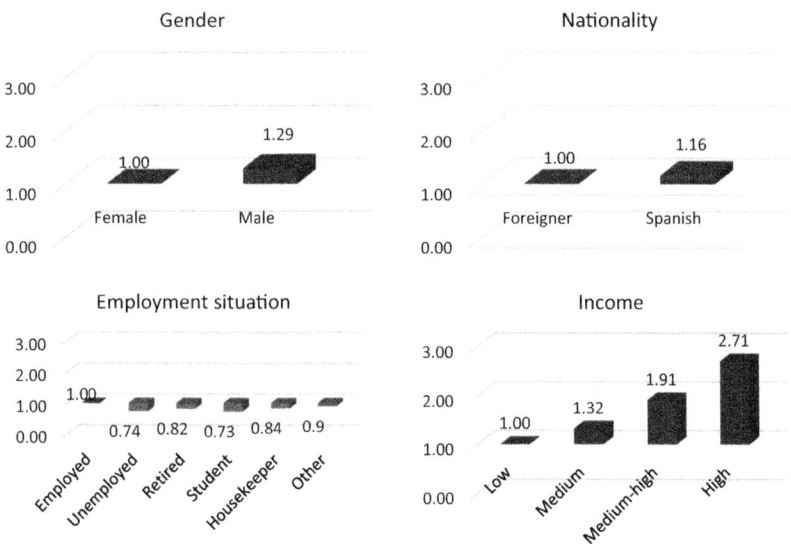

Fig. 4.7 Odds ratios of e-commerce. Pool (2008–2017) (*Source* Authors' creation from estimation results, Table 4.2)

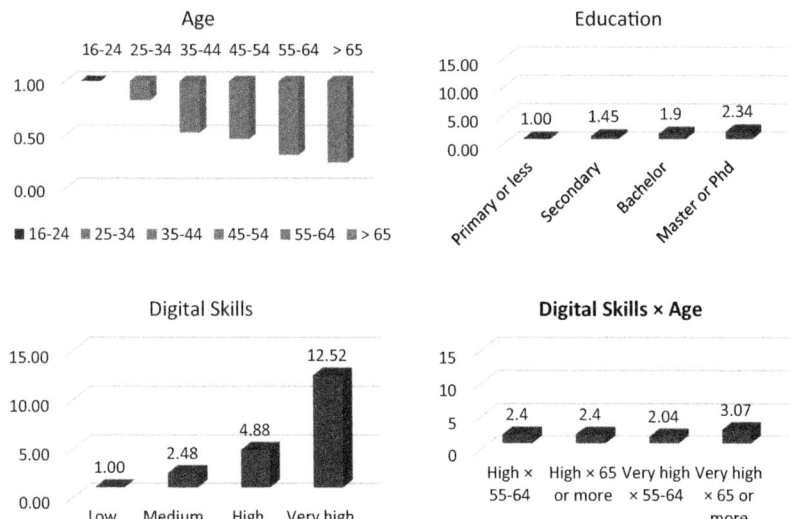

Fig. 4.8 Odds of e-commerce. Pool (2008–2017) (*Source* Authors' creation from estimation results, Table 4.2)

References

Ai, C., and E.C. Norton. 2003. Interaction Terms in Logit and Probit Models. *Economics Letters* 80 (1): 123–129. https://doi.org/10.1016/S0165-1765(03)00032-6.

Ajzen, I. 1991. The Theory of Planned Behavior. *Organizational Behavior and Human Decision Processes* 50 (2): 79–211.

Ajzen, Icek, and M. Fishbein. 1977. Attitude-Behavior Relations: A Theoretical Analysis and Review of Empirical Research. *Psychological Bulletin* 84 (5): 888–918. https://doi.org/10.1037/0033-2909.84.5.888.

Ajzen, Icek, and M. Fishbein. 1980. *Understanding Attitudes and Predicting Social Behavior*. Englewood Cliffs, NJ: Prentice-Hall.

Alonso, J., and A. Arellano. 2015. Heterogeneity and Diffusion in the Digital Economy: Spain's Case. https://www.bbvaresearch.com/wp-content/uploads/2015/11/15-28_WP-Ec_Digital_e.pdf. Accessed 21 Nov 2018.

Cerno, L., and T. Pérez-Amaral. 2006a. Demand for Internet Access and Use in Spain. In *Governance of Communication Networks: Connecting Societies and Markets with IT*, ed. B. Preissl and J. Müller, pp. 333–353. Heidelberg: Physica-Verlag HD. https://doi.org/10.1007/3-7908-1746-5_18.

Cerno, L., and T. Pérez-Amaral. 2006b. *Medición y determinantes de la brecha tecnológica en España*. Retrieved from https://eprints.ucm.es/7909/1/0601.pdf.

Cerno, L., and T. Pérez-Amaral. 2009. E-Commerce Use in Spain. In *Telecommunication Markets: Drivers and Impediments*, ed. P. Curwen, J. Haucap, and B. Preissl, 157–172. Heidelberg: Physica-Verlag HD. https://doi.org/10.1007/978-3-7908-2082-9_9.

Correa, M., J.R. García, and A. Tabanera. 2015. Comercio electrónico y hábitos de consumo en España: la importancia de la banca on-line. *Observatorio de Economía Digital, BBVA Research*. Retrieved from https://www.bbvaresearch.com/publicaciones/comercio-electronico-y-habitos-de-consumo-en-espana-la-importancia-de-la-banca-on-line/.

Davis, Fred D. 1989. Perceived Usefulness, Perceived Ease of Use, and User Acceptance of Information Technology. *MIS Quarterly* 13 (3): 319–340. https://doi.org/10.2307/249008.

Davis, Fred D. 1993. User Acceptance of Information Technology: System Characteristics, User Perceptions and Behavioral Impacts. *International Journal of Man-Machine Studies* 38 (3): 475–487. https://doi.org/10.1006/imms.1993.1022.

Davis, F.D., and P.R. Warshaw. 1989. User Acceptance of Information Technology: System Characteristics, User Perceptions, and Behavioral Impacts. *Management Sciences* 35 (8): 982–1003.

Davis, F.D., R.P. Bagozzi, and P.R. Warshaw. 1989. User Acceptance of Computer Technology: A Comparison of Two Theoretical Models. *Management Sciences* 35 (8): 982–1003.

Demoussis, M., and N. Giannakopoulos. 2006. Facets of the Digital-Divide in Europe: Determination and Extent of Internet Use. *Economics of Innovation and New Technology* 15 (3): 235–246. https://doi.org/10.1080/10438590500216016.

DiMaggio, P., and E. Hargittai. 2001. From the "Digital Divide" to "Digital Inequality": Studying Internet Use as Penetration Increases. Working Papers No. 47. Princeton University, Woodrow Wilson School of Public and International Affairs, Center for Arts and Cultural Policy Studies. Retrieved from https://ideas.repec.org/p/pri/cpanda/workpap15.html.html.

European Commission. 2018a. Digital Economy and Society Index (DESI) 2018 Country Report Spain. https://ec.europa.eu/digital-single-market/en/countries-performance-digitisation. Accessed 11 Nov 2018.

European Commission. 2018b. The Digital Economy and Society Index (DESI). https://ec.europa.eu/digital-single-market/en/desi. Accessed 11 Nov 2018.

Eurostat. 2018a. Glossary: E-commerce—Statistics Explained. https://ec.europa.eu/eurostat/statistics-explained/index.php/Glossary:E-commerce. Accessed 7 Oct 2018.

Eurostat. 2018b. ICT Usage in Households and by Individuals. https://ec.europa.eu/eurostat/cache/metadata/en/isoc_i_esms.htm. Accessed 6 Oct 2018.

Garín-Muñoz, T., and T. Pérez-Amaral. 2011. Factores Determinantes del Comercio Electrónico en España. *Boletín Económico Del ICE* 3016: 51–65.

Garín-Muñoz, T., R. López, T. Pérez-Amaral, I. Herguera, and A. Valarezo. 2019. Models for Individual Adoption of eCommerce, eBanking and eGovernment in Spain. *Telecommunications Policy* 43 (1): 100–111. https://doi.org/10.1016/j.telpol.2018.01.002.

Hargittai, E. 2001. Second-Level Digital Divide: Mapping Differences in People's Online Skills. ArXiv:Cs/0109068. Retrieved from http://arxiv.org/abs/cs/0109068.

Heckman, J.J. 1977. Sample Selection Bias as a Specification Error (with an Application to the Estimation of Labor Supply Functions). Working Paper No. 172. National Bureau of Economic Research. https://doi.org/10.3386/w0172.

Helsper, E.J., and A.J.A.M. van Deursen. 2015. The Third-Level Digital Divide: Who Benefits Most from Being Online? In *Communication and Information Technologies Annual*, vol. 10, 29–52. Emerald Group Publishing Limited. https://doi.org/10.1108/S2050-206020150000010002.

INE. 2018. Encuesta sobre equipamiento y uso de tecnologías de información y comunicación en los hogares. Documentos metodológicos y cuestionarios.

http://www.ine.es/dyngs/INEbase/es/operacion.htm?c=Estadistica_C&cid=1254736176741&menu=metodologia&idp=1254735976608. Accessed 7 Oct 2018.

Lera-López, F., M. Billon, and M. Gil. 2011. Determinants of Internet Use in Spain. *Economics of Innovation and New Technology* 20 (2): 127–152. https://doi.org/10.1080/10438590903378017.

Norris, P. 2001. *Digital Divide: Civic Engagement, Information Poverty, and the Internet Worldwide*. Cambridge University Press.

NTIA (National Telecommunications and Information Administration). 1995. Falling Through the Net: A Survey of the "Have Nots" in Rural and Urban America. https://www.ntia.doc.gov/ntiahome/fallingthru.html. Accessed 27 Nov 2018.

OECD. 2001. Understanding the Digital Divide. https://doi.org/10.1787/236405667766.

Oliver, R.L. 1980. A Cognitive Model of the Antecedents and Consequences of Satisfaction Decisions. *Journal of Marketing Research* 17 (4): 460–469. https://doi.org/10.2307/3150499.

Pérez-Hernández, J., and R. Sánchez-Mangas. 2011. To Have or Not to Have Internet at Home: Implications for Online Shopping. *Information Economics and Policy* 23 (3): 213–226. https://doi.org/10.1016/j.infoecopol.2011.03.003.

Rogers, E.M. 2003. *Diffusion of Innovations*, (Edición: 5th) 5 ed. New York: Simon & Schuster.

Srinuan, C., and E. Bohlin. 2011. Understanding the Digital-Divide: A Literature Survey and Ways Forward. Budapest: ITS. Retrieved from https://www.econstor.eu/handle/10419/52191.

Taylor, S., and P.A. Todd. 1995. Understanding Information Technology Usage: A Test of Competing Models. *Information Systems Research* 6 (2): 144–176. https://doi.org/10.1287/isre.6.2.144.

Valarezo, Á., T. Pérez-Amaral, T. Garín-Muñoz, I. Herguera García, and R. López. 2018. Drivers and Barriers to Cross-Border e-commerce: Evidence from Spanish Individual Behavior. *Telecommunications Policy* 42 (6): 464–473. https://doi.org/10.1016/j.telpol.2018.03.006.

Varian, H.R. 2002. The Demand for Bandwidth: Evidence from the INDEX Project. In *Broadband: Should We Regulate High-Speed Internet Access?* ed. R.W. Crandall and J.H. Alleman, 39–56. Washington, DC: AEI-Brookings Joint Center for Regulatory Studies, American Enterprise Institute.

Venkatesh, V., and F.D. Davis. 2000. A Theoretical Extension of the Technology Acceptance Model: Four Longitudinal Field Studies. *Management Science* 46 (2): 186–204. https://doi.org/10.1287/mnsc.46.2.186.11926.

Venkatesh, V., M.G. Morris, G.B. Davis, and F.D. Davis. 2003. User Acceptance of Information Technology: Toward a Unified View. *MIS Quarterly* 27 (3): 425–478. https://doi.org/10.2307/30036540.

PART II

Forecasting

CHAPTER 5

Predictive Accuracy Tests for Prediction of Economic Growth Based on Broadband Infrastructure

Walter Mayer, Gary Madden, and Xin Dang

INTRODUCTION

The development of broadband infrastructure is regarded as important by policymakers in many countries. A key motive is the belief that increased broadband deployment will stimulate economic growth. Three studies in the literature provide empirical support for the notion that broadband infrastructure is an important source of economic growth. For 15 European Union countries for the years 2003–2006, Koutroumpis (2009) estimates that a 1% increase in the penetration rate

---

In memory of late Gary Madden. The majority of the paper is published as a conferences paper at the 20th ITS Biennial Conference, Rio de Janeiro, 2014, where Gary Madden presented the paper.

W. Mayer · X. Dang (✉)
University of Mississippi, Oxford, MS, USA
e-mail: xdang@olemiss.edu

G. Madden
Curtin University, Perth, WA, Australia

for a given year increases GDP by approximately 3%. For 24 OECD member countries during the years 1996–2007, Czernich et al. (2011) estimate that a 10% increase in broadband penetration increases per capita GDP of between 0.9 and 1.5% per annum. Using a dynamic panel model and a sample of 29 OECD countries, Madden et al. (2016) investigate the impact of network speed as well as penetration. They estimate that a given increase in (the level of) network speed has a greater impact on economic growth in countries with lower penetration rates, and that, on average, for all countries, a 10% increase in broadband speed initially increases per capita GDP by about 0.13% and by a total of about 0.32% in the long-run, of which 97% is accomplished after four quarters.

In view of this evidence, the present paper investigates whether standard measures of broadband infrastructure are useful for out of sample predictions of future economic growth. The investigation is carried out by comparing the predictive accuracy of dynamic panel models of GDP with and without measures of broadband infrastructure. The data are for 29 OECD countries for the period 2008–2012. Following previous studies, the broadband infrastructure measures are broadband penetration, speed and the number of years since broadband was introduced. Various goodness-of-fit statistics are available for making such comparisons but cannot determine if differences in predictive accuracy are purely sampling errors. To control sampling errors, formal hypothesis tests are required. Formal hypothesis tests of predictive accuracy have been proposed and developed by Granger and Newbold (1977), Diebold and Mariano (1995), West (1996), Harvey et al. (1997) and others. Two that have become standard are the Morgan–Granger–Newbold (MGN) and Diebold–Mariano (DM) tests (Clark and McCracken 2001; Clark and McCracken 2014; McCracken 2007). Recently, Mayer et al. (2017) propose modifications of the DM test that improve asymptotic power by exploiting available sample information more fully to estimate the tested parameters. In the same spirit, we propose more powerful versions of the DM test and the MGN test by estimating the test parameter more efficiently. Those tests are employed to test the hypothesis that measures of broadband infrastructure can improve predictions of GDP growth after controlling for standard growth determinants.

Prediction Model

Following Madden et al. (2016), we specify a dynamic panel model with fixed effects.[1] The model has the form:

$$y_{it} = \sum_{h=1}^{2} \delta_h y_{i,t-h} + x_{it}\beta + z_{it}\gamma + u_i + \varepsilon_{it} \quad (5.1)$$

where y_{it} denotes the logarithm of national GDP per capita for country i ($=1, \ldots, n$) at time t ($1=1, \ldots, T$), x_{it} is a vector of growth *determinants and* z_{it} a vector of measures of broadband infrastructure. The lagged values $y_{i,t-h}$ allow changes in the explanatory variables and errors to impact GDP in future periods. The variables in x_{it} include the ratio of capital formation to GDP (*CAPGDP*), an education index (*EDU*), the growth rate of the labor force (*LABOR*) and a share price index (*SHAREPRICE*). The latter variable controls for the Global Financial Crisis which occurred during our sample period (2008–2012), while the former variables are standard determinants of steady-state growth. Also included in x_{it} is a time trend (*TREND*) and interactions with regional dummy variable (*TREND*REGION*), which capture region-specific growth for selected regions. Three measures of broadband infrastructure in z_{it} are broadband penetration (*PEN*), broadband speed (*SPEED*) and the number of years since broadband was introduced (*YEAR*). All measures are expected to be positively related to economic growth. To allow for the possibility that the impact of increased broadband speed on economic growth depends on the existing level of penetration, and the interaction term, PEN*SPEED is also included in z_{it}. The fixed effect α_i reflects time-invariant country-specific factors such as initial conditions that influence growth and may be correlated with the observed explanatory variables. The error term ε_{it} captures shocks that vary over both i and t. Table 5.1 provides definitions, means and standard deviations of

---

[1] Equation (5.1) is a simplified version of the model estimated by Madden et al. (2016). The model in that study allows the coefficients to vary by income group. This additional source of endogeneity raises new issues for the present prediction problem which have yet to be addressed in the literature. The purpose of the present paper is to assess the predictive content of broadband infrastructure using a standard dynamic panel data model. Hence, (5.1) suffices for the present paper.

Table 5.1 Variable definitions and descriptive statistics

Variable	Definition	Mean	SD
Dependent variable			
GDPPC	Real GDP per capita (quarterly)	9009.689	3350.091
Economic and demographic conditions			
LABOR	Growth change in working-age (15+) population	−0.000214	0.0250718
EDU	Years of schooling	10.8707	1.502812
CAPGDP	Gross fixed capital formation/GDP (quarterly)	0.2028873	0.035515
SHAREPRICE	All-share price indices (quarterly) (2005=100)	102.4766	52.63776
Broadband developments			
PEN	Fixed broadband subscribers per 100 inhabitants	26.21154	8.180294
SPEED	Real broadband download speed	11316.35	6739.085
YEAR	Years since broadband are introduced	10.24138	1.907266

Source Authors' computation
Note For estimation, log is used for GDPPC, EDU, CAPGDP, and SHAREPRICE

the variables for the quarterly data for 29 OECD Member Countries[2] from 2008:1 to 2012.

To investigate whether standard measures of broadband infrastructure are useful for predicting future economic growth, predictions were generated from two models. The first (Model 1) includes all of the broadband variables: PEN, SPEED, PEN*SPEED and YEAR. The second (Model 2) excludes all of the broadband variables. For each model, predictions of $y_{i,t+h}$ for $h=1, 2$ were computed conditional on the assumed information set. For Model 1, the information set is: $\Omega_{i,t+h|t} = \{(y_{i,r}, x_{i,s}, z_{i,s}) : r = 1, \ldots, t; s = 1, \ldots, t+h\}$.

For Model 2 the information set excludes z_{it} but is otherwise the same. Let $X_{it}\Gamma = x_{it}\beta + z_{it}\gamma$, $V_{it} = u_i + \varepsilon_{it}$ and $\hat{\delta}_1, \hat{\delta}_2, \hat{\Gamma}$ denote estimates. The predictors have the form:

$$\hat{y}_{i,t+1|t} = \hat{\delta}_1 y_{i,t} + \hat{\delta}_2 y_{i,t-1} + X_{i,t+1}\hat{\Gamma} + \hat{V}_{i,t+1|t} \quad (5.2)$$

---

[2]Chile, Estonia, Israel and Slovenia are not included as they have only joined the OECD in 2010, and this does not cover our period from 2008 to 2012.

5 PREDICTIVE ACCURACY TESTS FOR PREDICTION OF ECONOMIC ... 141

$$\hat{y}_{i,t+2|t} = \hat{\delta}_1 \hat{y}_{i,t+1|t} + \hat{\delta}_2 y_{i,t} + X_{i,t+2}\hat{\Gamma} + \hat{V}_{i,t+2|t} \tag{5.3}$$

The predicted values of the first three terms on the RHS of the above equations are straightforward. To predict $V_{i,t+h}$, we use a linear projection of $V_{i,t}$. The linear projection of $V_{i,t+h}$ is

$$E^*(V_{i,t}|\Omega_{i,t+h|t}) = \pi_0 + \sum_{s=1}^{t+h} \pi_{1ts} x_{is} + \sum_{s=1}^{t+h} \pi_{2ts} Y_{is} + \sum_{s=1}^{t} \pi_{3ts} y_{is} \tag{5.4}$$

Consistent estimates of the coefficients can be obtained by applying OLS to:

$$y_{it} - \hat{\delta}_1 y_{i,t-1} - \hat{\delta}_2 y_{i,t-2} - x_{it}\hat{\beta} - Y_{it}\hat{\gamma} = \pi_0 + \sum_{s=1}^{t+h} \pi_{1ts} x_{is} + \sum_{s=1}^{t+h} \pi_{2ts} Y_{is} + \sum_{s=1}^{t} \pi_{3ts} y_{is} + \eta_{it} \tag{5.5}$$

for $i=1, \ldots, n$; $t=1, \ldots, T$, where η_{it} denotes the projection error. The coefficients generally depend on t as functions of the covariances: $\text{cov}(x_{is}, u_i + \varepsilon_{it})$, $\text{cov}(Y_{is}, u_i + \varepsilon_{it})$ and $\text{cov}(y_{is}, u_i + \varepsilon_{it})$. The problem this poses is that a large number of coefficients must be estimated. If we assume that the covariances do not depend on i, x_{is} is uncorrelated ε_{it} and that the vector $(Y_{it}, \varepsilon_{it})$ is covariance-stationary, then $\pi_{1st} = \pi_{1s,t-s}$, $\pi_{2st} = \pi_{2s,t-s}$ and $\pi_{3st} = \pi_{3s,t-s}$.[3] However, in our application $n=29$ and $T=16$. Consequently, even under these restrictions the number of coefficients generally exceeds the number of observations available to estimate (5.4). To resolve the problem, instead of $E^*(V_{it}|\Omega_{it})$ we use the linear projection $E^*(V_{it}|\bar{x}_i)$ where $\bar{x}_i = (t+h)^{-1} \sum_{i=1}^{t+h} x_{it}$. By iterated projections, $E^*(V_{it}|\bar{x}_i) = E^*[E^*(V_{it}|\Omega_{it})|\bar{x}_i]$ and, consequently, the linear projection $E^*(V_{it}|\bar{x}_{it})$ can be viewed as a best linear predictor of $E^*(V_{it}|\Omega_{it})$. We report Arellano–Bond GMM estimators of (5.1) in Table 5.2 and discuss the significance of those coefficients in the Results Section.

---

[3] Under these assumptions, $\text{cov}(x_{is}, u_i + \varepsilon_{it})$ depends only on s, while $\text{cov}(Y_{is}, u_i + \varepsilon_{it})$ and $\text{cov}(y_{is}, u_i + \varepsilon_{it})$ depend only on s and $t-s$.

Table 5.2 Arellano-Bond GMM estimates of Log GDP per capita growth, 2008–2012

Number	Variable	Model 1		Model 2	
1	Log(GDPPC(−1))	0.7444 (17.80)	****	0.6344 (13.82)	****
2	Log(GDPPC(−2))	−0.0226 (−0.58)		0.0177 (0.47)	
3	PEN	0.0009 (1.31)			
4	SPEED	$2.2*10^{-6}$ (2.72)	***		
5	PEN*SPEED	$-7.12*10^{-8}$ (−3.17)	***		
6	YEAR	−0.002 (−1.34)			
7	Log(CAPGDP)	0.0195 (2.19)	**	0.0111 (1.00)	
8	EDU	−0.0141 (0.13)		0.2411 (1.73)	*
9	LABOR	0.056 (0.23)		−0.0187 (0.81)	
10	Log(SHAREPRICE)	0.0383 (10.47)	****	0.0434 (5.42)	****
11	TREND	0.0018 (2.97)	***	0.0011 (5.42)	****
12	NORTH AMERICA TREND	−0.0004 (1.23)		−0.0001 (0.36)	
13	PACIFIC TREND	0.0002 (0.73)		0.0005 (1.08)	

Source Authors' computation

Note The numbers in parentheses are Z statistic values. ****significant to 0.1%; ***significant to 1%; **significant to 5%; *significant to 10%

HYPOTHESIS TESTS OF PREDICTIVE ACCURACY

Given a sample of P predictions for country i, let $\hat{y}_{ip(j)}$ and $e_{ip(j)}$ denote, respectively, the pth ($p=1, \ldots, P$) prediction and prediction error from the Model $j=1, 2$. For the two models, we will test the null hypothesis of equal predictive accuracy:

$$H_0: \theta = 0 \text{ verse } H_1: \theta < 0 \tag{5.6}$$

where $\theta = E\left(e_{ip(1)}^2 - e_{ip(2)}^2\right)$. The alternative is one-sided. If the null is rejected, we are able to conclude that Model 1 with broadband variables yields a more accurate prediction on GDP growth than Model 2 does.

Let $d_{ip} = e_{ip(1)}^2 - e_{ip(2)}^2$. The original Diebold–Mariano (1995) (DM) test is designed for time series and is based on the sample mean of the difference between the squared prediction errors. An analogous test for panel data is:

$$DM_e = \hat{\theta}_e \sqrt{N/\hat{V}_e}, \tag{5.7}$$

where

$$\hat{\theta}_e = \frac{1}{NP}\sum_{i=1}^{N}\sum_{p=1}^{P} d_{ip} = \frac{1}{N}\sum_{i=1}^{N}\frac{1}{P}\sum_{p=1}^{P} d_{ip} = \frac{1}{N}\sum_{i=1}^{N}\bar{d}_{i.} = \bar{d}_{..}$$

and

$$\hat{V}_e = \frac{1}{N-1}\sum_{i=1}^{N}(\bar{d}_{i.} - \bar{\bar{d}}_{..})^2.$$

The covariance estimator $\hat{V}_e$ assumes that the prediction errors are uncorrelated across countries but allows serial correlation. Under this assumption and the null hypothesis (5.6), (5.7) converges to the standard normal distribution as N approaches infinity for fixed P. The null hypothesis also implies that the mean of $\bar{d}_i$ is zero. We consider a version of (5.7) with this restriction imposed on $\hat{V}_e$ and denote it by DMV_e. We expect it performs better than DM_e since it incorporates more information under the null hypothesis and hence yields higher power.

Like the original DM test, the MGN) test is designed for time series but is based on the sample correlation between the difference and sum of the prediction errors. Unlike the DM test, the MGN test assumes normally distributed prediction errors. An MGN-type test for panel data can be devised in terms of the country means:

$$\bar{e}_{i(j)} = \frac{1}{P}\sum_{p=1}^{P} e_{ip(j)} \qquad (5.8)$$

The MGN test for panel data is

$$MGN_e = \hat{\rho}_e\sqrt{(N-1)/(1-\hat{\rho}_e^2)} \qquad (5.9)$$

where $\hat{\rho}_\varepsilon$ is the sample correlation between $\bar{e}_{i(1)} + \bar{e}_{i(2)}$ and $\bar{e}_{i(1)} - \bar{e}_{i(2)}$. Under (5.6) and normally distributed prediction errors that are uncorrelated across countries, (5.9) also converges to the standard normal distribution.

The above DM and MGN tests depend on the predictions $\hat{y}_{ip(j)}$ only through the prediction errors $e_{ip(j)}$. We propose alternative versions of the DM and MGN tests based directly on the predictions.

The panel data analog of the DM test is based on the following alternative estimator of θ:

$$\hat{\theta}_y = \frac{1}{NP} \sum_{i=1}^{N} \sum_{p=1}^{P} (\hat{y}_{ip(1)} - \hat{y}_{ip(2)})(\hat{\alpha}_1^* \hat{y}_{ip(1)} + \hat{\alpha}_2^* \hat{y}_{ip(2)})$$
$$= \frac{1}{NP} \sum_{i=1}^{N} \sum_{p=1}^{P} g_{ip} \qquad (5.10)$$

where $\hat{\alpha}_j^* = 2\hat{\alpha}_j - 1$, $j = 1, 2$, and $\hat{\alpha}_j$ are estimated coefficients of the linear projection:

$$E^*(y_{it}|\hat{y}_{ip(1)}, \hat{y}_{ip(2)}) = \alpha_0 + \alpha_1 \hat{y}_{ip(1)} + \alpha_2 \hat{y}_{ip(2)}$$

We assume that the number of observations used to compute the $\hat{\alpha}_j$, M, is such that $p/M \to \infty$ so that they are asymptotically irrelevant in the distribution of $\hat{\theta}_y$. To satisfy this assumption for the application in the next section, the prediction (29 times 4 observations) and estimation (29 times 16 observations) samples are combined to compute the $\hat{\alpha}_j$. Using a GMM framework, it is expected $\hat{\theta}_y$ is more efficient than $\hat{\theta}_\varepsilon$ because it exploits available sample information more fully, and argue that tests based on $\hat{\theta}_y$ have greater power. Note that the closely related but different estimator is the one studied in Mayer et al. (2017). They formulate it as the restricted GMM estimator based on the different constraints and study asymptotical properties of the estimator. Using the estimator (5.10), our DM test for the panel data is

$$DM_y = \hat{\theta}_y \sqrt{N/\hat{V}_y} \qquad (5.11)$$

where

$$\hat{V}_y = \frac{1}{N-1} \sum_{i=1}^{N} (\bar{g}_{i.} - \bar{g}_{..})^2, \quad \bar{g}_{i.} = \frac{1}{P} \sum_{p=1}^{P} g_{ip}, \quad \bar{g}_{..} = \frac{1}{N} \sum_{i=1}^{N} \bar{g}_{i..}$$

The panel data analog of the MGN test is given by

$$MGN_y = \hat{\rho}_y \sqrt{(N-1)/(1-\hat{\rho}_y^2)} \qquad (5.12)$$

where $\hat{\rho}_y$ denotes the sample correlation between $\bar{\hat{y}}_{i(2)} - \bar{\hat{y}}_{i(1)}$ and $\alpha_1^* \bar{\hat{y}}_{i(1)} + \alpha_2^* \bar{\hat{y}}_{i(2)}$. The average is taken over the P predictions for each country.

In addition, we can have generalized versions of the MGN test that simultaneously addresses the problems of serially correlated and heavy-tailed prediction errors. The proposed tests include as special cases the DM and MGN tests, and the modified MGN test proposed by Harvey et al. (1997). These generalized tests can also be adapted to panel data and have the form:

$$MGNV_a = DM_a(1 - \hat{\rho}_a^2)^{-1/2} N^{-1/2}(1-N)^{1/2}, \ a = e, \bar{y} \quad (5.13)$$

The generalized versions $MGNV_a$ have a correction term to address the serial correlation. Pooling across P predictors is avoided and hence they are more powerful.

Results

For the two models, one and two-step-ahead prediction errors were generated. Recall that Model 1 includes all broadband variables, but Model 2 excludes all broadband variables. The sample consists of quarterly data from the first quarter of 2008 to the fourth quarter of 2012. The four quarters of 2012 were used to compare the predictive accuracy of the three models. The Arellano–Bond estimates for (5.2) were computed using a recursive scheme to generate successive observations. Table 5.2 reports the estimates for the full sample period of 2008–2012.

The estimated coefficients for capital formation (CAPGDP), the Price Share Index (SHAREPRICE) and time trend (TREND) have the expected positive signs in two models. However, the CAPGDP coefficient is significant only in Model 1, while Price Share Index and trend coefficients are significant in both. The EDU coefficient is insignificant in Model 1 but is significant at 5% level in Model 2. The estimated LABOR coefficients are insignificant in both models. The pattern of statistically significant CAPGDP coefficient and insignificant EDU and LABOR coefficients accords with the findings of Czernich et al. (2011). SPEED and the PEN*SPEED interaction coefficients are significant, while the separate variable coefficients for PEN and YEAR are insignificant. As discussed in Madden et al. (2016), the statistical significance of

Table 5.3 AIC and BIC of both models with and without Luxembourg

		One-step Prediction		Two-step Prediction	
		Model 1	Model 2	Model 1	Model 2
With	AIC	14.481	14.512	15.236	15.218
	BIC	14.790	14.726	15.544	15.432
Without	AIC	12.831	13.237	13.717	14.060
	BIC	13.147	13.455	14.032	14.279

Source Authors' computation

the positive coefficient sign for broadband speed and the negative sign for interaction term imply that increased speed has more impact in countries with less broadband penetration.

After the estimates were obtained, the models were applied to predict the GDPPC of the four quarters of 2012. Predictors and prediction errors of the two models were then calculated. Before applying the hypothesis tests described in the last section to compare prediction accuracy, we first consider the relative performance of the models in terms of fit. The AIC and BIC criteria were calculated for the prediction errors of each model. The results with and without Luxembourg are reported in Table 5.3 for a reason explained below. For both cases, the differences in the values of the AIC and BIC across models are fairly small. When all 29 countries are included in the calculations, the AIC favors Model 1 for one-step ahead predictions and favors Model 2 for two-step ahead predictions, while BIC favors Model 2 for both one and two-step ahead predictions.

To investigate this further, the distributions of the prediction errors for the two models were plotted. The plots are shown in Fig. 5.1 and reveal outliers in the right tail. The outliers in the right tail correspond to Luxembourg which has the largest GDP per capita in the sample. Apart from the outlier, the distributions of the prediction errors appear to be approximately normal centered at zero. The plots suggest that Model 1 has a smaller variance than Model 2, and, therefore, support the notion that the broadband variables are important predictors of economic growth. Recalculating the AIC and BIC after deleting Luxembourg, we now find that Model 2 is now ranked inferior to the Model 1 in all cases. With and without Luxembourg, different conclusions of AIC and BIC

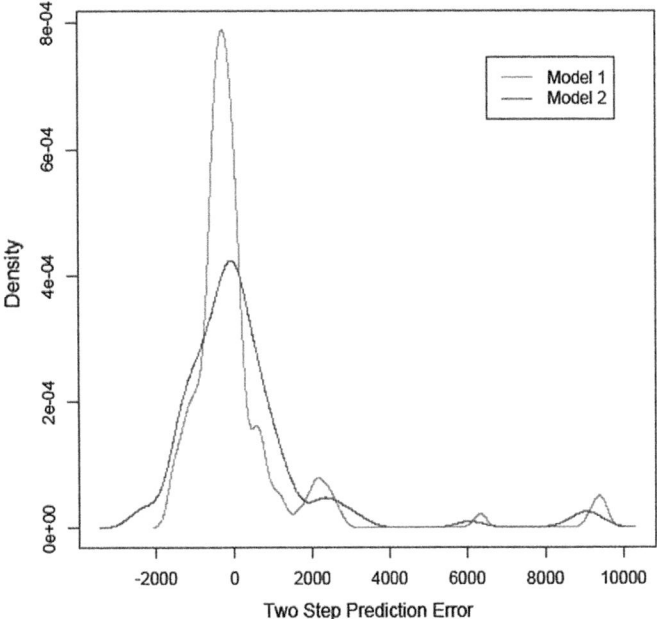

Fig. 5.1 Density estimation of two-step ahead prediction error for the models (*Source* Authors' computation)

can be explained by the fact that AIC and BIC are normal log-likelihood based and hence are sensitive to outliers.

To determine if the differences in predictive accuracy are statistically significant, we next apply the tests described in the Results section. The eight tests (four based on prediction errors, four based on predictors) were computed for the one-step and two-step predictions. The tests were carried out with and without Luxembourg. Table 5.4 reports the p-values of the tests. The tests clearly favor Model 1 over Model 2 as only 2 of the 16 tests are not significant at conventional levels significance with Luxembourg included. Without Luxembourg, all tests unambiguously claim that Model 1 has greater predictive accuracy than Model 2, and therefore measures of broadband infrastructure can improve predictions of economic growth.

Table 5.4 P-values of various tests for model forecasting accuracy with and without Luxembourg included

		DM		DMV		MGN		MGNV	
		e	y	e	y	e	y	e	y
With	1-step	0.0096	0.0023	0.0159	0.0060	0.1440	0.0032	0.0000	0.000
	2-step	0.0912	0.0000	0.0978	0.0005	0.5082	0.0000	0.0038	0.000
WO	1-step	0.0047	0.0017	0.0099	0.0053	0.0032	0.0005	0.0000	0.000
	2-step	0.0000	0.0000	0.0002	0.0000	0.0011	0.0005	0.0000	0.000

Source Authors' computation

We also note that the tests based on the predictors are generally more significant than the tests based on the prediction errors. This is not surprising given that the tests based on the predictors are asymptotically more powerful because they exploit available sample information more fully and are better able to detect small predictive differences between two models. MGNV tests are most significant among four tests on all cases.

With Luxembourg included, the MGN test based on prediction errors is the only test insignificant for the one-step and two-step predictions. It becomes significant when the outlier Luxembourg is deleted. This can be explained by the normally distributed assumption of the MGN test, which is obviously not satisfied if Luxembourg is included. As expected, tests on the predictors are more significant in two-step than in one-step ahead predictions. Recall that the two-step ahead predictor depends on the one-step ahead predictor in Eq. (5.3). If a test is significant in one-step ahead predictions, then we would expect it to be also significant in two-step ahead predictions. Such results only hold for tests based on the prediction errors with Luxembourg excluded. Tests based on the prediction errors for two-step ahead predictions, however are less significant than that for one-step ahead predictions with outliers included. In this sense, tests on the predictors are more robust than tests on the prediction errors.

Concluding Remarks

This paper investigated whether predictions of future economic growth can be improved by using standard measures of broadband infrastructure. The measures of broadband considered in the paper are penetration, speed and the number of years since broadband was introduced. The investigation was carried out by comparing the predictive accuracy of dynamic panel models of economic growth estimated with and without those measures of broadband infrastructure using the data of 29 OECD countries for the period 2008–2012. One-step and two-step ahead predictions were considered and compared for the two models. Adapted from tests for time-series data, we devised eight tests for panel data and performed them to compare predictive accuracy. Although the outliers contained in the data set complicate the problem, the tests clearly favor the model with broadband variables. Among those eight tests, tests based on the predictors are more powerful and more robust than those based on the prediction errors. Also, the generalized MGNV tests are recommended.

References

Clark, T.E., and M.W. McCracken. 2001. Test of Equal Forecast Accuracy and Encompassing for Nested Models. *Journal of Econometrics* 105: 85–110.

Clark, T.E., and M.W. McCracken. 2014. Tests of Equal Forecast Accuracy for Overlapping Models. *Journal of Applied Econometrics* 29: 415–430.

Czernich, N., O. Falck, T. Kretschmer, and L. Woessmann. 2011. Broadband Infrastructure and Economic Growth. *Economic Journal* 121: 505–532.

Diebold, F.X., and R.S. Mariano. 1995. Comparing Predictive Accuracy. *Journal of Business and Economic Statistics* 13: 253–263.

Granger, C.W.J., and P. Newbold. 1977. *Forecasting Econometric Time Series*. Orlando, FL: Academic Press.

Harvey, D., S. Leybourne, and P. Newbold. 1997. Testing the Equality of Prediction Mean Squared Errors. *International Journal of Forecasting* 13: 281–291.

Koutroumpis, P. 2009. The Economic Impact of Broadband on Growth: A Simultaneous Approach. *Telecommunications Policy* 33: 471–485.

Madden, G., W.J. Mayer, and C. Wu. 2016. *Broadband and Economic Growth: A Reassessment*. Technical Report, Bankwest Curtin Economics Centre, Curtin University. https://bcec.edu.au/publications/broadband-economic-growth/.

Mayer, W.J., F. Liu, and X. Dang. 2017. Improving the Power of the Diebold-Mariano-West Test for Least Squares Predictions. *International Journal of Forecasting* 33: 618–626.

McCracken, M.W. 2007. Asymptotics for Out of Sample Tests of Granger Causality. *Journal of Econometrics* 140: 719–752.

West, K.D. 1996. Asymptotic Inference About Predictive Ability. *Econometrica* 64: 1067–1084.

CHAPTER 6

Modelling Internet Diffusion Across Tourism Sectors

Miriam Scaglione and Jamie Murphy

INTRODUCTION

Everett M. Rogers' (1962) book, Diffusion of Innovations (DOI), the most cited work in innovation research, discusses innovation adoption and implementation at the individual and organisational level (Jeyaraj et al. 2006; Zhu and Kraemer 2005). Rogers (2003, p. 12) defines an innovation as "an idea, practice, or object that is perceived as new by an individual or other unit of adoption". Rogers' book proposes that diffusion follows a normal distribution and therefore, fixed percentages for his five—innovators, early adopters, early majority, late majority and

---

This chapter draws on a pedagogical explanation of Bass modelling of Swiss accommodation in (Scaglione, forthcoming), and a first draft of this paper presented at the Annual Conference of International Association of Scientific Experts in Tourism—AIEST 2012 held at Khon Kaen (Thailand).

M. Scaglione (✉)
Institute of Tourism, University of Applied Sciences and Arts Western Switzerland Valais, Sierre, Switzerland
e-mail: miriam.scaglione@hevs.ch

J. Murphy
University of Eastern Finland, Joensuu, Finland

laggards—adopter categories. These fixed percentages, however, rarely hold across countries and innovations.

Similar to the diffusion of many innovations, internet adoption follows an abnormal distribution due to complex and diverse economic, political, cultural, infrastructural and geographical factors (Andrés et al. 2010; see Kim 2011). Rather than such macro-factors, this study adds communication channels (Bass 1969) and heterogeneous adopter categories (Bemmaor 1994) to improve modelling innovation adoption.

Rogers proposes that at the beginning of the adoption process a few innovators influence others using intrapersonal communication such as word-of-mouth. But if the role of communication channels, such as word-of-mouth, have the same importance along with the adoption process is an open question in the literature (Easingwood et al. 1983).

Incorporating adopter category differences, abnormal diffusion rates and differing adoption drivers improves the diffusion calculations. These calculation results are critical for industry to predict market size and sales of an innovation. This study takes two small steps in modelling the organisational adoption of innovations, adding a threshold due to a critical mass of adopters and heterogeneity adoption propensities. These two factors help compare innovations across countries and tourism sectors.

This modelling story begins by reviewing two improvements to Roger's diffusion model. Bass' (1969) model predicts specific—industry, product and country—new product adoption rates and market sizes. Bemmaor (1994) complements Bass' innovation-specific adoption model by incorporating heterogeneous rather than homogenous adoption factors at the actor level. The results of these two models, across 13 tourism sector datasets, add to a growing body of innovation-specific adoption research and key factors related to internet diffusion in the tourism sector. Based on these two analyses, this paper proposes future research streams related to modelling the DOI.

Literature Review

The tourism sector has been at the forefront of innovation adoption and implementation at least since the dawn of airline reservation systems. A chance meeting of IBM and American Airlines executives in 1953 led to the development of Sabre, an automated system to help manage the link between airline inventory and customer reservations (Sabre, n.d.).

Since that moment, information and communication technologies (ICTs) have been a key tool for marketing and distribution in tourism

(O'Connor 1999). ICTs offer opportunities information dissemination (24*7*365) on goods and services, pricing strategies and promotions (last-minute, location-based offers, etc.), as well as broaden selling opportunities for tourism suppliers including getting customer feedback from social media sites, no matter the size of the company (O'Connor and Murphy 2004). Therefore, tourism organisations and individuals have been at the forefront of using ICTs such as websites, email and social media.

Understanding and predicting tourism ICT adoption rates, though, remains an ongoing challenge.

Innovation and Imitation

Bass (1969) expanded Rogers' (1962) adoption model by quantifying two factors that drive individual and organisational adoption, *innovation (p)* driven by external channels such as mass communication and *imitation (q)* driven by intrapersonal communication such as word-of-mouth. Bass' model (1969) calculates innovation-specific and abnormally shaped distribution and adoption curves. This non-cumulative distribution curve has three important values that set the adopter categories. The inflexion point $T1$ separates the early adopters and early majority categories. Next, the peak of absolute adoption T demarcates the early majority and late majority categories. Finally, the second inflexion point $T2$ delineates the late majority and laggards.

Equation (6.1) below, a modified exponential and logistic function (Meade and Islam 2006), shows the Bass function (1969) for adopting an innovation. $N(t)$ is the cumulative number of adopters at time t, m is the total market adoption. Parameters p and q, the coefficients of innovation and imitation, respectively, fuel the adoption.

$$\frac{dN(t)}{dt} = \underbrace{p(m - N(t))}_{\text{adoption due to external influence or independent adoption}} + \underbrace{\frac{q}{m}N(t)(m - N(t))}_{\text{adoption due to internal influence or internal adoption}}$$

(6.1)

The coefficients p and q, respectively, suggest the propensity to adopt driven by external information and interpersonal communication channels (Mahajan et al. 1990). With pure innovation, $q=0$ and $p>0$ and $p=0$ and $q>0$ with pure imitation.

Apart from these extreme cases, the sum and ratio of q and p delineate the diffusion curves. Their sum controls the scale of the S-shaped cumulative and non-cumulative adoption curve (respectively Figs. 6.1 and 6.2), and the bell-shaped non-cumulative adoption curve. The q/p ratio indicates the shape of the cumulative and non-cumulative adoption curves. The bigger the q/p ratio, the more prominent the S-shape and slower penetration rates are as shown in Fig. 6.1 (Bemmaor and Lee 2002; Meade and Islam 2006). Comparing the p and q parameters helps examine adoption across countries and the adoption of different innovations by the same population.

Van den Bulte and Stremersch's (2004) study of individual-level q/p ratios for 52 consumer durables across 28 countries found, among other aspects, a positive relationship between the q/p ratio and income inequality. The bigger a country's income inequality, the more that internal rather than external influences—respectively imitation and innovation—drove adoption. These authors argue that the adoption of the same product in different countries will show different q/p ratios.

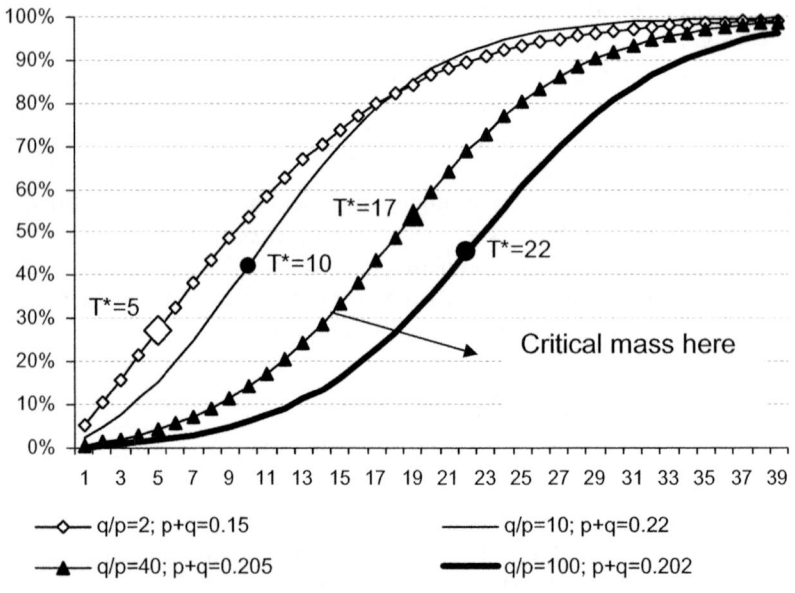

Fig. 6.1 Cumulative bass distribution models with different q/p ratios; T^* is the peak rate of adoption, namely the inflexion point of the cumulative distribution (*Source* Authors' computation)

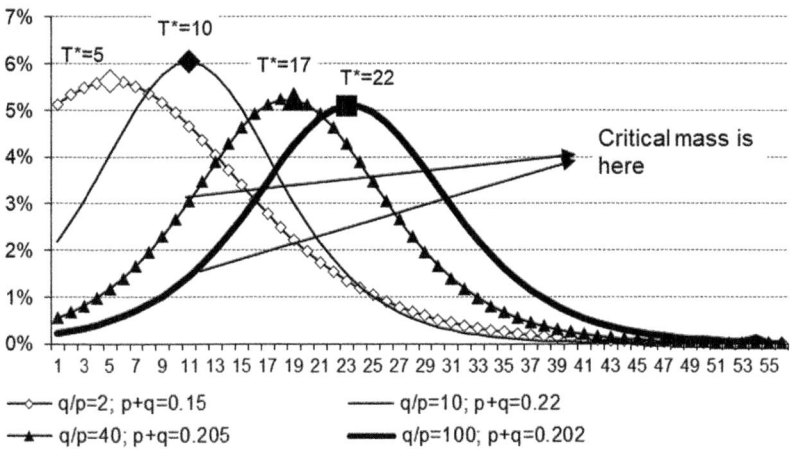

Fig. 6.2 Non-cumulative bass distribution models with different q/p ratios; T^* is the peak rate of adoption, namely the inflexion point of the cumulative distribution (*Source* Authors' computation)

Homogenous and Heterogeneous Adoption

The Bass model as in Eq. (6.1) assumes external communication—e.g. price, advertising and promotion—and intrapersonal communication such as word-of-mouth drive adoption. An extension of Bass' distribution curves helps bridge Rogers' (2003) five adopter categories into the Bass/Rogers or BR model (Mahajan et al. 1990). Rather than using the timing of adoption, the BR categories assume a homogenous likelihood to adopt across the five adopter categories (Bemmaor 1994, p. 203).

Bemmaor (1994) proposes a general diffusion model with a heterogeneous propensity for the first-adoption (purchase) in order to address these limitations. At the individual adopter level, the propensity to adopt an innovation at time t follows a shifted Gompertz distribution (SG) whereby early adopters take up the innovation in a more random fashion than latter adopters. Under this assumption, Bemmaor and Lee (2002) propose the Gamma/shifted Gompertz (G/SG) model of Eq. (6.2) below,

$$F(t) = \left[1 - e^{-bt}\right] \Big/ \left[1 + \beta e^{-bt}\right]^{\alpha} \tag{6.2}$$

The β parameter of Eq. (6.2) is its scale parameter; b is the scale parameter of the shifted Gompertz prior distribution of the individual propensity model. Bass model parameters, $b = p+q$ and $\beta = q/p$, can replace b and β. For fixed values of b and β, the shape of the Eq. (6.2) gamma distribution function depends on the parameter α, which measures adopter population heterogeneity. The coefficient of variation of a Gamma function is $\alpha^{-1/2}$.

Homogeneity has a direct relationship with α, and when α tends towards ∞, the population is homogenous. The propensity to adopt is roughly the same across potential adopters. When $\alpha \approx 1$, the model is the Bass model (Eq. 6.1). When $\alpha \approx 0$, potential adopters' acceptance rates differ across strata population. Bemmaor (1994, p. 220) suggests that the propensity to adopt differs across the BR adopter categories.

Critical Thresholds and Exogenous Variables

In addition to ignoring heterogeneous adoption propensities among the population, the Bass model as expressed by Eq. (6.1) ignores the critical threshold necessary to adopt (Van den Bulte and Stremersch 2004). This threshold reflects the requisite critical mass of adopters. Analysis of the q/p ratio gives insights on critical mass's importance for comparing diffusion across countries or comparing innovations across the same adopter population. For example, a qualitative study of German banks showed that relative to non-interactive innovations, the diffusion of interactive innovations (i.e. Electronic Funds Transfers and home banking for private customers) was slow until reaching a critical mass of adopters (Mahler and Rogers 1999). If this finding is the case, then the S-Shape for the non-interactive innovations should be less pronounced and the q/p ratio smaller than for interactive innovations as shown in Fig. 6.1.

The critical threshold can also depend on exogenous, external or environmental factors. At the organisational or aggregate-level, perceived innovation profitability (Jensen 1982) and risk-averse behaviour (Oren and Schwartz 1988) may accelerate or delay adopting an innovation. At the individual level, Chatterjee and Eliashberg (1990) used a micro-modelling approach to show that initial adopter value perceptions, aversion and perceived information reliability were both heterogeneity sources and threshold factors (see Van den Bulte and Stremersch 2004, p. 534).

The literature review shows that the Bass model (Eq. 6.1) cannot capture heterogeneous adoption dispositions across a given population of adopters or a given sector. Nevertheless, the q/p ratio helps capture critical mass differences across sectors. These two main facts, namely heterogenous adoption and critical mass, with an innovation's final penetration rate allow comparing adoption processes of an innovation across sectors.

This research compares internet diffusion across tourism sectors/countries, firstly by analysing the Bass parameters (Eq. 6.1), particularly the q/p ratio, in order to grasp the requisite critical mass for each sector/country and the different adoption curve shapes. Next, the analysis compares heterogeneity across these sectors/countries via the G/SG model.

Data and Methodology

Table 6.1 shows characteristics of the 13 datasets such as the sample and gathering the time of first adoption. Ten studies used the internet Archives' Wayback Machine (archive.org) to gather the website age (Murphy et al. 2007). The other three studies, all related to Swiss accommodation, used a customised softbot, or software robot (Steiner 1999). The softbot gathered the initial domain name registration date, registrant name and location for all Swiss (.ch) domain names registered from December 1995 to February 2004, which contained the keywords "auberge, gasthaus or hotel".

The datasets range in size from 96 to 2467 organisations and range geographically across the globe, Europe and four countries—Austria, Germany, Malaysia and Switzerland. The six tourism sectors in the datasets comprise cable cars/ski lifts, destination management organisations (DMOs), one- to five-star hotels, guesthouses, restaurants and tour operators, The analysis for both the Bass and G/SG model parameters draw on the SAS V9.2 Proc Model procedure (SAS Institute Inc. 2011).

Table 6.1 Study sample

Acronym	Sector	Age source	Market size (base year) (1)	Source	Website age/n (URL)	Observed penetration rate respect to (1)
DMO CH	Swiss DMOs	WM	155 (2005)	myswitzerland.ch	149/155	100%
DMO AU	Austrian DMOs	WM	N/A	Klimek et al. (2011)	96/96	N/A
DMO DE	German DMOs	WM	N/A	Klimek et al. (2011)	182/188	N/A
Restaurant CH	Swiss restaurants	WM	18867 (2005)**	Swisscom Directories	1573/1858	10%
Cable car CH	Swiss cable car companies	WM	370 (2010)	Seilbahnen Schweiz (seilbahnen.org)	190/190	51%
Hotel Chain	International hotel chains	WM	325 (2010)	hotelsmag.com (July 2006)	267/276	85%
European tour operators	European tour operators	WM	N/A	etoa.org, european-travel-market.com	117/121	N/A
Malaysian hotels	Malaysian hotels	WM	530	Hachim et al. (2012)	305/315	60%
No affiliated hotel CH	Unaffiliated Swiss hotels	Softbot	1133 (2003)*	Swisscom Directories	780/780	69%
Hotelleriesuisse CH	Affiliated Swiss hotels	Softbot	2122 (2003)*	Scaglione et al. (2010)	1677/1733	82%
Hotel CH	Swiss hotels	WM	3255 (2003)*	Scaglione et al. (2010)	2467/2513	77%
Guesthouse CH	Swiss guesthouses	Softbot	3463 (2003)*	Scaglione et al. (2004)	2250/2269	65%
Travel agencies CH	Swiss travel agencies	WM	272 (2008)	Schweizerischer Reisebüro-verband (www.srv.ch)	244/232	90%

Source **Recensements des entreprise (2005) Office fédéral de la *statistique* (OFS); *Calculated from statistique de l'hébergement touristique (HESTA, OFS)

RESULTS

Table 6.2 shows the model estimation results, such as the coefficients of innovation and imitation.

Figures 6.4, 6.5, 6.6, 6.7, 6.8, 6.9, 6.10, 6.11, 6.12, 6.13, 6.14, 6.15, and 6.16 in the Appendix show the following cumulative series for each sector: *observed*, i.e. the actual value, *estimated* values namely one-step-ahead estimation and *forecasted* values for the last 24 months of the sample using Bass model for each of the sector under study. Table 6.3 in the Appendix shows the goodness of fit measures for the Bass model estimation and forecasts observations.

Three Swiss sectors—cable cars, guesthouses and DMOs—have a logistic distribution. Only imitation, influence via intrapersonal communication, drives their internet adoption. Both innovation and imitation, respectively, external and intrapersonal communication, drive internet adoption in the ten other sectors with a 10.42 q/p median. The q/p ratio, intrapersonal communication divided by external communication, shows each channel's relative strength in the adoption process.

A high q/p ratio shows that imitation relative to innovation drives adoption. Similar to the three Swiss sectors with a logistic distribution, three additional Swiss sectors have the highest q/p ratios, ranging from 26 to 35. Using two measures of age, domain and website, these six Swiss tourism sectors adopt due to imitation more so than innovation. There seems a general parsimony of Swiss sectors, sensible to the critical mass of adopters and a wait and see behaviour.

At the innovation end of the communication channel ratio, Austrian DMOs are just under three with a $q/p=2.94$. Hotels chains and European TOs also had q/p ratios under six. More so than the other ten sectors, these sectors adopted due to innovation, that is, persuasion via mass communication. The median group—Malaysian hotels, Swiss restaurants, Swiss travel agents and German DMOs—had imitation/innovation ratios towards 10. The effects of channel communication accelerated the adoption process. As an example, Table 6.2 shows that Austrian DMOs had the lowest q/p (2.94) and the time to generate the early majority group was only 21 months against 48 months for the late majority. In contrast, Swiss hotels with the highest (35.03) ratio, took 42 months to generate the early majority against only 17 months for the late majority.

Table 6.2 Estimated Bass and G/SG α parameter with its standard deviations in (parentheses)

	Type of model	Drivers	Final market size	Final market size (respect to 1) Table 6.1	Initial & final obs.' dates	Bass model (Eq. 6.1)							G/SG
						T1 Inflection point	Peak Maximum	T2 Inflection point	p Innovation	q Imitation	q/p		α
DMO CH	Bass $p=0$ (Logistic)	imitation	150.1 (18.72)	96.77%	Dec-96–Feb-05	Feb-98	Jul-99	Jan-01	0 (na)	0.0661 (0.0113)			0.230 (0.023)
DMO AU	Bass	innov./imitat.	96.78 (17.02)	N/A	Dec-96–July-10	Jun-97	Sep-99	Jun-03	0.0079 (0.0021)	0.0232 (0.0171)	2.94		1.97E8 (1.34E7)
DMO DE	Bass	innov./imitat.	181.63 (21.96)	N/A	Nov-96–Aug-10	Dec-97	Dec-00	May-04	0.0050 (0.0022)	0.0503 (0.0153)	10.14		0.81 (0.09)
Restaurant CH	Bass	innov./imitat.	1647.35 (70.04)	8.73%	Nov-96–Feb-08	Apr-99	Nov-01	Jun-04	0.0035 (0.0011)	0.0392 (0.0067)	11.19		171.6 (3.37)
Cable car CH	Bass $p=0$ (Logistic)	imitation	194.5 (1.30)	52.70%	Dec-96–Feb-05	Jan-00	Jan-01	Jan-02	0	0.0875 (0.0018)			0.61 (0.07)
Hotel chain	Bass	innov./imitat.	269.2 (38.56)	82.77%	Dec-96–Nov-06	Feb-99	Jan-99	Apr-01	0.0099 (0.0101)	0.0376 (0.0165)	3.80		0.51 (0.056)
European tour operators	Bass	innov./imitat.	119.18 (9.39)	N/A	Oct-96–Feb-06	Aug-98	Aug-99	Nov-01	0.0076 (0.0049)	0.0399 (0.0205)	5.28		0.98 (0.28)
Malaysian hotel	Bass	innov./imitat.	316.9 (12.8)	59.81%	Nov-96–Aug-05	Oct-98	Jul-01	Apr-04	0.0037 (0.0015)	0.0364 (0.0093)	9.85		0.35 (0.02)
No affiliated hotel CH	Bass	innov./imitat.	796.4 (12.0)	70.26%	Oct-96–Feb-04	Sep-98	Dec-99	Apr-01	0.0029 (0.0014)	0.0799 (0.0054)	27.39		0.74 (0.04)
Hotelleriesuisse CH	Bass	innov./imitat.	1716.6 (22.0)	81.93%	Jul-96–Feb-04	Feb-98	Jul-99	Jan-01	0.0028 (0.0007)	0.0718 (0.0035)	25.56		1.16 (0.07)
Hotel CH	Bass	innov./imitat.	2525.5 (25.2)	77.57%	Dec-95–Feb-04	Jan-97	Jul-99	Nov-01	0.0022 (0.0007)	0.0775 (0.0030)	35.03		1.27 (0.08)
Guesthouses CH	Bass $p=0$ (Logistic)	imitation	3131.0 (235.5)	90.41%	Dec-95–Dec-03	Jul-01	Oct-03	May-06[a]	0	0.0521 (0.0022)			No convergence
Travel agencies CH	Bass	innov./imitat.	235.13 (39.04)	98.72%	Nov-96–May-06	Sep-00	Apr-02	Dec-03	0.0063 (0.0033)	0.0676 (0.0563)	10.70		0.60 (0.08)

Source Authors' calculations

Market Penetration

The median forecasted internet penetration rate (m/total population) for the 10 sectors possible to fix the total market size was 80%. The penetration rate ranges from 9% for Swiss restaurants to over 90% for Swiss travel agents, Swiss guesthouses and Swiss DMOs.

Below the median are Swiss cable cars, Malaysian hotels and non-affiliated Swiss hotels. These sectors tend to comprise small organisations and have low levels of internationalisation, which can help explain their low level of final penetration. The low penetration of the accommodation sector could also be a *leapfrog effect*, whereby late adopters may forego traditional websites in favour of a new technology such as social networks (Hashim et al. 2012).

Close to the median, Swiss hotels (overall and affiliate) and international chain hotels, show a consistent accommodation sector behaviour. In the highest level are Swiss guesthouses, DMOs and travel operators. The Swiss government's push towards internet adoption could help explain the strong DMO adoption (Scaglione and Schegg 2015). Swiss travel agencies seem to follow the general trend of e-tourism. The increasing role of the internet in holiday planning by consumers and businesses may have spurred travel agencies to follow this tendency in order to meet their market.

Homogeneity and Heterogeneity

The α values in the G/SG models reflect homogeneity, with Austrian DMOs followed by Swiss restaurants at the highest levels. Swiss hotels, affiliated and unaffiliated, and European tour operators are $\alpha \approx 1$. The sectors, listed by increasing are German DMOs, Swiss unaffiliated hotels, cable cars, travel agents, and chain hotels, Malaysian hotels and finally Swiss DMOs. Only sectors showing α near 1 follow a Bass model, thus having a similar propensity of first adoption across companies and also among BR categories.

The other Swiss accommodation sector, non-affiliated hotels, shows a greater heterogeneity than overall Swiss hotels and affiliated hotels, in line with the assumption that unaffiliated hotels are smaller than affiliated hotels. The high α value reflects differences within these sectors.

Sectors showing the greatest homogeneity are Swiss restaurants and Austrian DMOs. The underlying reason for Austrian DMOs are

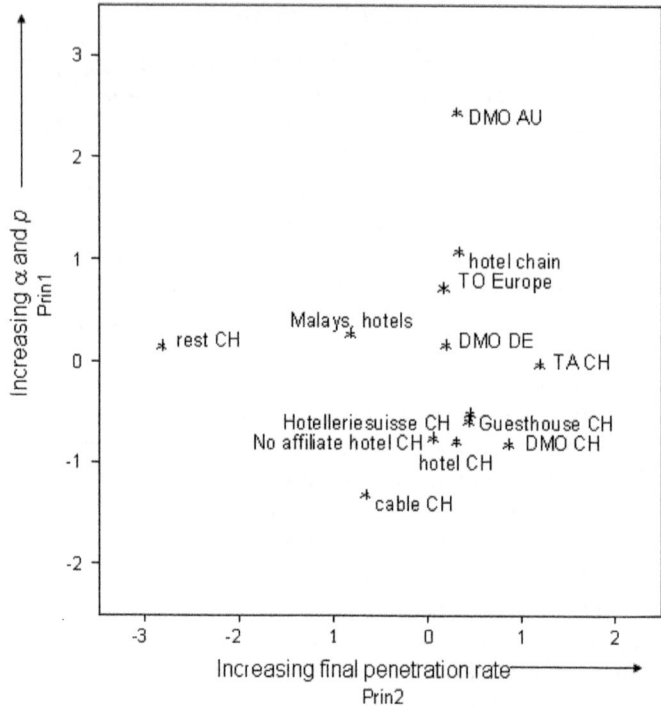

Fig. 6.3 Principal components analysis of Bass parameters, α level of heterogeneity and final penetration rate across tourism sectors/countries (*Source* Authors' computation)

beyond the understanding of the authors but the homogeneity of Swiss restaurants could be explained by the lowest internet penetration level of 9%. A careful study of the restaurant data shows that its homogeneity could be due to location. Swiss regions with high tourism intensity like the cantons of Bern, Vaud and Valais, have more restaurants in comparison to all others cantons.

To shed light on the mixed role of the four values—p, q, penetration rate and α—a principal components analysis using the Proc Princomp of SAS V9.2 yields the graph in Fig. 6.3. The two first components explain a total variation of 77%. Prin1, on the *y*-axis, shows a 0.59 positive correlation with p, a 0.52 positive correlation with α and a 0.62 negative correlation with *q*. This first component seems to capture the diffusion models parameters. Prin2 on the *x*-axis reflects the final penetration rate, showing a positive 0.96 correlation with predicted market penetration. The Swiss sectors clustered in the graph, excluding Swiss restaurants, suggest a possible influence of culture/national factors. Another clustering of international hotels chain and European tour operators suggests a possible influence of internalisation and company size.

Discussion and Conclusion

The analysis shows both the importance of innovation and imitation parameters, homogeneity, critical mass and final market size, and that the 13 tourism sectors varied. Almost all Swiss sectors reflect a high sensibility to critical mass. Possible organisational and cultural factors include resistance to innovation and aversion to risk. Unlike the other Swiss sectors, Swiss DMOs show high heterogeneity. This diverseness aligns with the country's different linguistics/cultural factors and organisational differences such as the size—local, regional or national—budget and intensity of the tourism they represent.

The two international datasets illustrate the importance of testing homogeneity. With near homogenous adoption across categories, European tour operators follow the Bass model of adoption. International hotels chains, however, illustrate the importance of modelling heterogeneity.

As future research streams, quantitative and qualitative studies should continue ferreting out the drivers and barriers to both technology adoption and organisational implementation of that same technology. Qualitative findings can lead to improvements in modelling the diffusion of innovations. The empirical research will enlighten comprehension of the tourism sector and improve diffusion forecasts, particularly the take-off phase that many models ignore.

Limitations of this research are mostly related to the gathering of data process as it uses two different methods, one of them (softbot) only collects the ".ch" domain. Hence the comparison across Swiss accommodations sectors deserves a cautious attitude. Another limitation is that the only parameter reported of G/SG model is heterogeneity (α). This report does not fully address the latter model and future research should make a deeper results analysis.

Appendix

Cumulative series, Observed, estimated (one-step ahead) and forecasted for the last 24 months using Bass model for all sectors under study.

See Figs. 6.4, 6.5, 6.6, 6.7, 6.8, 6.9, 6.10, 6.11, 6.12, 6.13, 6.14, 6.15, and 6.16.

Residual and forecasting errors statistics.

See Table 6.3.

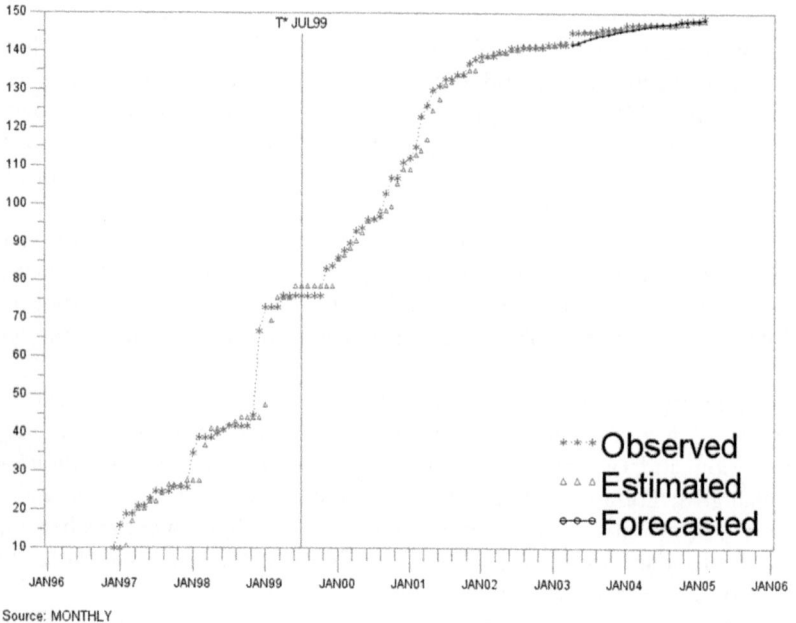

Fig. 6.4 Swiss Destination Managements Organizations (DMOs) (*Source* Authors' computation)

6 MODELLING INTERNET DIFFUSION ACROSS TOURISM SECTORS 165

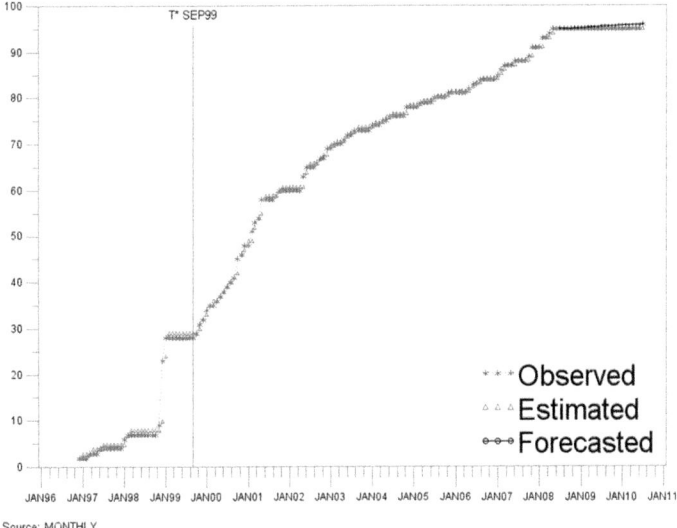

Fig. 6.5 Austrian DMOs (*Source* Authors' computation)

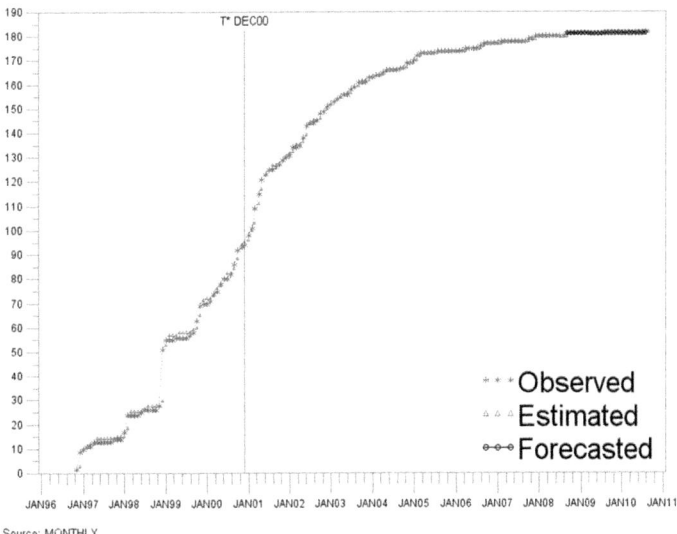

Fig. 6.6 German DMOs (*Source* Authors' computation)

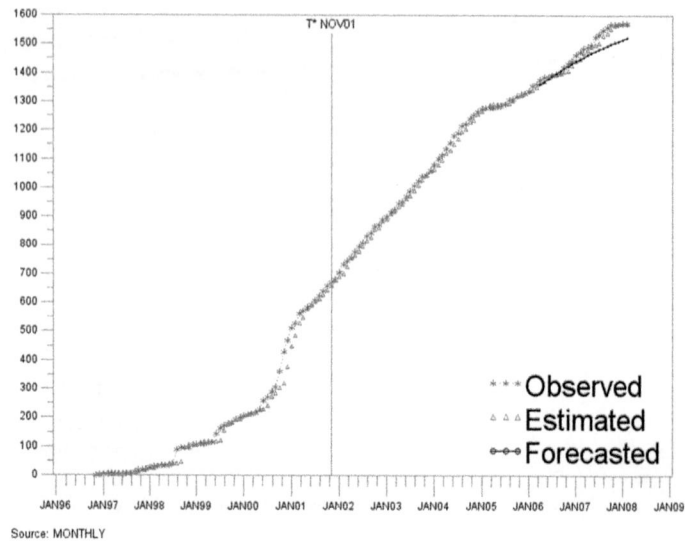

Fig. 6.7 Swiss restaurants (*Source* Authors' computation)

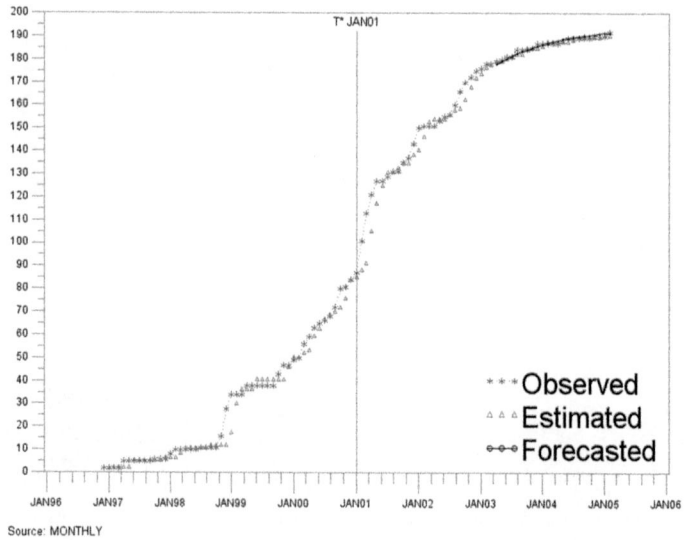

Fig. 6.8 Swiss cable cars companies (*Source* Authors' computation)

6 MODELLING INTERNET DIFFUSION ACROSS TOURISM SECTORS 167

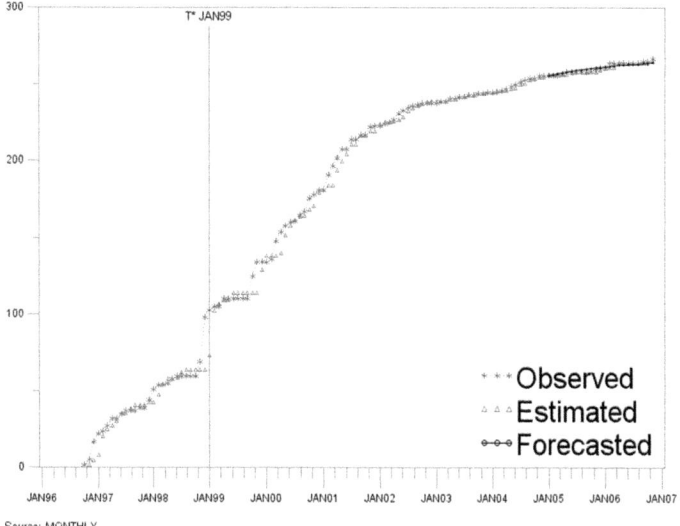

Fig. 6.9 International hotel chains (*Source* Authors' computation)

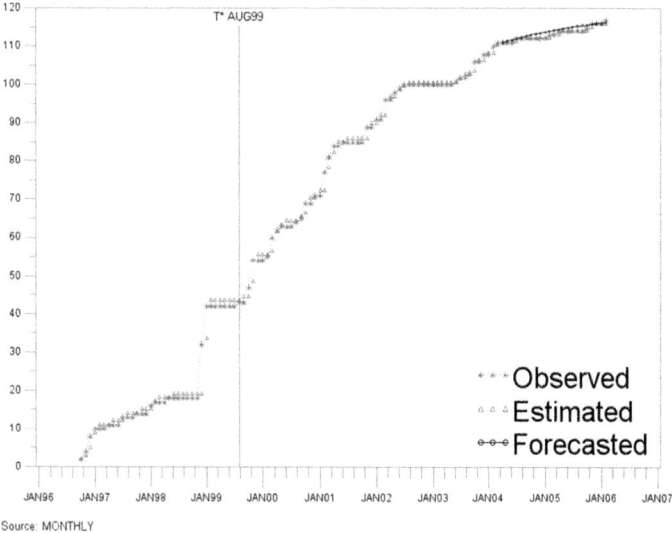

Fig. 6.10 European tour operators (TOs) (*Source* Authors' computation)

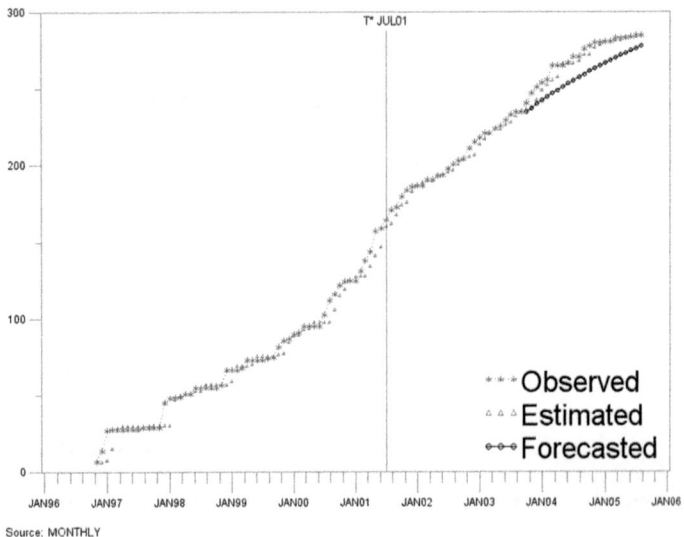

Fig. 6.11 Malaysian hotels (*Source* Authors' computation)

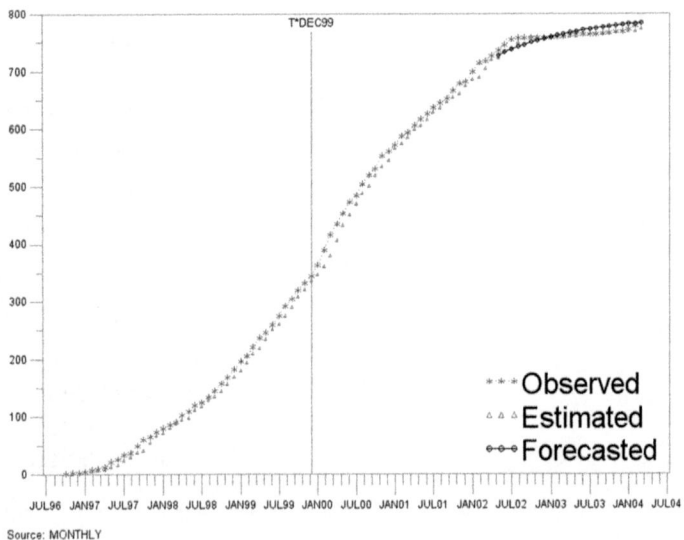

Fig. 6.12 Swiss non-affiliated hotels (*Source* Authors' computation)

6 MODELLING INTERNET DIFFUSION ACROSS TOURISM SECTORS 169

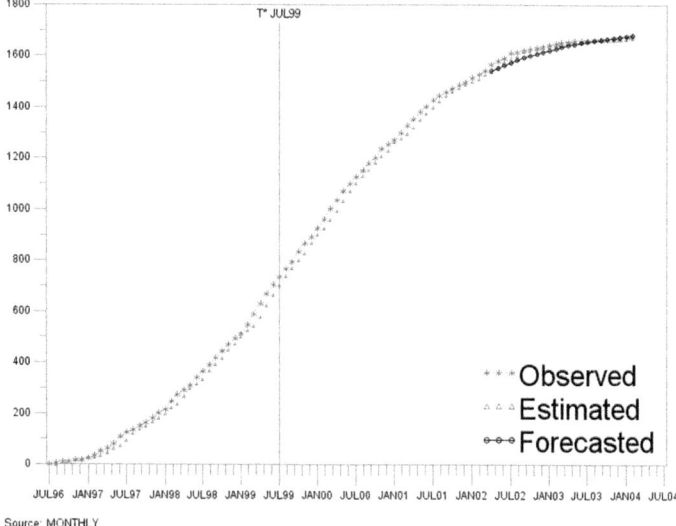

Fig. 6.13 Swiss affiliated hotels (*Source* Authors' computation)

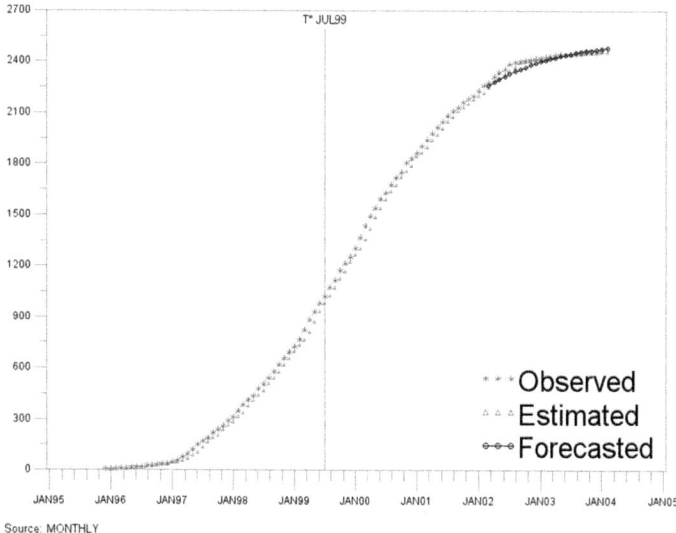

Fig. 6.14 Swiss hotels (*Source* Authors' computation)

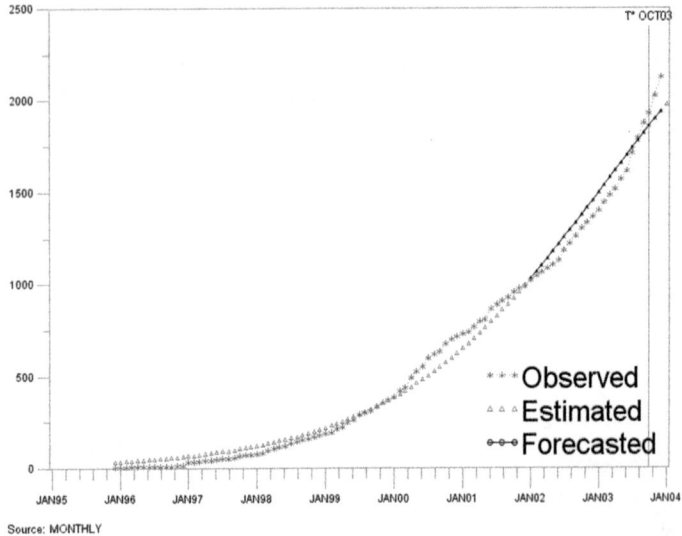

Fig. 6.15 Swiss *gasthouses* (*Source* Authors' computation)

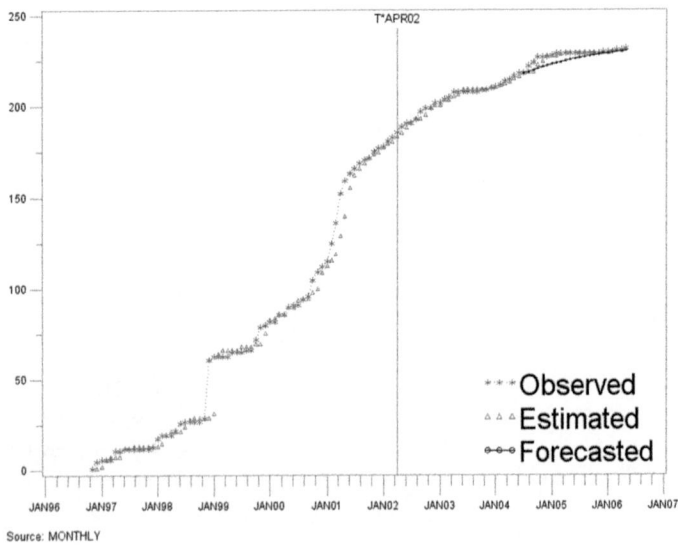

Fig. 6.16 Swiss travel agencies (*Source* Authors' computation)

Table 6.3 Residual and forecasting error statistics

	Residual error statistics: estimation model Bass reported in Table 6.2			Forecasting errors statistics-24 last months				
	MSE	Root MSE	Adj R²	Range of the sample (24 last observations)	Mean Abs Error	Mean Abs % Error	RMS Error	RMS % Error
DMO CH	7.3937	2.7191	0.9966	Mar03–Feb05	0.6861	0.4700	0.9199	0.6322
DMO AU	1.754	1.3244	0.9981	Aug08–Jul10	0.4610	0.4852	0.5309	0.5589
DMO GE	4.6745	2.162	0.9987	Sep08–Aug10	0.3447	0.1904	0.3856	0.213
Restaurant CH	105.5	10.2729	0.9997	Mar06–Feb08	26.4983	1.7302	34.5525	2.228
Cable car CH	6.022	2.454	0.9989	Mar03–Feb05	0.8279	0.4426	0.9744	0.5199
Hotel chain	12.3322	3.5117	0.9983	Dec04–Nov06	1.1229	0.4312	1.305	0.5013
European tour operators	4.0498	2.0124	0.9973	Mar04–Feb06	0.9542	0.8447	1.1094	0.9831
Malaysian hotel	7.5925	2.7554	0.9993	Sep03–Aug05	10.1686	3.7449	10.8393	3.9852
No affiliated hotel CH	14.2494	3.7748	0.9998	Mar04–Feb04	7.2274	0.9499	8.4162	1.1046
Hotelleriesuisse CH	29.2468	5.408	0.9999	Mar02–Feb04	11.5324	0.7077	13.5408	0.8346
Hotel CH	54.6344	7.3915	0.9999	Mar02–Feb04	16.1906	0.6734	19.8001	0.8263
Gasthouses CH	3922	62.6261	0.9892	Jan02–Dec03	67.9966	4.5786	78.837	5.0285
Travel agencies CH	13.1253	3.6229	0.9981	Jun04–May06	2.2008	0.9664	2.7593	1.2122

Source Authors' calculations

References

Andrés, L., D. Cuberes, M. Diouf, and T. Serebrisky. 2010. The Diffusion of the Internet: A Cross-Country Analysis. *Telecommunications Policy* 34 (5–6): 323–340.

Bass, F.M. 1969. A New Product Growth Model for Consumer Durables. *Management Science* 15 (5): 215–227.

Bemmaor, A.C. 1994. Modeling the Diffusion of New Durable Goods: Word-of-Mouth Effect Versus Consumer Heterogeneity. In *Research Traditions in Marketing*, ed. G. Laurent, G.L. Lilien, and B. Pras, 201–223. Boston, MA: Kluwer Academic Publish Group.

Bemmaor, A.C., and J. Lee. 2002. The Impact of Heterogeneity and Ill-Conditioning on Diffusion Model Parameter Estimates. *Marketing Science* 21 (2): 209–220.

Chatterjee, R., and J. Eliashberg. 1990. The Innovation Diffusion Process in a Heterogeneous Population: A Micromodeling Approach. *Management Science* 36 (9): 1057–1079.

Easingwood, C.J., V. Mahajan, and E. Muller. 1983. *A Nonuniform Influence Innovation Diffusion Model of New Product Marketing Science* 2 (3): 273–295.

Hashim, N.H., M. Scaglione, and J. Murphy. 2012. Modelling and Comparing Malaysian Hotel Website Diffusion. In *Information and Communication Technologies in Tourism 2012, Proceedings of ENTER 2012*, ed. M. Fuchs, F. Ricci, and L. Cantoni, 167–178. Wein and New York: Springer-Verlag.

Jensen, R. 1982. Adoption and Diffusion of an Innovation of Uncertain Profitability. *Journal of Economic Theory* 27 (1): 182–193.

Jeyaraj, A., J.W. Rottman, and M.C. Lacity. 2006. A Review of Predictors, Linkages, and Biases in IT Innovation Adoption. *Journal of Information Technology* 21: 1–23.

Kim, S. 2011. The Diffusion of the Internet: Trend and Causes. *Social Science Research* 40: 602–613.

Klimek, K., M. Scaglione, R. Schegg, and R. Matos-Wasem. 2012. Marketing and Sustainable Tourism in Alpine Destinations. In *New Challenges for Tourism Promotion: Tackling High Competition and Multimedia Changes*, vol. 6, ed. P. Keller and C. Laesser, 155–169. Berlin: Erich Schmidt Verlag.

Mahajan, V., E. Muller, and R. Srivastava. 1990. Determination of Adopter Categories by Using Innovation Diffusion Models. *Journal of Marketing Research* XXVII (February): 37–50.

Mahler, A., and E.M. Rogers. 1999. The Diffusion of Interactive Communication Innovations and the Critical Mass: The Adoption of Telecommunications Services by German Banks. *Telecommunications Policy* 23: 717–740.

Meade, N., and T. Islam. 2006. Modelling and Forecasting the Diffusion of Innovation-A 25-Year Review. *International Journal of Forecasting* 22 (3): 519–545.

Murphy, J., N.H. Hashim, and P. O'Connor. 2007. Take Me Back: Validating the Wayback Machine. *Journal of Computer-Mediated Communication* 13 (1). http://jcmc.indiana.edu/vol13/issue11/murphy.html.

O'Connor, P. 1999. *Electronic Information Distribution in Tourism and Hospitality.* Wallingford: CAB international.

O'Connor, P., and J. Murphy. 2004. A Review of Research on Information Technology in the Hospitality Industry. *International Journal of Hospitality Management* 23 (5): 473–484.

Oren, S.S., and R.G. Schwartz. 1988. Diffusion of New Products in Risk-Sensitive Markets. *Journal of Forecasting* 7 (4): 273–287.

Rogers, E.M. 1962. *Diffusion of Innovations*, 1st ed. New York: The Free Press.

Rogers, E.M. 2003. *Diffusion of Innovations*, 5th ed. New York: The Free Press.

Sabre. n.d. *The Sabre Story.* Retrieved from https://www.sabre.com.website: https://www.sabre.com/files/Sabre-History.pdf.

SAS Institute Inc. 2011. *SAS/STAT® 9.22 User's Guide.* Cary, NC: SAS Institute Inc.

Scaglione, M. Forthcoming. The Diffusion of Information and Communication Technologies (ICT) in the Tourism Sector. In *Handbook of E-Tourism*, ed. Z. Xiang, M. Fuchs, U. Gretzel, and W. Höpken. Cham: Springer International Publishing.

Scaglione, M., C. Johnson, and J.-P. Trabichet. 2010. How the Swiss Tourism Sector Is Managing the Change to Web 2.0. In *Managing Change in Tourism Creating Opportunities—Overcoming Obstacles*, ed. P. Keller and T. Bieger, 101–118. Berlin: ERICH SCHMIDT VERLAG.

Scaglione, M., and R. Schegg. 2015. The Case of Switzerland During the Last 20 Years. In *Tourism and Leisure: Current Issues and Perspectives of Development*, ed. H. Pechlaner and E. Smeral, 175–202. Berlin, Germany: Springer Verlag.

Scaglione, M., R. Schegg, T. Steiner, and J. Murphy. 2004. The Diffusion of Domain Names by Small and Medium-Sized Swiss Hotels. In *The Future of Small and Medium Sized Enterprises in Tourism 54th AIEST Congress*, vol. 46, ed. P. Keller and T. Bieger, 259–271. St. Gallen: International Association of Scientific Experts. in Tourism.

Steiner, T. 1999. Distributed Software Agents for WWW-Based Destination Information System. PhD Thesis. Lausanne: University of Lausanne.

Van den Bulte, C., and S. Stremersch. 2004. Social Contagion and Income Heterogeneity in New Product Diffusion: A Meta-Analytic Test. *Marketing Science* 23 (4): 530–544.

Zhu, K., and K.L. Kraemer. 2005. Post Adoption Variations in Usage and Value of E-Business by Organizations: Cross-Country Evidence from the Retail Industry. *Information Systems Research* 16 (1): 61–84.

CHAPTER 7

Exploring Drivers of Online Political Participation in the European Union

María Rosalía Vicente and Hiroaki Suenaga

Introduction

The internet has provided citizens with a wide array of options to participate into political activities: people can express their opinions in forums and on their social network profiles, read official documents online, send emails to their public representatives, and sign petitions or even start them. Governments and political parties have also utilized the internet to interact closely with citizens (Gerl et al. 2018).

Political scientists have been trying to unveil the effects of the internet on citizen's political participation. In particular, they have focused on whether the internet breaks prevalent link between individuals' socioeconomic resources and political participation or whether it reproduces or even exacerbates that association. Empirical results are rather mixed: while some studies suggest that the internet has mitigated the traditional socioeconomic resources gap (Anduiza et al. 2010a; Krueger 2002),

M. R. Vicente (✉)
Applied Economics, University of Oviedo, Oviedo, Spain
e-mail: mrosalia@uniovi.es

H. Suenaga
School of Economics, Finance and Property, Curtin University, Perth, WA, Australia

© The Author(s) 2020
J. Alleman et al. (eds.), *Applied Economics in the Digital Era*,
https://doi.org/10.1007/978-3-030-40601-1_7

others find no significant change (Best and Krueger 2006; Hansen and Reinau 2006). The only conclusive evidence refers to the importance of digital skills for participation (Anduiza et al. 2010a, b; Best and Krueger 2006; Krueger 2002).

Additionally to the socioeconomic factors and digital skills, two other factors are argued to influence citizens' online political participation: (i) online social networks and (ii) the online institutional environment, in particular, the availability of online tools developed by public institutions to promote citizens' political participation. Nonetheless, neither of these factors has been investigated from an integrated approach. Rather, their impacts are usually examined separately from other traditionally considered participation resources.

This paper bridges this gap in the literature and provides new evidence on the determinants of citizens' online political participation. In particular, we focus on participation through official channels, i.e., the online tools and resources implemented by public institutions to foster citizens' political involvement. The contributions of this paper are twofold. First, we analyze online political participation through an integrated resources approach which simultaneously considers both the traditional political participation-related resources and three online-specific resources, namely, digital skills, online group resources, and online institutional environment. Second, while previous research has focused on individual countries, we provide for the 28 member states of the European Union (EU). The use of cross-country data allows us to examine whether the development of the online institutional environment affects citizen's online political participation.

Literature Review

Political participation can be defined as the collection of actions with "the intent or effect of influencing government action—either directly by affecting the making or implementation of public policy or indirectly by influencing the selection of people who make those policies" (Verba et al. 1995, p. 38). Individuals' decisions to conduct these actions have long been studied in the field of political science, where a common approach is to focus on the availability of the required resources.[1] Following this

---

[1] Specifically, the resources required for political participation are classified into 4 groups: (1) individuals' socioeconomic attributes and skills, which determine abilities to afford

resources approach, previous studies have commonly identified that factors such as income, education, employment, living in urban areas, age, and being male are all positively associated with political participation in an offline setting. Other factors, such as individuals' interests in political issues (Armingeon 2007), feeling of citizen duty (Dalton 2008; Vráblíková 2010), self-efficacy (Pattie and Seyd 2003; Vráblíková 2010), ideological position (Holm and Robinson 1978; Miller and Shanks 1982; Robinson and Fleishman 1988), and (dis)satisfaction with politics and politicians (Armingeon 2007), can also influence individual's political participation.

A rapid diffusion of the internet has generated optimistic views about its potential to expand political participation by lowering many of the barriers faced by citizens in the offline environment. In particular, the internet is thought to enhance citizens' political involvement in several ways. First, it allows faster and easier access to information about political and public issues than traditional media (Anduiza et al. 2010a; Di Gennaro and Dutton 2006; Quintelier and Vissers 2007; Schlozman et al. 2010). Second, it fosters participation through lowering the cost of involvement in certain activities. For example, a petition can be easily signed from anywhere and anytime through any devices connected to the internet (computer, mobile phone, tablet, etc.) with less social pressure (Anduiza et al. 2010a; Di Gennaro and Dutton 2006; Quintelier and Vissers 2007; Schlozman et al. 2010). Third, it widens the array of options for political interactions among citizens and also between citizens and public authorities (Bimber 1999; Bimber and Davis 2003; Castells 2009, 2012; Davis 2005; Di Gennaro and Dutton 2006; McAdam et al. 2001; Quintelier and Vissers 2007; Schlozman et al. 2010; van Laer and van Aelst 2009). It also overcomes the unidirectionality of traditional mass communication and enables multidirectional communication (Castells 2009). Fourth, it allows individuals to foster other people's participation through posting information on their online social networks or generating petitions through e-petition platforms

participation-related costs and the access to relevant information, respectively; (2) individuals' political attitudes and opinions; (3) group resources, which help individuals to engage in various group activities and gain information through social networks; and (4) institutional and political environment, which directly influence the availability of channels for citizens' participation.

(Vicente and Novo 2014). Finally, it enables easier mobilization actions at a global level (Etzioni 2003; McAdam et al. 2001; Norris 2001; Schlozman et al. 2010; Schuler 1996; van Laer and van Aelst 2009).[2]

These potentials have led to a mobilization hypothesis which claims that the internet helps to break the link between individuals' socioeconomic characteristics and political activity, and facilitates the participation of individuals who otherwise have been traditionally excluded from political activities (Barber 1989; Dahl 1989; Schlozman et al. 2010). Opposing to this view, the reinforcement or normalization hypothesis states that the internet reinforces the traditional socioeconomic divides in political participation through intensifying the participation activities of those already engaged (Bimber 1999, 2001; Gibson et al. 2005; Murdoc and Golding 1989; Norris 2001, 2003; Schlozman et al. 2010). Other studies (Hirzalla et al. 2011; Nam 2012) argue that the two hypothesized effects are not mutually exclusive; rather the internet plays a dual role: it mobilizes certain groups of individuals who are usually not involved in political activities and, at the same time, it reinforces existing divides in offline participation.

These mixed views on the role of the internet in political participation have led the political science community to explore empirically the patterns of citizens' participation over the internet (Boulianne 2009; Lutz et al. 2014). As one of the first studies on this topic, Bimber (1999) analyzed citizens' online contacts with public officials and reported that education is not significant in explaining online political participation. Krueger (2002) confirmed Bimber's results on education and also observed that citizens with low income utilize the internet more often for political participation than those with higher income. Gibson et al. (2005) similarly report that the internet mitigates the traditionally observed income divide. In contrast, Anduiza et al. (2010a) found that income has no significant effects on online political participation while education exhibits a significant positive effect on online donations. Additionally, Cantijoch et al. (2013) observed that social class explains online political participation. Schlozman et al. (2010) extended this result and identified the same pattern of socioeconomic stratification in both online and offline environments. This result contrasts with the findings of Jensen et al. (2007) that the association between individuals'

---

[2] Mobilization actions refer to citizens' collective actions, such as an organized boycott, or the recent example of the "me too" movement.

socioeconomic characteristics and political engagement is significant only in the offline setting but not in the online environment. Other studies similarly report no conclusive evidence for gender and age (Anduiza et al. 2010b; Best and Krueger 2006; Krueger 2002; Saglie and Vabo 2009).

To summarize, the evidence provided by the previous empirical studies are mixed or inconclusive about the effects of socioeconomic factors on online political participation, with the only consensus being on the positive association of online participation with political interest and efficacy (Anduiza et al. 2010a, b; Best and Krueger 2006; Hargittai 2002; Krueger 2002; Novo and Vicente 2019; Saglie and Vabo 2009; Vicente and Novo 2014). These mixed or inconclusive results imply the presence of factors that are relevant to individuals' political participation specifically in an online environment. Among these, three resources are of particular importance: digital skills, online social networks, and online institutional environment.

With regard to the first of these factors, the use of the internet itself requires specific resources, broadly termed as digital literacy, which refers to individuals' ability to operate computers, smartphones, or other devices to connect to the internet, search for information, and evaluate its quality. The empirical evidence consistently suggests that digital skills are a key resource for online participation and individuals with high levels of digital skills are more likely to participate online (Anduiza et al. 2010a, b; Beam et al. 2018; Krueger 2002; Vicente and Novo 2014).

Second, the roles of online group resources in political participation have received increased attention, particularly since the first Obama campaign utilized social media extensively to mobilize young voters (Boulianne 2015; Jensen 2013; Parikh 2012). However, the empirical evidence is mixed, with Casteltrione (2015) reporting that about half of 60 studies she reviewed find online social networks reinforcing the unequal patterns of political participation, while the other half reports the opposite and highlights mobilization effects.

Finally, over the last years, public institutions have substantially increased the online availability of public services to foster their interactions with citizens and facilitate their engagement in public policy.[3]

---

[3] For example, European Commission's website "Your voice in Europe" provides access to several online tools which allow Europeans to join online discussions on public issues, to contact political representatives, and to provide their feedback to European institutions on specific matters.

The empirical evidence of the effectiveness of such efforts is still limited (Garret and Jensen 2011; Saglie and Vabo 2009). Previous research reports that the success of these actions depends on the type of online participation activities (Vicente and Novo 2014), the local government structure (Zheng et al. 2014), and the effectiveness of public sector governance (Gulati et al. 2014). Furthermore, the online institutional environment has been hardly integrated into the resources approach.

DATA

Data examined in this study are sourced from a survey conducted by the European Commission on the use of online services across Europe (European Commission 2012b, 2013). The population in the survey referred to internet users, i.e., individuals aged from 16 to 74 years old who accessed the internet in the last three months, across the 28 members of the EU. For each country, a representative sample was collected, according to Eurostat's figures on the national composition of the internet users by age, gender, and regional distribution at NUTS1 level (European Commission 2012a). Individuals were contacted by email and invited to participate in the online survey over the last two weeks of November 2012 (European Commission 2012a). Sample size per country is about 1000 respondents except for 200 respondents in Croatia, Cyprus, Luxembourg, and Malta. This leads to a total sample of 24,961 internet users.

By including only internet users in the sample, the analysis reveals the determinants of online participation conditional on the use of the internet. While we cannot extend the results to infer participation behaviors of non-internet users, we can still discuss how individuals' socioeconomic characteristics would affect online participation, were the barriers to the internet access completely eliminated.[4]

The survey asked questions on different online political activities undertaken by internet users. In particular, it asked internet users whether they had contacted any political representatives through email, read official policy documents and decisions on government websites, and participated in online policy-related consultations. These three

---

[4] This is the same approach as Krueger (2002) who analyzed a sample of Internet users to examine the link between online participation and socioeconomic characteristics in a hypothetical situation where equal access were fully achieved.

activities correspond to different types of political participation; according to Ekman and Amna's (2012) typology, reading policy documents is classified as a latent form of civic engagement, while sending an email and taking part in consultation are classified as manifest expressions of political participation. They also differ in terms of the level of commitments required from each side. For any public institution, publishing their email contact online is much easier than creating an online consultation. At the other end, email allows citizens to express their opinions on any matters of their interest, whereas an online consultation allows them to provide their feedbacks only on specified topics. Uploading policy documents is relatively easy, but it implies a passive role of the citizen. In contrast, allowing direct feedback through email entails proactivity by citizens.

ECONOMETRIC MODEL AND VARIABLES

Model

The variables of main interest are citizens' participation in three online political activities: (1) contacting political representatives by email; (2) reading policy documents or decisions online; and (3) participating in online consultations. For each of these activities, we have data recorded in binary terms in the dataset (have done it versus have not). Following Fairlie (2004), an individual's decision to participate online can be modeled in terms of the utility gained from participation relative to the utility from not doing it. Let U_{i1} denote the utility that an internet user i receives from online participation and U_{i0} be i's utility when not participating. These utilities are specified as linear functions of i's resources, $\mathbf{X}$:

$$U_{i0} = \mathbf{X}_i \boldsymbol{\beta}_0 + \varepsilon_{i0} \tag{7.1}$$

$$U_{i1} = \mathbf{X}_i \boldsymbol{\beta}_1 + \varepsilon_{i1} \tag{7.2}$$

where an additive error term, ε_{ik}, is assumed to be normally distributed $\forall i$ and $k = 0, 1$.

We define a dichotomous variable, Y_i, so that $Y_i = 1$ if the user i engages in online political activity and $Y_i = 0$ if he does not. Since the user participates only if U_{i1} exceeds U_{i0}, the probability of user i's participation is expressed as:

$$P(Y_i = 1) = P(U_{i1} > U_{i0}) = F[X_i(\beta_1 - \beta_0)] \tag{7.3}$$

where F is the cumulative normal distribution function of the error term $\varepsilon_i = \varepsilon_{i0} - \varepsilon_{i1}$ with $E(\varepsilon_i)=0$ and $V(\varepsilon_i)=1\ \forall i$.

Variables

The objective is to examine how individuals' decisions to participate in each of the three activities are affected by the set of explanatory variables identified in the literature review section. Table 7.1 provides descriptions of these variables as well as three dependent variables.

The model includes variables representing a wide range of individuals' socio-demographic features (age, gender, employment status, and educational attainment), their level of digital skills,[5] and use of online social networks. The cross-country variations in the institutional and political environment are controlled for by the inclusion of the e-participation index developed by the United Nations (2012). This index evaluates the extent to which citizens access public information, interact with public stakeholders, and participate in decision-making in each country (United Nations 2012). Finally, the model includes regional Gross Domestic Product (GDP), which aims to control for unobserved common factors that are specific to the region where the respondent lives. These data come from Eurostat (2015b).

Table 7.2 presents the descriptive statistics of the full set of variables. The first three rows of the table show that European internet users exhibit moderate levels of online political participation, with an average participation rate ranging from 27.6% for sending email to a political representative to 41.2% for reading policy documents online. As naturally expected, the average participation rate is lower for an activity that requires active participation (sending email to political representative) than for the one that requires passive participation (reading policy documents).

The sample covers individuals aged from 16 to 74 years old and represents roughly equal gender balance. With regard to respondents' employment status, the majority of the sample (58.2%) are employed, while others are either unemployed (8.9%), inactive (8.7%), or students (14.2%).

---

[5] Following the convention in the literature (Anduiza et al. 2010a; Best and Krueger 2006; Eurostat 2015a; Krueger 2002; Vicente and Novo 2014), we approximate the level of digital skills by the number of online activities conducted out of 12 activities. This measure is then divided by 12 to convert into 0–1 scale.

Table 7.1 Description of variables

Variables	Description
Dependent variables	
E-mail	Dummy variable that takes value 1 if respondent has contacted by email any political representatives in the last 12 months (0 otherwise)
Policy documents	Dummy variable that takes value 1 if respondent has consulted policy documents or decisions on government websites in the last 12 months (0 otherwise)
Consultations	Dummy variable that takes value 1 if respondent has participated in online consultations on policy issues in the last 12 months (0 otherwise)
Independent variables	
Age	Age of the respondent
Gender	Dummy variable that takes value 1 if respondent is a woman (0 otherwise)
Network	Dummy variable that takes value 1 if respondent has utilized online social networks in the last 12 months (0 otherwise)
Digital skills	Index for respondent's level of digital skills. It is the number of online activities respondents conduct on a daily (or almost daily) basis, out of the following 12 activities: purchase personal consumer goods/services; purchase tickets or book cultural events; travel bookings; buy/sell goods/services through auction sites; administer a bank account; contribute to blogs; download, watch/listen to music, films, video files, radio or television; download games or online gaming; make calls; check professional email; download/upload documents for professional purposes; search the web for information for professional purposes. It is then divided by 12 to convert into 0–1 scale
Employment situation	Categorical variable on current employment situation: employed; unemployed; inactive; student
Educational attainment	Categorical variable on highest educational attainment: primary/lower secondary school or no formal education; upper secondary school; higher education (e.g., university, college)
Regional GDP per capita pps	Regional Gross Domestic Product (GDP) per capita in purchasing power standards (pps) (in Euros)
UN e-participation index	Index developed by the United Nations (2012) to measure 'the quality and usefulness of information and services provided for the purpose of engaging its citizens in public policy-making through the use of e-government programs'

Source Authors'

Table 7.2 Descriptive statistics

Variables	Mean	Standard deviation	Minimum	Maximum
Email	0.276	0.447	0	1
Policy documents	0.412	0.492	0	1
Consultations	0.311	0.463	0	1
Age	38.75	14.01	16	74
Gender (women)	0.508	0.500	0	1
Network	0.847	0.360	0	1
Digital skills	0.134	0.153	0	1
Employed	0.582	0.49	0	1
Unemployed	0.089	0.285	0	1
Inactive	0.187	0.390	0	1
Students	0.142	0.349	0	1
Primary education	0.147	0.355	0	1
Upper secondary education	0.437	0.496	0	1
Higher education	0.416	0.493	0	1
Regional GDP per capita pps	22,902.64	8868.71	7800	65,200
UN e-participation index	0.419	0.270	0.026	1
n	24,961			

Source Authors' calculations using data from European Commission (2013), Eurostat (2015b) and United Nations (2012)

For education, the majority of sample has achieved either upper secondary or higher education (43.7 and 41.6%, respectively) with the remaining 14.7% having primary education only. As to respondents' online-specific resources, a large share (84.7%) of respondents report that they use online social networks. In contrast, respondents have done on average only 1.6 activities out of 12 online activities considered. Finally, regional GDP per capita ranges from 7800 (Bulgaria) to 65,200 (Luxembourg) Euros. The UN e-participation index varies between 0.026 (Bulgaria) and 1 (the Netherlands).

Results

Table 7.3 shows the results of estimating the three probit equations, one for each of the three online participation activities. The first three columns show the results when the cross-country variations in the institutional and political environment are controlled by country dummies

Table 7.3 Results of the probit regression on online political participation

	(1) Email	(2) Policy documents	(3) Consultations	(4) Email	(5) Policy documents	(6) Consultations
Age	−0.024***	−0.007	−0.028***	−0.023***	−0.005	−0.026***
	(−4.09)	(−1.39)	(−5.21)	(−3.99)	(−0.87)	(−4.87)
Age^2	0.000***	0.000*	0.000***	0.000**	0.000	0.000***
	(3.48)	(2.24)	(5.21)	(3.26)	(1.31)	(4.65)
Women	−0.243***	−0.209***	−0.185***	−0.246***	−0.211***	−0.195***
	(−12.27)	(−10.95)	(−9.75)	(−12.83)	(−10.68)	(−9.53)
Unemployed	−0.151***	−0.176***	−0.069**	−0.137***	−0.110**	−0.074
	(−4.21)	(−5.90)	(−2.24)	(−3.70)	(−2.67)	(−1.79)
Inactive	−0.096**	−0.155***	−0.067*	−0.096**	−0.119***	−0.050
	(−3.12)	(−5.61)	(−2.52)	(−2.86)	(−3.51)	(−1.63)
Student	−0.162***	−0.037	−0.067	−0.184***	−0.054	−0.118**
	(−4.72)	(−1.00)	(−1.62)	(−5.08)	(−1.48)	(−2.59)
Upper secondary education	0.124***	0.200***	0.129***	0.164***	0.273***	0.215***
	(4.76)	(7.16)	(4.17)	(6.10)	(7.85)	(5.39)
Higher education	0.299***	0.509***	0.222***	0.333***	0.566***	0.239***
	(11.51)	(18.45)	(6.86)	(11.88)	(26.03)	(6.24)
Network	0.337***	0.288***	0.390***	0.341***	0.289***	0.368***
	(12.36)	(11.18)	(13.62)	(12.39)	(10.27)	(11.52)
Digital skills	1.329***	1.297***	1.322***	1.355***	1.387***	1.389***
	(18.94)	(18.10)	(18.37)	(18.07)	(19.36)	(18.10)
Regional GDP per capita	0.086	0.116**	−0.002	0.210**	0.063	0.122
	(1.82)	(2.74)	(−0.03)	(3.12)	(0.53)	(1.01)
UN e-participation index				−0.403***	−0.797***	−0.273
				(−3.22)	(−5.40)	(−1.47)
Country dummies	Included	Included	Included			
Constant	−1.354**	−2.205***	0.173	−2.837***	−1.769	−1.858
	(−2.66)	(−4.72)	(0.29)	(−4.04)	(−1.47)	(−1.60)

Source Authors' calculations

Notes The table reports the estimated coefficients and the t-statistic (the latter in brackets). In estimations (1)–(3) country-specific effects are controlled by country dummies (the latter are not reported due to space limitations); in estimations (4)–(6) the UN e-participation index is used instead of the country dummies. ***, **, * indicate statistically significant at the one, five and ten percent levels, respectively. The reference categories are men, employed, primary studies and not users of online networks. The UN index is at country-level whereas the last three columns show the results when they are controlled by the United Nations' e-participation index.

Starting with the age variables, the estimated coefficients are significant negative and positive, respectively, for a linear and quadratic terms in two of the three activities (email and consultations), indicating that the probability of online participation decreases with age at decremental rates. These results are in line with previous findings that youngsters are more involved in online political participation than older counterparts (Albrecht 2006; Best and Krueger 2006; Jensen 2013; Quintelier and Vissers 2007). For the policy document equation, both linear and quadratic coefficients of age are not significant.

In all three activities, coefficients are significant negative for women, whereas they are significant positive for two education variables (upper secondary and tertiary), as well as for Network, and Digital Skills. Coefficients are significant negative for two labor condition variables (unemployed and inactive) in email and policy document equations while it is significant negative for being student only in the email equation. These results indicate that being male, employed, with higher education, better digital skills, and belonging to an online social network all contribute positively to the probability of online political participation. Education has strong effects particularly on the probability of reading policy documents online. Yet, individuals' digital skills have an even greater effect on all three activities, indicating that abilities to operate in the online environment are important for online political participation. The significance of online networks might suggest some mobilization effects and reflects their potential to foster citizens' interaction with political agents (Borge et al. 2012; Krueger 2002; Gibson et al. 2005, 2008; Xenos and Moy 2007).

The importance of an individual's educational attainment and employment in online participation indicates that the traditional offline participation divides are reproduced in the online environment. That is, the socioeconomically inactive or disadvantaged population, both in terms of their status in labor market and educational attainment, are significantly less likely to participate in any online initiatives.

At the regional level, the results show only limited association between income and the three participation activities. At country level, the coefficient is significant negative for the United Nations e-participation index in email and policy document equations. The estimated negative sign might be an indication that citizens are engaged in online political participation activities at a slower rate than the efforts made by public authorities to facilitate population's involvement in policy-making through the internet.

DISCUSSION OF RESULTS AND CONCLUDING REMARKS

Over the last years, the role of the internet in political participation has been discussed extensively in the field of political science. A key question commonly asked in the literature is whether the internet reproduces or ameliorates the socioeconomic divides traditionally observed for various political activities in the offline environment. Previous studies have provided no clear empirical evidence on this debate. Hence, we have tried to contribute to this line of research by applying an integrated resources approach, which incorporates into the analysis the four types of traditional resources as well as a set of resources specifically relevant for online participation, namely, digital skills, online social networks and the availability of officially implemented online tools for citizens' political participation.

The analysis has considered citizens' participation in three types of online political activities through channels implemented by public authorities. These activities vary both in terms of the type and level of commitment required from each side: sending emails to politician implies an active role by the citizens; joining online consultations is also an interactive activity but limited to the specific topic of the public request; and reading policy documents is characterized as a non-interactive activity.

The analysis reveals that the internet replicates the unequal patterns of political participation traditionally observed in offline environment. Specifically, women, the eldest, low-educated and unemployed individuals are significantly less likely to engage in any of the three online political activities than their counterparts. These results obtained for a sample of internet users indicate that, even after controlling for individuals' digital skills and online networking, socioeconomic characteristics are still important in determining political participation online. The results imply that any policies targeting an increased access to the internet and/or increased availability of official online channels will not alleviate inequality in political participation by itself.

These findings raise a very serious concern: even if all the barriers against the internet access were removed, political participation would be still unequal and determined by individuals' socioeconomic background. Indeed, participation divides are substantial and further complicated when combined with digital divides. Traditionally, socioeconomically disadvantaged individuals are hindered from political activities in an offline environment. They are also less likely to utilize the internet, and thus, their opportunity for online political participation is even further limited, compared to internet users who generally come from better socioeconomic background. Additionally, even among internet users, individuals' socioeconomic backgrounds still matter for their online political participation, with those coming from better backgrounds (in particular, with higher educational achievement and being active in labor market) are more likely to participate.

The analysis has also found a negative association between public development of online participation tools and their use by citizens. The negative association indicates underutilization of these official online tools. If this is caused by citizens' unawareness of the services implemented by public institutions, the situation will improve as these online public services become widely acknowledged. To verify this scenario requires further investigation. It would be also interesting to compare non-official and official participatory tools by their usage levels as well as drivers. In particular, it will be necessary to assess whether people are less aware of official channels than non-official channels, whether they find it more difficult to use official channels, or whether it is a matter of preference; they underutilize official channels because they feel less freedom to express themselves in official environment, have less trust with official channels, or feel that these public tools are of no use because in the end their opinions would not be listened to, etc. Future research should address these questions to assess the efforts currently devoted by public institutions and design better policy tools to foster citizens' participation.

References

Albrecht, S. 2006. Whose Voice Is Heard in Online Deliberation? A Study of Participation and Representation in Political Debates on the internet. *Information, Communication & Society* 9: 62–82.

Anduiza, E., M. Cantijoch, A. Gallego, and J. Salcedo. 2010a. *Internet y Participación Política en España*. Madrid: Centro de Investigaciones Sociológicas.

Anduiza, E., A. Gallego, and M. Cantijoch. 2010b. Online Political Participation in Spain: The Impact of Traditional and Internet Resources. *Journal of Information Technology and Politics* 7: 356–368.

Armingeon, K. 2007. Political Participation and Associational Involvement. In *Citizenship and Involvement in European Democracies: A Comparative Analysis*, ed. J. Van Deth, J. Montero, and A. Westholm, 358–384. London: Routledge.

Barber, B.R. 1989. *Strong Democrazy: Participatory Politics for a New Age*. Berkeley: University of California Press.

Beam, M.A., J.D. Hmielowski, and M.J. Hutchens. 2018. Democratic Digital Inequalities: Threat and Opportunity in Online Citizenship from Motivation and Ability. *American Behavioral Scientist* 62: 1079–1096.

Best, S.J., and B.S. Krueger. 2006. Online Interactions and Social Capital: Distinguishing Between New and Existing Ties. *Social Science Computer Review* 24: 395–410.

Bimber, B. 1999. The Internet and Citizen Communication with Government: Does the Medium Matter? *Political Communication* 16: 409–428.

Bimber, B. 2001. Information and Political Engagement in America: The Search for Effects of Information Technology at the Individual Level. *Political Research Quarterly* 54: 53–67.

Bimber, B., and R. Davis. 2003. *Campaigning Online*. New York: Oxford University Press.

Borge, R., A.S. Cardenal, and C. Malpica. 2012. El Impacto de internet en la Participación Política: Revisando el Papel del Interés Político. *Arbor* 188: 733–750.

Boulianne, S. 2009. Does Internet Use Affect Engagement? A Meta-Analysis of Research. *Political Communication* 26: 193–211.

Boulianne, S. 2015. Social Media Use and Participation: A Meta-Analysis of Current Research. *Information, Communication & Society* 18: 524–538.

Cantijoch, M., D. Cutts, and R. Gibson. 2013. *Internet Use and Political Engagement: The Role of e-Campaigning as a Pathway to Online Political Participation*. Retrieved June 26, 2016 from https://escholarship.org/uc/item/538243k2.

Castells, M. 2009. *Communication Power*, 2nd ed. Oxford: Oxford University Press.

Castells, M. 2012. Autocomunicación de Masas y Movimientos Sociales en la Era de Internet. *Anuari Del Conflicte Social 2011* 1: 11–19. Retrieved October 13, 2015 from http://revistes.ub.edu/index.php/ACS/article/view/6235/7980.

Casteltrione, I. 2015. The Internet, Social Networking Web Sites and Political Participation Research: Assumptions and Contradictory Evidence. *First Monday* 20. Retrieved June 26, 2016 from http://firstmonday.org/article/view/5462/4403.

Dahl, R. 1989. *Democracy and Its Critics.* New Haven: Yale University Press.
Dalton, R. 2008. Citizenship Norms and the Expansion of Political Participation. *Political Studies* 56: 76–98.
Davis, R. 2005. *Politics Online: Blogs, Chatrooms, and Discussion Groups in American Democracy.* New York: Routledge Taylor and Francis Group.
Di Gennaro, C., and W. Dutton. 2006. The Internet and the Public: Online and Offline Political Participation In the United Kingdom. *Parliamentary Affairs* 59: 299–313.
Ekman, J., and E. Amna. 2012. Political Participation and Civic Engagement: Towards a New Typology. *Human Affairs* 22: 283–300.
Etzioni, A. 2003. Are Virtual and Democratic Communities Feasible? In *Democracy and New Media*, ed. H. Jenkins and D. Thorburn, 85–100. Cambridge, MA: MIT Press.
European Commission. 2012a. *eGovernment Benchmark Framework 2012–2015.* Brussels: European Commission.
European Commission. 2012b. *Public Services Online: Assessing User Centric eGovernment Performance in Europe—eGovernment Benchmark 2012. Digital Agenda for Europe.* Brussels: European Commission.
European Commission. 2013. *Public Services Online Database.* Brussels: European Commission.
Eurostat. 2015a. *Infomation Society Statistics: E-skills of Individuals.* Retrieved July 29, 2015 from http://ec.europa.eu/eurostat/web/information-society/data/main-tables.
Eurostat. 2015b. *Regional Statistics.* Retrieved July 29, 2015, from http://ec.europa.eu/eurostat/web/regions/data/main-tables.
Fairlie, R.W. 2004. Race and the Digital Divide. *Contributions in Economic Analysis & Policy* 3: 1–40.
Garrett, R.K., and M.J. Jensen. 2011. e-Democracy Writ Small. *Information, Communication & Society* 14: 177–197.
Gerl, K., S. Marschall, and N. Wilker. 2018. Does the Internet Encourage Political Participation? Use of an Online Platform by Members of a German Political Party. *Policy & Internet* 10: 87–118.
Gibson, R., W. Lusoli, and S. Ward. 2005. Online Participation in the UK: Testing a Contextualised Model of Internet Effects. *British Journal of Politics and International Relations* 7: 561–583.
Gibson, R.K., W. Lusoli, and S. Ward. 2008. Nationalizing and Normalizing the Local? A Comparative Analysis of Online Candidate Campaigning in Australia and Britain. *Journal of Information Technology and Politics* 4: 15–30.
Gulati, G.J.J., C.B. Williams, and D.J. Yates. 2014. Predictors of Online Services and e-Participation: A Cross-National Comparison. *Government Information Quarterly* 31: 526–533.

Hansen, H.S., and K.H. Reinau. 2006. The Citizens in e-Participation. *Lecture Notes in Computer Science* 4084: 70–82.
Hargittai, E. 2002. Second-Level Digital Divide: Differences in People's Online Skills. *First Monday* 7. Retrieved October 13, 2015 from http://firstmonday.org/ojs/index.php/fm/article/view/942.
Hirzalla, F., L. van Zoonen, and J. de Ridder. 2011. Internet Use and Political Participation: Reflections on the Mobilization/Normalization Controversy. *The Information Society* 27: 1–15.
Holm, J., and J. Robinson. 1978. Ideological Identification and the American Voter. *The Public Opinion Quarterly* 42: 235–246.
Jensen, J.L. 2013. Political Participation Online: The Replacement and the Mobilisation Hypotheses Revisited. *Scandinavian Political Studies* 36: 347–364.
Jensen, M.J., J.D. Danziger, and A. Venkatesh. 2007. Civil Society and Cyber Society: The Role of the Internet in Community Associations and Democratic Politics. *The Information Society* 23: 39–50.
Krueger, B.S. 2002. Assessing the Potential of Internet Political Participation in the United States: A Resource Approach. *American Politics Research* 30: 476–498.
Lutz, C., C.P. Hoffmann, and M. Meckel. 2014. Beyond Just Politics: A Systematic Literature Review of Online Participation. *First Monday* 19. Retrieved October 13, 2015 from http://firstmonday.org/ojs/index.php/fm/article/view/5260.
McAdam, D., C. Tarrow, and C. Tilly. 2001. *Dynamics of Contention*. Cambridge: Cambridge University Press.
Miller, W., and J. Shanks. 1982. Policy Directions and Presidential Leadership: Alternative Interpretations of the 1980 Presidential Election. *British Journal of Political Science* 12: 299–356.
Murdoc, G., and P. Golding. 1989. Information Poverty and Political Inequality: Citizenship in the Age of Privatized Communications. *Journal of Communication* 39: 180–193.
Nam, T. 2012. Dual effects of the Internet on Political Activism: Reinforcing and Mobilizing. *Government Information Quarterly* 29: S90–S97.
Norris, P. 2001. *Digital Divide: Civic Engagement, Information Poverty and the Internet*. Cambridge: Cambridge University Press.
Norris, P. 2003. *Democratic Phoenix: Reinventing Political Activism*. Cambridge: Cambridge University Press.
Novo, A., and M.R. Vicente. 2019. Exploring the Determinants of e-Participation in Smart Cieies. In *E-Participation in Smart Cities: Technologies and Models of Governance for Citizen Engagement*, ed. M.P. Rodríguez-Bolívar and L. Alcaide-Múñoz, 157–178. Cham: Springer International.

Parikh, K.H. 2012. *Political Fandom in the Age of Social Media: Case Study of Barack Obama's 2008 Presidential Campaign.* LSE. Retrieved October 13, 2015 from http://www.lse.ac.uk/media@lse/research/mediaWorkingPapers/MScDissertationSeries/2011/64.pdf.

Pattie, C., and P. Seyd. 2003. Citizenship and Civic Engagement: Attitudes and Behavior in Britain. *Political Studies* 51: 443–468.

Quintelier, E., and S. Vissers. 2007. The Effect of Internet Use on Political Participation: An Analysis of Survey Results for 16-Year-Olds in Belgium. *Social Science Computer Review* 26: 411–427.

Robinson, J., and J. Fleishman. 1988. A Report: Ideological Identification: Trends and Interpretations of the Liberal-Conservative Balance. *The Public Opinion Quarterly* 52: 134–145.

Saglie, J., and S.I. Vabo. 2009. Size and e-Democracy: Online Participation in Norwegian Local Politics. *Scandinavian Political Studies* 32: 382–401.

Schlozman, K.L., S. Verba, and H.E. Brady. 2010. Weapon of the Strong? Participatory Inequality and the Internet. *Perspectives on Politics* 8: 487–509.

Schuler, D. 1996. *New Community Networks: Wired for Change.* Reading, MA: Addison-Wesley.

United Nations. 2012. *e-Goverment Survey 2012: e-Government for the People.* New York: United Nations.

van Laer, J., and P. van Aelst. 2009. Cyber-Protest and Civil Society: The Internet and Action Repertoires in Social Movements. In *Handbook on Internet Crime*, ed. M.Y. Yvonne Jewkes, 230–254. New York: Willan Publishing.

Verba, S., K.L. Schlozman, and H. Brady. 1995. *Voice and Equality: Civic Voluntarism in American Politics.* Cambridge: Cambridge University Press.

Vicente, M.R., and A. Novo. 2014. An Empirical Analysis of e-Participation. The Role of Social Networks and e-Government over Citizens' Online Engagement. *Government Information Quarterly* 31: 379–387.

Vráblíková, K. 2010. *Contextual Determinants of Political Participation in Democratic Countries.* Retrieved September 25, 2013 from http://epubs.surrey.ac.uk/2577/1/Masaryk1_PIDOP_Barret.pdf.

Xenos, M., and P. Moy. 2007. Direct and Differential Effects of the Internet on Political and Civic Engagement. *Journal of Communication* 57: 704–718.

Zheng, Y., H.L. Schachter, and M. Holzer. 2014. The Impact of Government Form on e-Participation: A Study of New Jersey Municipalities. *Government Information Quarterly* 31: 653–659.

CHAPTER 8

Machine Learning and Forecasting: A Review

Petrus H. Potgieter

INTRODUCTION

Forecasting is an ancient art and science already used by the ancient Egyptians to estimate tax revenue from their measurements of the level of the Nile.[1] The advent of modern electronic computers and the growth in both the availability of data and in computing power put forecasting at the forefront of much of modern management and enabled planning on a scientific basis. Since the 1990s

- a vast increase in easily available data and computer power (including cloud computing, which makes available extraordinary computer resources) and

---

[1] This great African river is surprisingly influential in time series analysis, as in civilization in general, recently because of H.E. Hurst's studies of the previous mid-century (Sutcliffe et al. 2016; Gneiting and Schlather 2004) and the discovery of self-affinity and fractal dimension.

P. H. Potgieter (✉)
University of South Africa, Pretoria, South Africa
e-mail: potgiph@unisa.ac.za

© The Author(s) 2020
J. Alleman et al. (eds.), *Applied Economics in the Digital Era*,
https://doi.org/10.1007/978-3-030-40601-1_8

- the rise of business requirements such as customizing the customer experience in real time,

have made the use of machine learning in forecasting inevitable. One can observe how the serious use of computer-generated foresight has spread from manufacturing and retail into the services industry.

Awareness of the use and utility of forecasting predates the computer era, e.g., White (1928) who pointed out the increase in available business data and described the use of forecasting at the Eastman Kodak Company in the 1920s. As all good things have their dark side, incidentally, the same era saw the rise of scientism[2] in government with Herbert Hoover's ominously named Waste Committee emphasizing the use of forecasting and business statistics in government (Metcalf 1975) in the United States. In the academic world, Einhorn (1972) warned of "an assortment of techniques in search of a substantive problem" in combination with even the computing power of almost half a century ago producing spurious results in abundance.

Nevertheless, in the second decade of the twenty-first century, vast numbers of people are confronted with forecasts in their everyday life. They see content that they were forecast to like, check the weather prediction, use computer estimates of travel time and much more.

Overview

The paper conducts a brief survey of the machine learning technique and its application to forecasting of various types, aimed at readers with a higher than intermediate but not necessarily higher form level of numeracy. It starts with an overview of machine learning and its relation to and application in forecasting in the subsequent section. The main body of work consists of sections devoted to an analysis of theoretical questions arising and practical considerations that relate specifically to forecasting (without and with) machine learning, respectively. The latter reports in summary on selected research findings on concrete applications. An extensive list of references is supplied.

---

[2] Q.v. Thomas Burnett at the American Association for the Advancement of Science (AAAS) https://www.aaas.org/programs/dialogue-science-ethics-and-religion/what-scientism for example.

Background

On the surface, machine learning and forecasting—especially when casting both of these as "computational statistics"—resemble each other quite closely. First, a fixed data set is employed to train an algorithm, by which we mean that parameters are determined that generate a good fit to the training data. In econometric type forecasts such as regression analysis, this can be an explicit mathematical method that might be known to converge to an optimal solution. In machine learning, it is perhaps more usual for there to be an explicit method, the mathematical properties of which are not well understood (in other words, a heuristic).

In any case, the aim of the activity is always to predict or estimate an unknown data point. Forecasting is mainly thought of in terms of time series (e.g., next month's consumer inflation) whereas machine learning practitioners tend to envision the system extracting general rules about an environment and then being used to optimize the behavior of some system (e.g., deciding which product to display on the screen of an online customer in order increase sales). The facility with which human language expresses volition and intention, obscures the prediction which might underly the latter example. It could consist simply of forecasting, for various possible items, the probability that a consumer will make a purchase if that item is shown on the screen and then ranking the possibilities.

Machine Learning

For machine learning, many applications are based on the automated recognition of patterns that allow for the classification of data about hitherto unobserved objects and individuals. A typical application is facial recognition where the most basic application in the area is the ability to pick out faces from a given photograph. Textbook examples include the classification of pictures of objects from a small class (such as items of clothing) and labeling them (as shoes or dresses, for instance). These are applications that are quite far from the practice of forecasting.

A somewhat closer fit to forecasting can be found in the problem that was the subject of the Netflix Prize competition which was based on a training set of over 100 million ratings given by nearly half a million users to some 18,000 films. The data was given as a set of quadruplets and participants in the competition were asked to predict ratings for a test set of nearly 3 million actual ratings which were known only to the organizers. The root mean square error (RMSE) test was used to evaluate the results.

The prize money in the Netflix competition was a million dollars and it awarded smaller prizes for substantial progress over the RMSE performance of Netflix' own algorithm. The prize was awarded in 2009 to one of two essentially equivalently scoring teams and the competition. Privacy concerns (Narayanan and Shmatikov 2006) about the training data lead to the competition not being pursued. The winning team used ensemble methods, including random forests and restricted Boltzmann machine. In the problem, given historical numerical data was used to predict numerical ratings given in the future but because of the nature of the idea, the conventional statistician would perhaps rather approach the problem by first doing a factor or principal component analysis. In machine learning, it might not be necessary to consciously and explicitly make this kind of analysis (Lázaro et al. 2018).

The method used to train the machine learning object is often thought to be an inherent part of the paradigm and not a secondary consideration, for example steepest descent for certain types of neural network (Ahmed et al. 2010). Machine learning methods tend to be more computationally demanding than the older statistical methodologies (Makridakis et al. 2018a).

Forecasting

The usual time series forecasting problem, consists of predicting the value one step into the future by considering a window of n past values and possibly a time lag, $d \geq 1$. For machine learning, the problem is set up as an input data matrix consisting of the available historical windows (as the rows of the matrix) and the target output vector consists of the available historical values corresponding to the windows (Bontempi et al. 2013). Any suitable machine learning algorithm can of course be applied to this simplified problem and numerous software libraries exist for doing so.

In computational statistics, the chief objective of the training exercise is to find mathematically well-defined parameters (that best fit the training data, for instance regression coefficients). The method employed in estimating these parameters (and it is always an estimate) is often thought to be of secondary importance even though being mathematically determined does not necessarily imply that solutions can be found effectively (Brattka et al. 2016) and certainly not efficiently.

Theoretical Approaches

In this section, we consider some theoretical issues affecting both general forecasting and machine learning. The Takens theorem (Bontempi et al. 2013) provides strong theoretical support for the approach of forecasting from a finite set of data but it is in essence non-constructive as it provides no way to picking the method that does well at forecasting future values. In the domains where forecasting has traditionally been used, practitioners are advised to be conservative and to use all available domain-specific knowledge (Armstrong et al. 2015). This practice is appropriate where data is plentiful and there is perhaps some underlying theory. In these cases, one can hope to construct a *knowledge model* (Armstrong and Green 2019) which might include causal relationships that are based on science or experience.

Machine learning is most frequently seen as a tool for *data models* where the entire model is derived purely from the data provided and where no specific theory underlies the whole. In the case of the Netflix competition data mentioned above, this is an example. It is often thought that machine learning could expose or hint at an underlying theory to be discovered through its ability to expose patterns in large data sets. This is especially true when the data sets involve unusual or qualitative data such as internet search histories (for example, Aziz and Dowling 2019). In this sense, machine learning borders on heuristics when it is used as a tool for suggestion relationships that are later investigated and developed into a possible knowledge model.

Forecasting Trouble

Naturally, practitioners need to prepare the fitness of forecasting methodologies and for this a suitable measure is required. The choice of measure itself (in the Netflix competition, it was root mean square deviation) is sometimes made without regard to the eventual application of the forecast so errors with a severe effect (the temperature being 20 degrees lower the forecast) are not distinguished from those with a generally less severe effect (the temperature being 20 degrees higher than forecast).

Mean absolute percentage error (MAPE) has been frequently recommended (Hyndman and Koehler 2006) as a measure of fit but it poses severe problems around values of zero, for example. Using MAPE, a forecast temperature of 5 degrees (Celsius) is as great a forecast error as one of 15 degrees in case the actual temperature is measured as 12 degrees. The same is true of a measured Fahrenheit temperature of 50 degrees compared to forecasts of 25 and 100 degrees respectively. This is a general problem that effects forecasting competitions and applies equally to machine learning techniques and this paper returns to the topic below, when discussing the empirical evidence.

Einhorn (1972) emphasizes the tendency of data analysis techniques in general to produce type I errors (rejecting the null hypothesis). This is actually a general problem when considering possible relations since even (actually, especially) a random process will generate patterns with high probability (almost surely in the case of an infinite random series, Fouché and Potgieter 1998).

Machine Learning Difficulties

The naïve application of machine learning to time series data can easily result in a neural network (for example) applied[3] to random walk data, discovering a persistence model. Although similar problems arise in forecasting using more traditional statistical methods, they can be much less obvious and more difficult to detect in the machine learning context due to the "black box" functioning of the methodology.

Ben-David et al. (2019) have investigated the mathematical foundations of machine learning employed for the estimation of the maximum problem. They correctly summarize the machine learning problem as to "[a]pproximate a target concept given a bounded amount of data about it." They then formulate it as a general set theoretic problem and derive theoretical impossibility theorems based on mathematical set theory. Nevertheless, that study concludes that inter alia that the results "may not be directly related to practical machine learning applications."

---

[3] Vegard Flovik writing on Medium at https://towardsdatascience.com/how-not-to-use-machine-learning-for-time-series-forecasting-avoiding-the-pitfalls-19f9d7adf424 for example.

Missing Data

Madden et al. (2017) have demonstrated the difficulties generated by missing data, a ubiquitous phenomenon in practice (Moon et al. 2017 provide on example). They demonstrated the importance of being able to distinguish absent values in data sets as being generated by a process where the values are either

- missing completely at random,
- missing at random, or
- missing not at random.

Using machine learning, this process can be automated (thereby saving time and non-computing resources) and the risk of modeling error by imposing an incorrect and rigid assumption about missing data might be avoided. Although econometricians might dislike admitting it, real datasets almost always feature missing data and quite possibly frequently of the third kind (above). This is true not only of surveys but also of other kinds of data where coding errors, capturing errors, incorrect readings (a zero from an instrument where a reading is in fact missing) all play a role. (Guyon et al. 1994) pioneered the modern use of machine learning in cleaning data, including the use of the Vapnik–Chervonenkis criterion (discussed in the next subsection).

Over-Fitting and Vapnik–Chervonenkis Theory

Over-fitting is an ever-present danger for all methods. Statistical forecasting has however developed information criteria that are relatively well understood and can be used to avoid the problem. So far, equivalent methods for machine learning are not well developed (Makridakis et al. 2018a; Ahmed et al. 2010). Actually, it appears to be worth considering suboptimal determination of parameters during training and there is evidence to support the idea that suboptimal parameter determination need to materially influence forecast accuracy (Nikolopoulos and Petropoulos 2018). This is similar to results in other areas where approximations to optimal choices are not in practice highly distinguishable from theoretical optima (Howell and Potgieter 2018). In the machine learning environment, the *explore-exploit tradeoff* is the explicit name given to the relationship between the cost of gathering training data and exploiting the trained model (Amat et al. 2018).

For classifier problems in machine learning, the Vapnik–Chervonenkis dimension provides a useful tool (Abu-Mostafa 1989) for determining whether the training set is sufficiently large for the model to be effective and whether the model will be efficient. Here we consider classification in two classes only. The Vapnik–Chervonenkis inequality implies that if the set of classifiers for n observations associated with a model has Vapnik–Chervonenkis dimension S_n then the probability that a model classifier misses on n data points differs from the global probability that it misclassifies by at most

$$2\sqrt{\frac{\log S_n + \log 2}{n}}$$

in other words: if S_n grows sufficiently slowly, it is possible to know how many data points is needed in order to be able to judge the fit of the available classifiers to the data. This means that the training set will identify a good universal classifier.

Armed with the Vapnik–Chervonenkis theory and the related Perles–Sauer–Shelah theorem (Buzaglo et al. 2012), classifier problems in machine learning appear to be on a solid footing and it might be conjectured that this is the class of prediction problems (including those in banking, mentioned below, where a forecast indicator is used as a classifier) where machine learning techniques outperform statistical forecasting methods. A word of caution is nevertheless not out of place. A textbook example is the classification of photographs that have trees in them. One can use the theory to derive an error bound and proceed to train a classifier with a sufficiently large training set. However, we do not know the probability of an arbitrary photograph contains a tree. If this probability is low, a classifier might produce many errors (of both type I and type II) relative to the number of photographs with trees in them. Using the Vapnik–Chervonenkis theory involves the implicit assumption that the distribution over the two classes in the population is more-or-less known to the modeler!

Empirical Evidence

In this section we consider some of the empirical evidence around the use of machine learning in forecasting problems. No specific methodology has been employed to select the examples but a significant correlation with the interests of the author is to be assumed.

Case Studies

We look at case studies about the use of machine learning for forecasting in a few important areas in this subsection: banking, economics, medicine and other applications. Given the success of machine learning in commercial applications of various kinds, it is not surprising that it is being applied in new fields, including medicine and public policy.

Banking

Bequé and Lessmann (2017) used artificial neural networks for credit scoring, a widely used tool in financial institutions which has as its goal the prediction of a client's loan repayment behavior and/or profitability. This is not ordinary time series forecasting as the prediction relates either to a borrower belonging to a certain class in the future or a probability of default for an individual, based on historical default rates for a class and some individual characteristics. The study finds that various machine learning techniques (notably extreme learning machines and regularized linear regression) perform well in terms of predictive accuracy, ease of use and computing resource complexity when compared to the current industry standard, linear regression. Earlier work by Lessmann et al. (2015) had speculated that the most recent machine learning appeared not to be particularly good at credit scoring but nevertheless confirmed that certain techniques (neural network classifiers, for example) performed better than linear regression. They argue that linear regression should no longer be used as the benchmark for assessing the performance of new credit scoring algorithms.

(Lázaro et al. 2018) apply machine learning in the problem of managing a bank's cash supply chain by using it to forecast demand for cash at branches (including ATMs). They combine the forecast demand with an optimization of the required cash transportation logistics and identity significant savings using their test data. This problem well illustrates a useful application of machine learning since it involves

- A complicated modeling problem, and
- A clear objective—cost saving—supported by forecast values that are used to determine acceptable incidences of stock-out days.

(Gogas et al. 2018) forecast bank failure using support vector machines (SVMs) and find their method to outperform the traditionally used Ohlson score—a bankruptcy forecasting model based on a linear combination of several scalars and ratios as well as two binary-valued indicator functions. It is instructive to consider their criteria however. In the out-of-sample data considered, the Ohlson method predicted insolvency for 381 banks, 114 of which turned out to be insolvent. Solvency was predicted to 898 banks, one of which in fact turned out to be insolvent. The false alarms generated are therefore high. The SVM method predicted only 119 insolvencies (of which 112 correctly) and 1160 solvent banks (of which 3 turned out to be insolvent). One's assessment of this method depends not on a statistical methodology but on the severity if missing an insolvency and the advantage of not incorrectly predicting insolvency for a bank that will turn out to be solvent. It is possible that the Ohlson method can be modified to produce similar results.

Economics

(Klimek et al. 2019) used linear response theory to forecast economic growth for 56 industrial sectors in 43 countries and found that the machine learning technique outperformed ARIMA forecasting in the years after the great financial crisis.

Butaru et al. (2016) predict credit card delinquency using three methods: logistic regression, decision trees, and random forests and find that both latter methods outperform logistic regression. Similarly, Kim and Swanson (2018) find hybrid methods using machine learning and factor methods to perform acceptably in macro-economic forecasting.

Medicine

Aging populations, declining medical workforces and the availability of patient data have propelled healthcare applications to prominence in the machine learning literature. Predicting the pathogenicity of genetic mutations is one example (Esteva et al. 2019). These are new areas of application wherein statistical forecasting has not had a prominent role and researchers find it natural to use machine learning. Electronic health records have been used to predict cardiovascular risk (Beam and Kohane 2018).

Other Applications

Moon et al. (2017) use principal component and factor analysis in combination with machine learning methods to forecast power consumption in higher education. In the public sector, machine learning has been used in situations where traditional forecasting has not played a big role, e.g., the predictions of the probability that police contacts will be involved in adverse incidents (Ackermann et al. 2018). Support vector machines have proven useful in several areas, including predicting permeability in a gas reservoir for example (Baziar et al. 2014).

The M3 and M4 Competitions

Ahmed et al. (2010) considered the third Madridakis (M3) competition data of the International Institute of Forecasters to compare the performance of eight machine learning techniques on roughly one thousand monthly time series of business data. The best performance reported was by the multilayer perceptron (MLP) and the Gaussian process models. They used symmetric MAPE to measure forecast accuracy. In order to identify the best models, they looked at best-performance fraction averages for several pre-processing methods as well as other rankings. However with the better pre-processing methods, the performance of Bayesian neural networks does not appear to be much worse than that of Gaussian processes. From the M4 Competition results, Makridakis et al. (2018b) report a poor performance by pure machine learning methods but much better results from hybrid methods.

Conclusion

Armstrong and Green (2019) consider pure data models (including machine learning) to be of dubious scientific value and there is a considerable literature to support the idea that simple knowledge-based models perform better than complicated models that require extensive computation. Nevertheless, machine learning is widely used in commercial applications and its versatility makes it useful for pre-processing and imputing (Mullainathan and Spiess 2017). In the banking industry in particular, machine learning methods have been reported to be successful. We have conjectured that where forecasting is used as a classifier, machine learning might be especially successful.

Opinions expressed in this work are those of the author at the time of writing and not necessarily those of any project, his social network persona(e), employer(s), employee(s), donors, clients or suppliers nor of any company of which he might be director and/or shareholder.

Acknowledgements This project has received funding from the European Union's Horizon 2020 research and innovation program under the Marie Skłodowska-Curie grant agreement No. 731143 for Computing with Infinite Data.

The author is grateful to the late Gary Madden for innumerable stimulating conversations on many topics that have influenced the author's thinking and also to Patricia Madden for much kindness and hospitality in Perth and to both of them for friendship and good cheer in many locations around the world.

References

Abu-Mostafa, Yaser S. 1989. The Vapnik-Chervonenkis Dimension: Information Versus Complexity in Learning. *Neural Computation* 1 (3): 312–317. https://doi.org/10.1162/neco.1989.1.3.312.

Ackermann, Klaus, Lauren Haynes, Rayid Ghani, Joe Walsh, Adolfo De Unánue, Hareem Naveed, Andrea Navarrete Rivera, et al. 2018. Deploying Machine Learning Models for Public Policy. In *Proceedings of the 24th ACM SIGKDD International Conference on Knowledge Discovery & Data Mining—KDD 18*. ACM Press. https://doi.org/10.1145/3219819.3219911.

Ahmed, Nesreen K., Amir F. Atiya, Neamat El Gayar, and Hisham El-Shishiny. 2010. An Empirical Comparison of Machine Learning Models for Time Series Forecasting. *Econometric Reviews* 29 (5–6): 594–621. https://doi.org/10.1080/07474938.2010.481556.

Amat, Fernando, Ashok Chandrashekar, Tony Jebara, and Justin Basilico. 2018. Artwork Personalization at Netflix. In *Proceedings of the 12th ACM Conference on Recommender Systems—RecSys 18.* ACM Press. https://doi.org/10.1145/3240323.3241729.

Armstrong, J. Scott, and Kesten C. Green. 2019. Why Didn't Experts Pick M4-Competition Winner? https://repository.upenn.edu/marketing_papers/431.

Armstrong, J. Scott, Kesten C. Green, and Andreas Graefe. 2015. Golden Rule of Forecasting: Be Conservative. *Journal of Business Research* 68 (8): 1717–1731. https://doi.org/10.1016/j.jbusres.2015.03.031.

Aziz, Saqib, and Michael Dowling. 2019. Machine Learning and AI for Risk Management. In *Disrupting Finance: FinTech and Strategy in the 21st Century,* ed. Theo Lynn, John G. Mooney, Pierangelo Rosati, and Mark Cummins, 33–50. Cham: Springer International Publishing. https://doi.org/10.1007/978-3-030-02330-0_3.

Baziar, Sadegh, Mehdi Tadayoni, Majid Nabi-Bidhendi, and Mohsen Khalili. 2014. Prediction of Permeability in a Tight Gas Reservoir by Using Three Soft Computing Approaches: A Comparative Study. *Journal of Natural Gas Science and Engineering* 21 (November): 718–724. https://doi.org/10.1016/j.jngse.2014.09.037.

Beam, Andrew L., and Isaac S. Kohane. 2018. Big Data and Machine Learning in Health Care. *JAMA* 319 (13): 1317–1318. https://doi.org/10.1001/jama.2017.18391.

Ben-David, Shai, Pavel Hrubeš, Shay Moran, Amir Shpilka, and Amir Yehudayoff. 2019. Author Correction: Learnability Can Be Undecidable. *Nature Machine Intelligence* 1 (2): 121. https://doi.org/10.1038/s42256-019-0023-6.

Bequé, Artem, and Stefan Lessmann. 2017. Extreme Learning Machines for Credit Scoring: An Empirical Evaluation. *Expert Systems with Applications* 86 (November): 42–53. https://doi.org/10.1016/j.eswa.2017.05.050.

Bontempi, Gianluca, Souhaib Ben Taieb, and Yann-Aël Le Borgne. 2013. Machine Learning Strategies for Time Series Forecasting. In *Business Intelligence: Second European Summer School, EBISS 2012, Brussels, Belgium, July 15–21, 2012, Tutorial Lectures,* ed. Marie-Aude Aufaure, and Esteban Zimányi, 62–77. Berlin, Heidelberg: Springer Berlin Heidelberg. https://doi.org/10.1007/978-3-642-36318-4_3.

Brattka, Vasco, Stéphane Le Roux, Joseph S. Miller, and Arno Pauly. 2016. The Brouwer Fixed Point Theorem Revisited. In *Pursuit of the Universal,* 58–67. Springer International Publishing. https://doi.org/10.1007/978-3-319-40189-8_6.

Butaru, Florentin, Qingqing Chen, Brian Clark, Sanmay Das, Andrew W. Lo, and Akhtar Siddique. 2016. Risk and Risk Management in the Credit Card Industry. *Journal of Banking & Finance* 72 (November): 218–239. https://doi.org/10.1016/j.jbankfin.2016.07.015.

Buzaglo, Sarit, Rom Pinchasi, and Günter Rote. 2012. Topological Hypergraphs. In *Thirty Essays on Geometric Graph Theory*, 71–81. New York: Springer. https://doi.org/10.1007/978-1-4614-0110-0_6.

Einhorn, Hillel J. 1972. Alchemy in the Behavioral Sciences. *Public Opinion Quarterly* 36 (3): 367. https://doi.org/10.1086/268019.

Esteva, Andre, Alexandre Robicquet, Bharath Ramsundar, Volodymyr Kuleshov, Mark DePristo, Katherine Chou, Claire Cui, Greg Corrado, Sebastian Thrun, and Jeff Dean. 2019. A Guide to Deep Learning in Healthcare. *Nature Medicine* 25 (1): 24–29. https://doi.org/10.1038/s41591-018-0316-z.

Fouché, W.L., and P.H. Potgieter. 1998. Kolmogorov Complexity and Symmetric Relational Structures. *The Journal of Symbolic Logic* 63 (03): 1083–1094. https://doi.org/10.2307/2586728.

Gneiting, Tilmann, and Martin Schlather. 2004. Stochastic Models That Separate Fractal Dimension and the Hurst Effect. *SIAM Review* 46 (2): 269–282. https://doi.org/10.1137/s0036144501394387.

Gogas, Periklis, Theophilos Papadimitriou, and Anna Agrapetidou. 2018. Forecasting Bank Failures and Stress Testing: A Machine Learning Approach. *International Journal of Forecasting* 34 (3): 440–455. https://doi.org/10.1016/j.ijforecast.2018.01.009.

Guyon, I., N. Matić, and V. Vapnik. 1994. Discovering Informative Patterns and Data Cleaning. In *Proceedings of the 3rd International Conference on Knowledge Discovery and Data Mining*, 145–156. AAAIWS'94. Seattle, WA: AAAI Press. http://dl.acm.org/citation.cfm?id=3000850.3000866.

Howell, Bronwyn E., and Petrus H. Potgieter. 2018. Bundles of Trouble: Can Competition Law Adapt to Digital Pricing Innovation? *Competition and Regulation in Network Industries* 19 (1–2): 3–24. https://doi.org/10.1177/1783591718801102.

Hyndman, Rob J., and Anne B. Koehler. 2006. Another Look at Measures of Forecast Accuracy. *International Journal of Forecasting* 22 (4): 679–688. https://doi.org/10.1016/j.ijforecast.2006.03.001.

Kim, Hyun Hak, and Norman R. Swanson. 2018. Mining Big Data Using Parsimonious Factor, Machine Learning, Variable Selection and Shrinkage Methods. *International Journal of Forecasting* 34 (2): 339–354. https://doi.org/10.1016/j.ijforecast.2016.02.012.

Klimek, Peter, Sebastian Poledna, and Stefan Thurner. 2019. Quantifying Economic Resilience from Input–output Susceptibility to Improve Predictions of Economic Growth and Recovery. *Nature Communications* 10 (1). https://doi.org/10.1038/s41467-019-09357-w.

Lázaro, Jorge López, Álvaro Barbero Jiménez, and Akiko Takeda. 2018. Improving Cash Logistics in Bank Branches by Coupling Machine Learning and Robust Optimization. *Expert Systems with Applications* 92 (February): 236–255. https://doi.org/10.1016/j.eswa.2017.09.043.

Lessmann, Stefan, Bart Baesens, Hsin-Vonn Seow, and Lyn C. Thomas. 2015. Benchmarking State-of-the-Art Classification Algorithms for Credit Scoring: An Update of Research. *European Journal of Operational Research* 247 (1): 124–136. https://doi.org/10.1016/j.ejor.2015.05.030.

Madden, Gary, Nicholas Apergis, Paul Rappoport, and Aniruddha Banerjee. 2017. An Application of Nonparametric Regression to Missing Data in Large Market Surveys. *Journal of Applied Statistics* 45 (7): 1292–1302. https://doi.org/10.1080/02664763.2017.1369498.

Makridakis, Spyros, Evangelos Spiliotis, and Vassilios Assimakopoulos. 2018a. Statistical and Machine Learning Forecasting Methods: Concerns and Ways Forward. Edited by Alejandro Raul Hernandez Montoya. *PLOS ONE* 13 (3): e0194889. https://doi.org/10.1371/journal.pone.0194889.

Makridakis, Spyros, Evangelos Spiliotis, and Vassilios Assimakopoulos. 2018b. The M4 Competition: Results, Findings, Conclusion and Way Forward. *International Journal of Forecasting* 34 (4): 802–808. https://doi.org/10.1016/j.ijforecast.2018.06.001.

Metcalf, Evan B. 1975. Secretary Hoover and the Emergence of Macroeconomic Management. *Business History Review* 49 (01): 60–80. https://doi.org/10.2307/3112962.

Moon, Jihoon, Jinwoong Park, Eenjun Hwang, and Sanghoon Jun. 2017. Forecasting Power Consumption for Higher Educational Institutions Based on Machine Learning. *The Journal of Supercomputing* 74 (8): 3778–3800. https://doi.org/10.1007/s11227-017-2022-x.

Mullainathan, Sendhil, and Jann Spiess. 2017. Machine Learning: An Applied Econometric Approach. *Journal of Economic Perspectives* 31 (2): 87–106. https://doi.org/10.1257/jep.31.2.87.

Narayanan, Arvind, and Vitaly Shmatikov. 2006. How to Break Anonimity of the Netflix Prize Dataset. *CoRR* abs/cs/0610105. http://arxiv.org/abs/cs/0610105.

Nikolopoulos, Konstantinos, and Fotios Petropoulos. 2018. Forecasting for Big Data: Does Suboptimality Matter? *Computers & Operations Research* 98 (October): 322–329. https://doi.org/10.1016/j.cor.2017.05.007.

Sutcliffe, John, Stephen Hurst, Ayman G. Awadallah, Emma Brown, and Khaled Hamed. 2016. Harold Edwin Hurst: The Nile and Egypt, Past and Future. *Hydrological Sciences Journal* 61 (9): 1557–1570. https://doi.org/10.1080/02626667.2015.1019508.

White, Charles P. 1928. Industrial Forecasting. *The ANNALS of the American Academy of Political and Social Science* 139 (1): 109–125. https://doi.org/10.1177/000271622813900115.

CHAPTER 9

Stated Consumer Behavior with D-Efficient Choice Set Designs: The Case of Mobile Service Bundles

Christian M. Dippon and Gary Madden

INTRODUCTION

Recently, competition, deregulation, and technological advances have radically shifted the boundaries of formerly distinct (and stable) telecommunication markets. The convergence of data, video, and voice markets foreshadowed that product pricing would increasingly result from market outcomes; that is, product innovations would be more frequent, and competition would be conducted within differentiated (by product, platform, and provider) markets. Consequently, the emphasis on establishing market-own-price-elasticity values changed to obtaining firms' cross-price elasticities among complementary and substitute products via

---

The authors acknowledge helpful comments from Kenneth Train (UC, Berkeley) and Jerry Hausman (MIT).

C. M. Dippon (✉)
NERA Economic Consulting, Washington, DC, USA
e-mail: christian.dippon@nera.com

G. Madden
Curtin University, Perth, WA, Australia

© The Author(s) 2020
J. Alleman et al. (eds.), *Applied Economics in the Digital Era*,
https://doi.org/10.1007/978-3-030-40601-1_9

demand simulations (Tardiff and Levy 2014). Thus, rather than examining presumably the outcomes of some market equilibrium, preferences for firms' products would need to be analyzed.

This search for firms' elasticity values for recently introduced products increased the dependence on data suitable to estimate the theoretical model underlying the elasticity estimates. There are two potential sources for data—stated preference (SP) data and revealed preference (RP) data. RP data are actual market data, whereas SP data are the result of an experiment, specifically a consumer survey.[1] SP experiments aim to provide data to model consumer choices when sales data are either not available or not practical. There are various trade-offs between RP and SP data. The main advantage of RP data is that subscribers actually made those choices and accepted the consequences. In practice, however, RP data present several difficulties and complexities. Chief among them is the fact that given the competitive environment of the U.S. mobile sector, price variations are minimal as mobile service providers are price takers and thus cannot widely experiment with changes in service plan offerings. The implication of this lack of variation is that a theoretical model cannot evaluate the interactions between demand determinants and sufficient information does not exist on how consumers react to attribute changes (Louviere et al. 2000). Consequently, the only feasible alternative is SP data.

SP data have often been applied in the analysis of various telecommunication service demands, for example, mobile services and regulatory policy (Leng 2016), terrestrial digital multimedia broadcast (Byun et al. 2006), broadband internet (Cardona et al. 2009; Ida and Kuroda 2006), mobile telephone (Ida and Kuroda 2009), international mobile telecommunications-2000 (Kim 2005), cable TV (Yoo 2002), and portable internet services (Yoo and Moon 2006). Operationally, combinations of attribute levels describe alternatives. Survey respondents trade off different attributes by indicating which of the alternatives they are most and least likely to purchase. Given the lack of variation in the actual market, the trade-off experiment must necessarily offer choices that are not present in the marketplace. Confraria et al. (2017) studied the joint instead of individual effect of switching costs and network effects (both local network and global effects)

---

[1] Perl (1978) first used survey data to model residential demand for basic telephone service in the United States. Madden et al. (1993) provide the first application of an orthogonal (fractional factorial) design of experimental price regimes.

in determining consumers' preferences when selecting their mobile telecommunication plans. Churchill (1995) found that attribute ranges should extend beyond what researchers typically observe in the market; however, they should not be so excessive as to make them unrealistic. Similarly, Bliemer and Rose (2009) found that a wide range of attribute levels yields parameter estimates with smaller standard errors. Evaluated at the sample mean, however, these elasticities reflect market realities rather than hypothetical choices.[2]

Because SP data provide information from experiments rather than actual choices, questions arise as to the reliability of inferences based on these responses. These questions are about both the experimental design (whether the internal structure of the experiment, that is, the combinations of the attributes and attribute levels, enable precise parameter estimates to be obtained) and the experiments' external validity (whether the responses really explain subsequent market behavior). This study focuses on the former question.

In designing an SP experiment, a researcher could include all possible combinations of attribute levels in the attribute matrix and present each of the resulting surveys to decision makers. This full factorial design would allow the researcher to obtain a specific response from each decision maker for each combination of attribute levels thereby ruling out the possibility of biasing the study results through the administration of a subset of these combinations. It further provides the researcher with full information on the effects of a change in the attributes on the dependent variable. While theoretically attractive, implementing a full factorial design is highly challenging due to the large number of attribute combinations. Consequently, full factorial designs are only practical for the simplest of SP experiments.

Considering these limitations, researchers have opted to present only a subset of all possible choice situations to each survey respondent, that is, fractional factorial design. The specific distribution of attributes across this matrix can have a material impact on the model outputs and the statistical power of the experiment, particularly when small numbers of observations are involved (Rose and Bliemer 2010). Thus, attribute levels within the incidence matrix must be selected with great care.

---

[2] As an additional measure to reflect market realities, Train (2009) describes a method to recalibrate the constants "to reflect the fact that unobserved factors are different for the forecast area or year compared to the estimation sample."

A common method for determining the attribute levels for fractional factorial design matrices is the orthogonal transformation (see Louviere et al. 2000). This transformation is primarily justified by its property to maximize the D-efficiency of linear models (e.g., Kuhfeld et al. 1994; Lusk and Norwood 2005).[3]

More recent research has raised several concerns as to whether orthogonal design yields the most accurate method for nonlinear models, such as the logit model (e.g., Huber and Zwerina 1996; Kanninen 2002; Kessels et al. 2006; Sandor and Wedel 2001, 2002, 2005; Vermeulen et al. 2011).[4] Given these concerns, Rose et al. (2008) proposed to attain D-efficiency through Monte Carlo simulation. In this method, the analyst initially populates the design matrix randomly. Using parameter priors from a properly specified pilot model, the analyst then calculates the choice probabilities for this design and then constructs the AVC matrix. To evaluate the statistical efficiency of this initial design, the initial D-error is calculated. This is simply the determinant of the AVC matrix from the initial design. In a next step, Rose et al. proposed a design change (i.e., changing one or more attributes in the SP design matrix) and then recalculating the D-error. If the D-error of the second run is smaller than the D-error of the initial run, the second design is retained; if not, it is rejected. This looping procedure is repeated R times up to the point where the researcher is satisfied with the D-error of the design. The objective of the present study is to assess the accuracy gains (measured in terms of D-efficiency) of this efficient design method relative to traditional design methods. On the methodological side, such comparisons have been made in previous research. For example, Huber and Zwerina (1996, Table 2) investigated percentage gains in expected D-error, similar to the results in Table 6 of this paper. Comparisons based on other metrics (such as RMSE or EMSE) have also been made

---

[3] Statistical efficiencies are measures of design "goodness," indicating the precision of the parameter estimates for a fixed sample size (Rose and Bliemer 2010). Among these efficiency measures is "D-efficiency," which measures the determinant of the variance-covariance matrix. Although D-efficiency seems to be the most prominent measure, there are several others. Rose and Bliemer (2005) and Kuhfeld et al. (1994) discuss A-efficiency, Scarpa and Rose (2008) analyze the C-error criterion, whereas Bliemer and Rose (2005) introduce S-efficiency.

[4] We note that this concern is not universal. For instance, Bliemer and Rose (2005) show that in unlabeled choice experiments where all parameter priors are zero, the orthogonal design will be D-optimal.

between D-efficient designs and traditional designs (e.g., Arora and Huber 2001; Sandor and Wedel 2001, 2002, 2005). Bliemer and Rose (2010) discuss the efficacy of D-efficient design in lowering standard errors and thereby yielding more efficient survey designs.[5] On the practical side, however, there is a clear void in the literature. Importantly, little research exists that examines whether D-efficiency gains are retained in the fitted model. The design matrix is optimized according to the coefficient priors. Are these efficiency gains retained in the fitted model? More specifically, is the D-error of a model that was based on an optimized design matrix smaller than the D-error of an identical model that relied on data from a non-optimized design? This study addresses this void and develops a test to address how much efficiency gains, if any, are retained in an optimized model. The findings from this research are consistent with those from past research. The benefits of D-efficient designs have been demonstrated by the aforementioned research. The reduction of design efficiency in circumstances when assumed values used in design construction deviate from true values has also been reported before (e.g., figures 3–5 in Sandor and Wedel 2002). To address this issue, Bayesian designs (Sandor and Wedel 2001; Kessels et al. 2006) have been proposed to account for the uncertainty of the prior specifications. Finally, this study is among the first applications of efficient design methods to the mobile telephony service markets.[6]

Traditional Design Methods

Full fractional designs contain all combinations of the prescribed attribute levels. However, potentially large numbers of attribute combinations limit reliable assessment by survey respondent comparisons. The alternatives from a full factorial design are:

$$T^{ff} = \prod_{j=1}^{J} \prod_{z=1}^{Z_j} l_{jz} = l^{Z \cdot J} \tag{9.1}$$

---

[5] Other studies that addressed the efficiency gains from experimental design include Toner et al. (1998, 1999), Viney et al. (2005), Tudela and Rebolledo (2006), Louviere et al. (2008), and Hess et al. (2008).

[6] A 2014 study by the Australian Government, Department of Communications, applied D-efficient design to measure consumer preferences with respect to broadband services in Australia.

where l_{jz} is the number of levels for attribute z in choice alternative j (Louviere et al. 2000).[7] Given the practical limitation of full factorial design, researchers have opted to present only a subset of all possible choice situations to each survey respondent, that is, fractional factorial design. Various methods exist from which a researcher can select the subset. A commonly used method is to select randomly a number of choice situations from the full factorial design and present this subset to survey respondents. Another method is to divide the number of choice situations in the full factorial design by the number of survey respondents and administer the choice situations in sequential blocks. Yet, another method creates a set of blocks that are then repeatedly administered to the survey respondents.

Traditionally, orthogonality defined a properly chosen fractional factorial design as the statistical property was thought to achieve D-efficiency. Orthogonal designs ensure that the attribute levels are not correlated and that they achieve D-efficiency by minimizing the value of the determinant of the asymptotic variance-covariance (AVC) matrix (minimum D-error). Smaller D-errors infer smaller standard errors for the estimated linear model coefficients (Kuhfeld et al. 1994; Lusk and Norwood 2005).

Efficient Design

Bunch et al. (1996) proved that the D-efficient property of orthogonal designs did not readily extend to nonlinear models, such as logit models. A further obstacle in applying non-orthogonal designs to nonlinear models is an inability to find closed-form solutions for D-efficiency when estimated by maximum likelihood (Bunch et al. 1996). Attempts to derive closed-form solutions for particular functional forms have also proved unsuccessful, for example, Chaloner and Larntz (1989) for logistic regressions.

With D-optimality criterion problematic for nonlinear models, an alternative concept, D-efficiency, was proposed (Rose et al. 2008). Specifically, Rose et al. proposed randomly assigning elements in the design matrix. From this initial design, an AVC matrix is constructed, and an initial

---

[7]For instance, in an SP experiment with three choice alternatives ($J = 3$), three attributes ($Z = 3$), and three attribute levels for each of the attributes ($l = 3$), there are $3^{3 \times 3} = 19{,}683$ choice situations. Full factorial design for this SP experiment would imply that all survey respondents state their preferences for 19,683 choice situations.

D-error (the determinant of the AVC matrix from the initial design) is calculated. Next, one or more attributes in the SP design matrix are changed, and the D-error is recalculated. Should this D-error be lower than the initial D-error, the second design is retained (alternatively rejected). This procedure is repeated until changes in the D-error are "small."

Several methods have been developed that guide sequential changes to the matrix designs. For instance, Cook and Nachtsheim (1980) proposed a modified Fedorov algorithm that starts with a full factorial design (simple choice situations) or a fractional factorial design (more complex choice situations) that iteratively replaces design matrix rows. The fractional matrix is selected so that attribute levels are distributed evenly across the design matrix. Monte Carlo simulations evaluate designs in the set and select the lowest D-error. Huber and Zwerina (1996) and Sandor and Wedel (2001) proposed the relabeling, swapping, and cycling (RSC) algorithm to replace attributes rather than rows in the matrix. Relabeling refers to replacing an attribute column by switching attribute levels. Swapping changes only selected attribute levels. Finally, cycling replaces all attribute levels (for all choice situations) with the next higher or lower value.

CHOICE CONTEXT, EXPERIMENT, AND SAMPLE

In the United States, mobile phone handsets (handsets) are often bundled by providers with mobile phone services. Bundles (or plans) typically prescribe an airtime allowance, overage charges, data download and upload speeds, and other data services such as short message service (SMS) and multimedia message service (MMS) options. This commercial practice raises the question of the extent to which bundle components, specifically handsets, affect consumer subscriptions. Thus, quantifying the sensitivity of a service bundle (bundle) and the bundle component demand to changes in plan structure (including prices) is important for commercial and policy input. An online survey of U.S. consumers' choice of mobile phone bundle enabled the implementation and testing of D-efficient design methods.[8]

---

[8] The questionnaire was comprised of a 14-page A4 document, and 31 responses were sought. In particular, three questions ascertained respondent age, current mobile phone subscription, and responsibility for billings; seven questions elicited current mobile phone use by service; another twelve questions formed the conjoint experiment and elicited first

The conjoint experiment required respondents to rank three (hypothetical) mobile service plans by preference from high to low.[9] Next, respondents indicated the least desirable of the non-preferred alternatives. The plans are potential discrete choices (alternatives). Three alternatives comprised a choice situation (set). An aspect of the experimental design is that respondents faced different choice sets. In total, survey respondents completed six independently drawn choice sets (with each containing three alternatives).

Table 9.1 lists the choice attributes viewed by respondents. Importantly, price and non-price attributes other than voice and SMS are considered. Further, Table 9.2 reports the attribute levels that define sample alternatives.

Table 9.3 illustrates three plans. Each plan has identical attributes, but their levels differ. For instance, Plan 1's monthly charge is $120 and Plan 2's $50. Respondents select the preferred plan based on the relative value placed on each attribute. For example, a low handset price is more important than the length of the term contract for some subscribers. However, other respondents prefer paying more for the handset and being free to switch providers.

Table 9.4 profiles the sampled respondent sociodemographic characteristic by the U.S. Bureau of Census population values. The sample means values are similar to that of the Bureau. Next, the reported mobile

---

and second preferences from the six choice sets; another question identified household landline subscriptions. Finally, eight questions elicited information concerning household location and respondent characteristics. Details of the respondent profile sought included age, education, gender, marital status, employment status, number of children, and income. SurveySavvy™ administered the survey in May 2010 via an online omnibus survey panel of 3.5 million persons. The full launch generated 653 responses. However, because of incomplete responses or short completion times (below 3.5 minutes), 164 questionnaires were eliminated. The study analyzed the remaining 489 completed questionnaires.

[9] In reality, there are many more mobile service plans from which to choose. Limiting the alternatives to three assumes that mobile subscribers narrow their selection to three plans before making a decision. It also takes into account that given the complexity of mobile service plans consumers might not be capable of comparing more than three plans at once.

Table 9.1 Definition of service bundle attributes

Variable	Attribute	Description	Unit
handset_price	Handset price	Price of the mobile phone	$
Mrc	Monthly recurring charge	Fixed-plan price component per month	$
v_allow	Monthly voice allowance	Number of voice minutes in monthly charge	minutes
d_allow	Monthly data allowance	Number of kilobytes of downloads and uploads in monthly charge	Kb
download	Speed	Speed in seconds that a file can be downloaded from the internet	kbps
v_over	Minutes penalty	Per-minute charge in excess of the monthly voice allowance	$/minute
d_over	Data penalty	Per-kilobyte charge in excess of the monthly data allowance	$/kb
text	SMS	Charge for each text message sent and received	$/message
Smartphone	Handset	=1, Smartphone (e.g., iPhone, Blackberry);=0, non-Smartphone	dummy
term_length	Term	Contract length in months with an ETF of $150	months

Source Author's computation

Table 9.2 Service bundle attribute levels

Handset price	MRC	Allowance		Speed	Penalty		SMS	Handset	Term
		Voice	Data		Minutes	Data			
0	20	400	0	250	0	0	0	1	0
50	40	800	50	500	0.10	0.10	0.05	0	6
100	60	1200	200	1000	0.15	0.15	0.10		12
200	80	1600	500	1500	0.20	0.20	0.20		18
300	100	2000	1000	2000	0.25	0.25	0.25		24
400	120	3000	5000	3000	0.30	0.30	0.30		30
500	160	9999	9999	6000	0.40	0.40	0.40		36

Source Authors' calculations

Table 9.3 An illustrative choice set

Attribute	Plan 1	Plan 2	Plan 3
Handset price	$200	$50	$400
Monthly recurring charge	$120	$60	$20
Monthly voice allowance	400	3000	2000
Monthly data allowance	5000	200	50
Speed	3000	500	2000
Minutes penalty	$0.40	$0.10	$0.10
Data penalty	$0.40	$0.25	$0.10
SMS	$0.30	$0.25	Free
Handset	Smart	Smart	Not smart
Term	30	18	12

Source Authors' calculations

communication profiles are compared to those reported by several industry sources. Overall, the mean values match well with the exception of plan minutes.[10]

Finally, a review of experimental choices was made to ensure that respondents considered carefully in making evaluations. Those reviews revealed that no mobile plan was always selected, viz., no plans were dominant. Second, the ordering of the alternatives responses did not influence decisions. Finally, there is no evidence of respondent fatigue or selections based on predetermined rules.

---

[10] Several internal consistency checks tested the accuracy of self-reported responses. First, respondents who had prepaid service plans should not have a term contract. Approximately 12% of respondents currently had prepaid mobile service. Of those, none indicated that they were under a term contract. Specifically, 98% responded in the negative with 2% stating that they did not know. Second, monthly data plan subscribers should use internet and email. Of the 41.7% that had data plans, 86.2% accessed the internet and 80.4% used email. Curiously, 17 data plan subscribers did not use their mobile phone to access the internet or email. Some service providers require the purchase of data plans with certain handsets. This might explain the seeming anomaly. Finally, respondents who did not subscribe to a mobile service plan at the time of the survey should not respond to questions about current consumption. As expected, none of those respondents answered questions about current consumption.

Table 9.4 Selected sample and population mean comparisons

Variable	Mean		Source	Unit of measurement
	Sample	Population		
Socio-demographic variables				
Age	44.4	45.2	US Bureau of Census	Years
Employed	62.6	65.0	US Bureau of Census	Percent
Female	52.9	50.8	US Bureau of Census	Percent
Married	50.7	50.3	US Bureau of Census	Percent
Children	1.8	1.9	US Bureau of Census	Number (households with children)
Annual income	$30–$50k	$30–$50k	US Bureau of Census	Modal income category
State of residence	25.4	24.8	US Bureau of Census	50 U.S. states and the District of Columbia numbered 1–51 by alphabetical order
Mobile communications profile				
	Sample	Other		
Wireless phone	94.3	93.0	CTIA (2011)	Percent
Plan minutes	900-1400	726	CTIA (2011)	Minutes per month
Data plan	41.7	41.3	Nielson (2010a)	Percent
SMS plan	46.3	43.0	Pew Research Center	Percent
Mobile Internet	40.1	47.0	comScore (2011)	Percent
Mobile email	38.4	30.5	comScore (2011)	Percent
Monthly expense	$50-$99	$47.09	FCC (2010b)	Sample median
Postpaid contract	69.8	70.9	FCC (2010b)	Percent
Landline communications profile				
Mobile only	28.0	26.6	CTIA (2011)	Percent

Source Authors' calculations based on cited data sources

Implementing the Experimental Design

The attribute matrix consists of ten choice attributes and three alternatives (or 30 columns). Each row contains three mobile service plans (choice sets or situations). The number of rows equals the number of conjoint exercises contained in the experiment.[11] With all attributes having seven or fewer levels, a design of seven blocks with each block containing six choice situations is the minimum number of rows to attain attribute balance. Hence, the attribute matrix subject to the D-efficient design is dimension (30 × 42). The matrix was determined D-optimized through *Ngene* by ChoiceMetrics using the RSC swapping procedure.

D-efficient designs require a priori information about model specification and coefficient values. Providing this information is critical because the optimized design matrix is based on these inputs.[12] The parameter priors were obtained from a pilot study in which 25 individuals responded to six choice situations each. At two observations per choice situation, this provided 300 observations. A standard exploded logit model was fitted to these data. Table 9.5 summarizes the results of the fitted *ex ante* model.

In the first iteration, the D-optimization routine generated a D-error of 0.000225. Through successive iterations, the RCS routine reduced this error by varying the attributes matrix in accordance with the RSC method. Table 9.6 shows that the routine ran 1,479,737 iterations and reduced the initial D-error by 40%, indicating that efficiency can be drastically improved through the use of efficient design. Stated differently, efficient design reduces the D-error of random attribute matrix by 40%.

---

[11] With repeating choice situations, the design matrix could consist of six rows defined as a single block, thus implying that all respondents would select from the same six choice situations. However, such a design would not allow attribute balance as each attribute level only has 14% chance of inclusion in the design.

[12] The corollary of this requirement is that the resulting D-efficient attribute matrix is only D-efficient with respect to the model that generated the a priori information. It follows that if the a priori information deviates from the final model specification, the attribute matrix may not be D-efficient with respect to the final (*ex post*) model. This chicken-or-egg problem creates a unique challenge for the analyst because the model's final specification is subject to the statistical significance (or absence thereof) of the attributes and its various interaction terms.

Table 9.5 Exploded logit estimates, pilot data

Variable	Coefficient	Std. Error	Z statistic
Phone	−0.0021	0.0005	−4.38
Mrc	−0.0167	0.0021	−7.76
v_allow (per 100)	0.0066	0.0042	1.58
d_allow (per 100)	0.0112	0.0038	2.98
download (per 100)	0.0034	0.0048	0.71
v_over	−0.0838	0.7544	−0.11
d_over	−0.3807	0.7668	−0.50
Text	−1.6333	0.6137	−2.66
Smartphone	1.0785	0.1869	5.77
term_length	−0.0032	0.0072	−0.45
Number of observations	750		
Number of groups	300		
Observations per group			
Minimum	2		
Average	2.5		
Maximum	3		
LR χ^2 (10)	137.56		
Probability > χ^2	0		
Log likelihood	−199.98		

Source Authors' computation

Table 9.6 Ngene optimization results

Statistics	Value
D-error start value	0.00022523557
D-error end value	0.00013464514
Percentage improvement	40
Total iterations	1,479,737

Source Authors' calculations

D-Optimality of the *Ex Post* Model

As per Table 9.6, subject to the specification and parameter priors of the *ex ante* model, D-efficient design promises to improve the accuracy of the parameter forecasts by 40%. However, knowledge of the exact model specification and coefficients is not usually available. This raises the question of whether the optimization gains are robust to deviations from this model specification. To test the sensitivity of D-efficient designs, sample data were fitted to various versions of the logit model. Tables 9.7 and 9.8 summarize the results of the *ex post* model.

Table 9.7 Mixed exploded logit estimates

Variable	Mean of log of coefficients			Standard deviation of log of coefficients		
	Estimate	Std. Error	Z statistic	Estimate	Std. Error	Z statistic
phone_neg	−5.7816	0.0674	−85.78	0.8043	0.0715	11.25
mrc_neg	−4.6736	0.1262	−37.03	1.0518	0.0967	10.88
v_allow	−5.9611	0.5215	−11.43	2.6188	0.3043	8.61
d_allow	−6.1022	0.5578	−10.94	1.4831	0.4493	3.30
download	−8.0862	1.4286	−5.66	1.9329	0.4253	4.54
text_neg	−1.3937	0.5098	−2.73	1.7141	0.2827	6.06
Smartphone	−1.8069	0.3046	−5.93	1.7144	0.2283	7.51
term_length_neg	−5.1307	0.2957	−17.35	1.5041	0.1881	8.00
dummy_high	−1.4487	0.3150	−4.60	0.9079	0.2642	3.44
Log likelihood at convergence = −4427.6						

Source Authors' computation

Table 9.8 Mixed exploded logit estimates, lognormal

Variable	Distribution	Median	Mean	Standard deviation	
phone_neg	Lognormal	−0.0031	−0.0043	0.0040	
mrc_neg	Lognormal	−0.0093	−0.0161	0.0224	
v_allow	Lognormal	0.0026	0.0913	1.6604	
d_allow	Lognormal	0.0022	0.0067	0.0183	
download	Lognormal	0.0003	0.0020	0.0110	
Text	Lognormal	−0.2482	−1.0786	4.1441	
Smartphone	Lognormal	0.1642	0.6997	2.6406	
term_length	Lognormal	−0.0059	−0.0182	0.0505	
dummy_high	Lognormal	0.2349	0.3553	0.4044	
		Coefficient	Standard error	Z	
age_int	Fixed	−0.4479	0.1085	−4.13	
gender_int	Fixed	1.0965	1.2696	0.86	
d_over	Fixed	−0.4371	0.196	−2.23	

Source Authors' computation

The *ex post* model differs from the *ex ante* model in two aspects. First, instead of a standard exploded logit model, the *ex post* model is a mixed logit model with lognormal distributed coefficients. Second, the coefficients estimated by the *ex post* model differ from the pilot study coefficient because the pilot study produced different estimates than the actual survey and due to the inclusion of three additional variables in the *ex post* model (e.g., dummy_high, age_int, and gender_int).

To test the robustness of the benefits from D-optimization, the D-error of the optimized design matrix is calculated using the specifications of the *ex post* model (rather than the *ex ante*—pilot—model). This error measure then is compared to the D-errors of 30 design matrices created from random draws from the full factorial matrix that are also evaluated using the specifications of *ex post* model.

Ngene allows mixed logit models in its optimization routine. However, at present, the software can only optimize mixed logit models for uniform and normal distributions. In the absence of lognormal distributions in *Ngene's* optimization routine, the D-errors of the design matrix and the 30 randomly designed matrices are evaluated under two model specifications. The first specification is the *ex post* model, as presented above with all variables distributed lognormal. The second specification is a derivate of the *ex post* model with all variables distributed normal. Although not used, this second specification normalizes for any optimality loss due to differences in the coefficient distribution. Table 9.9 presents the results of this comparison.

The average D-error of the random matrices evaluated under the lognormal version of the *ex post* model is 0.0870. This compares to the D-error of 0.0878 of the optimized design matrix evaluated under the same model. Similarly, as evaluated under the normal version of the *ex post* model, the average chance D-error is 0.0074 compared to the D-error of the optimized matrix of 0.0088. The D-error for the optimized matrix is larger when evaluated against the average of both benchmark models. This indicates that for the present study the potential benefits from D-optimization could not be retained due to differences between the *ex ante* model and the *ex post* model. Under the *ex post* model, a chance design would have generated equally accurate forecasts as the optimized design did.

Notwithstanding, Table 9.10 illustrates that if the model specification is known at the survey design stage, efficient survey design stands to significantly improve the D-error of the design matrix and thereby the accuracy of the study. Specifically, had the specification and parameter estimates of the *ex post* model been known at the survey design phase (i.e., the analyst has perfect foresight), D-efficient design could have improved the D-error by 83%.

Table 9.9 D-optimality comparisons

Design	D-error, Lognormal model	D-error, Normal model
Random1	0.1209	0.0075
Random2	0.0876	0.0076
Random3	0.0704	0.0078
Random4	0.0635	0.0075
Random5	0.0854	0.0078
Random6	0.0805	0.0073
Random7	0.0710	0.0071
Random8	0.0791	0.0080
Random9	0.0943	0.0080
Random10	0.0908	0.0068
Random11	0.1032	0.0077
Random12	0.1043	0.0071
Random13	0.0919	0.0068
Random14	0.0804	0.0070
Random15	0.0864	0.0072
Random16	0.0858	0.0067
Random17	0.0729	0.0072
Random18	0.0784	0.0071
Random19	0.0781	0.0071
Random20	0.0886	0.0067
Random21	0.0940	0.0074
Random22	0.0773	0.0082
Random23	0.0811	0.0077
Random24	0.0810	0.0075
Random25	0.0778	0.0075
Random26	0.0986	0.0076
Random27	0.0984	0.0070
Random28	0.0712	0.0076
Random29	0.0841	0.0075
Random30	0.1328	0.0078
Average random	0.0870	0.0074
Optimized	0.0878	0.0088

Source Authors' computation

Hence, for D-efficient design to be applied successfully to other studies, analysts must have accurate a priori information about the *ex post* model's specification and parameter estimates. As such, the method promises the most benefits in market research studies that are conducted on a regular basis (e.g., monthly, quarterly, or yearly). In fact, as Bliemer

Table 9.10 Ngene optimization results with perfect foresight

Statistics	Value
D-error start	0.135723
D-error end	0.022749
Total iterations	62,257
Last iteration with improvement	62,041
Improvement	83

Source Authors' calculations

and Rose (2010) show, efficient design methods promise best results in settings where studies are repeated frequently and thus priors become available on a continuous basis.

Conclusions

Testing for D-optimality revealed that the *ex post* model retained no benefits of D-optimization. A randomly derived design matrix yields on average a D-error of between 0.0074 and 0.0870. A design matrix that is optimized based on priors has an ultimate D-error of between 0.0088 and 0.078. This finding illustrates that D-optimization requires highly accurate a priori information of the model specification and its coefficients. However, if such information is available, it might question the need for conducting the study. Notwithstanding, with near perfect a priori information or in situation where serial efficient design is an option, the underlying D-error of the design matrix could have been improved by over 80%.

References

Arora, N., and J. Huber. 2001. Improving Parameter Estimates and Model Prediction by Aggregate Customization in Choice Experiments. *Journal of Consumer Research* 28 (September): 273–283.

Bliemer, M., and J. Rose. 2005. Efficiency and Sample Size Requirements for Stated Choice Studies. Working Paper ITLS-WP-5-08, Institute of Transport and Logistics Studies, University of Sydney.

Bliemer, M.C.J., and J.M. Rose. 2009. Efficiency and Sample Size Requirements for Stated Choice Experiments (Paper #09-2421). Presented at the Transportation Research Board 88th Annual Meeting, Washington, DC.

Bliemer, M.C.J., and J.M. Rose. 2010. Serial Choice Conjoint Analysis for Estimating Discrete Choice Models. In *Choice Modelling: State-of-the-Art and the State-of Practice: Proceedings from the Inaugural International Choice Modelling Conference*, ed. S. Hess and A. Daly, 139–161. Bingley, United Kingdom: Emerald Press.

Bunch, D., J. Louviere, and D. Anderson. 1996. A Comparison of Experimental Design Strategies for Multinomial Logit Models: The Case of Generic Attributes. Working Paper, Graduate School of Management, University of California Davis.

Byun, S., H. Bae, and H. Kim. 2006. A Contingent Valuation of Terrestrial DMB Services. In *Economics of Online Markets and ICT Networks*, ed. R. Cooper, G. Madden, A. Lloyd, and M. Schipp, 215–225. Heidelberg: Physica-Verlag.

Cardona, M., A. Schwarz, B. Yurtoglu, and C. Zulehner. 2009. Demand Estimation and Market Definition for Broadband Internet Services. *Journal of Regulatory Economics* 35: 70–95.

Chaloner, K., and K. Larntz. 1989. Optimal Bayesian Design Applied to Logistic Regression Experiments. *Journal of Statistical Planning and Inference* 21: 191–208.

Churchill, G.A., Jr. 1995. *Marketing Research, Methodological Foundations*, 6th ed.. Hinsdale, IL: Dryden Press. ISBN 0-03-098366-5.

Cook, R., and C. Nachtsheim. 1980. A Comparison of Algorithms for Constructing Exact D-optimal Designs. *Technometrics* 22: 315–324.

Confraria, J., T. Ribeiro, and H. Vasconcelos. 2017. Analysis of Consumer Preferences for Mobile Telecom Plans Using a Discrete Choice Experiment. *Telecommunications Policy* 41 (3): 157–169. https://doi.org/10.1016/j.telpol.2016.12.009.

Hess, S., J.M. Rose, and D.A. Hensher. 2008. Asymmetrical Preference Formation in Willingness to Pay Estimates in Discrete Choice Models. *Transportation Research* Part E, 44 (5): 847–863.

Huber, J., and K. Zwerina. 1996. The Importance of Utility Balance in Efficient Choice Designs. *Journal of Marketing Research* 33: 307–317.

Ida, T., and T. Kuroda. 2006. Discrete Choice Analysis of Demand for Broadband in Japan. *Journal of Regulatory Economics* 29: 5–22.

Ida, T., and T. Kuroda. 2009. Discrete Choice Model Analysis of Mobile Telephone Service Demand in Japan. *Empirical Economics* 36: 65–80.

Kanninen, B. 2002. Optimal Design for Multinomial Choice Experiments. *Journal of Marketing Research* 39: 214–217.

Kessels, R., P. Goos, and M. Vandebroek. 2006. A Comparison of Criteria to Design Efficient Choice Experiments. *Journal of Marketing Research* 43: 409–419.

Kim, Y. 2005. Estimation of Consumer Preferences on New Telecommunications Services: IMT-2000 Service in Korea. *Information Economics and Policy* 17: 73–84.

Kuhfeld, W., R. Tobias, and M. Garratt. 1994. Efficient Experimental Design with Marketing Research Applications. *Journal of Marketing Research* 31: 545–557.

Leng, Phirak. 2016. A Cost-Benefit Assessment of the Regulatory Policy in Cambodia's Mobile Telecommunications Market. *Asian Development Policy Review* 4 (1): 1–25.

Louviere, J., D. Hensher, and J. Swait. 2000. *Stated Choice Methods: Analysis and Application*. Cambridge, England: Cambridge University Press.

Louviere, J.J., T. Islam, N. Wasi, D. Street, and L. Burgess. 2008. Designing Discrete Choice Experiments: Do Optimal Designs Come at a Price? *Journal of Consumer Research* 35 (2): 360–376.

Lusk, J., and F. Norwood. 2005. The Effect of Experimental Design on Choice-based Conjoint Valuation Estimates. *American Journal of Agricultural Economics* 87: 771–785.

Madden, G., H. Bloch, and D. Hensher. 1993. Australian Telephone Network Subscription and Calling Demands: Evidence from a Stated-preference Experiment. *Information Economics and Policy* 5: 207–230.

Perl, L. 1975. *Economic and Demographic Determinants of Telephone Availability*. In AT&T filing before the Federal Communications Commission, Docket No. 20003, April 21.

Perl, L. 1978. *Economic and Social Determinants of Residential Demand for Basic Telephone Service*. White Plains: National Economic Research Associates.

Rose J., and M. Bliemer. 2005. Sample Optimality in the Design of Stated Choice Experiments. Working Paper ITLS-WP-5-13, Institute of Transport and Logistics Studies, University of Sydney.

Rose, J., and M. Bliemer. 2010. Stated Choice Experimental Design Theory: The Who, the What and the Why. Mimeo, University of Sydney.

Rose, J., M. Bliemer, D. Hensher, and A. Collins. 2008. Designing Efficient Stated Choice Experiments in the Presence of Reference Alternatives. *Transportation Research* 42: 395–406.

Sandor, Z., and M. Wedel. 2001. Designing Conjoint Choice Experiments Using Managers' Prior Beliefs. *Journal of Marketing Research* 38: 430–444.

Sandor, Z., and M. Wedel. 2002. Profile Construction in Experimental Choice Designs for Mixed Logit Models. *Marketing Science* 21: 455–475.

Sandor, Z., and M. Wedel. 2005. Heterogeneous Conjoint Choice Designs. *Journal of Marketing Research* 42: 210–218.

Scarpa, S., and J.M. Rose. 2008. Design Efficiency for Non-market Valuation with Choice Modelling: How to Measure It, What to Report and Why. *The Australian Journal of Agricultural and Resource Economics* 52 (3): 253–282.

Tardiff, T., and D. Levy. 2014. Prologue II: Lester Taylor's Insights. In *Demand for Communications Services – Insights and Perspectives: Essays in Honor of Lester D. Taylor*, ed. J. Alleman, Á. Ní-Shúilleabháin, and P. Rappoport, xxv–xxix. New York: Springer.

Toner, J.P., S.D. Clark, S.M. Grant-Muller, and A.S. Fowkes. 1998. Anything You Can Do, We Can Do Better: A Provocative Introduction to a New Approach to Stated Preference Design. *WCTR Proceedings, Antwerp* 3: 107–120.

Toner, J. P., M. Wardman, and G. A. Whelan. 1999. Testing Recent Advances in Stated Preference Design. Presented at the European Transport Conference, Cambridge, UK.

Train, K. 2009. *Discrete Choice Methods with Simulation*, 2nd ed. Cambridge, England: Cambridge University Press. ISBN 978-0-521-74738-7.

Tudela, A., and G. Rebolledo. 2006. Optimal Design of Stated Preference Experiments When Using Mixed Logit Models. European Transport Conference.

Vermeulen, B., P. Goos, and M. Vandebroek. 2011. Rank-Order Choice-Based Conjoint Experiments: Efficiency and Design. *Journal of Statistical Planning and Inference* 141: 2519–2531.

Viney, R., E. Savage, and J. Louviere. 2005. Empirical Investigation of Experimental Design Properties of Discrete Choice Experiments in Health Care. *Health Economics* 14: 349–62.

Yoo, S. 2002. Extending Dichotomous Choice Contingent Valuation Methods to Pre-test-Market Evaluation: The Case of a Cable Television Service. *Applied Economics Letters* 9: 315–318.

Yoo, S., and H. Moon. 2006. An Estimation of the Future Demand for Portable Internet Service in Korea. *Technological Forecasting and Social Change* 73: 575–587.

PART III

Governance of the Internet

CHAPTER 10

Platforms with Restrictive Licensing

Christiaan Hogendorn

INTRODUCTION

Two-sided platforms coordinate two types of users in order to increase the value of the whole system surrounding the platform. One side comprises the "components," such as software apps, that offer services on the platform, while the other side is usually end users or sometimes advertisers. Platforms take on complex roles as "facilitators" or "regulators" of their associated component systems. When platforms issue licenses to components, they often impose restrictive clauses that preclude certain activities off of the platform that would otherwise be profitable. A common example is shopping malls that limit the radius in which a tenant can build another store.

The restrictions we study are a subset of *exclusive dealing* where "the distributor agrees not to engage in any other business that competes directly with the manufacturer's activities (or even in any other business)" (Rey and Vergé 2008, p. 355). Our main point is that in the setting where the "manufacturer" is a platform, the various intra-platform externalities tend to make the special case of "even in any other business" more likely.

C. Hogendorn (✉)
Wesleyan University, Middletown, CT, USA
e-mail: chogendorn@wesleyan.edu

This paper focuses on an efficiency-based motivation for these restrictive clauses. Some component firms may have outside opportunities that conflict with the inside goals of the platform and the other components on the platform. Specifically, we consider a case where component firms have diseconomies of scope in developing high-quality components. Such diseconomies could arise because of scarce managerial attention, scarce creative talent, or scarce marketing resources. If the quality of platform components has spillovers to the platform as a whole, a situation we term a *quality commons*, then the platform may want to take steps to remove the diseconomies of scope via restrictive clauses.[1] While clearly beneficial to the platform, such clauses may or may not raise social welfare.

There are two other potential motivations for restrictions of the type we study. One is vertical foreclosure, which would typically involve the platform providing an integrated product that competes directly with the component provider (Gilbert and Katz 2011). The other is demand-side network effect where one platform would attempt to tip the market, or at least dominate the market, by denying components to another platform (Hogendorn and Yuen 2009). We do not suggest that either of these is implausible or even uncommon. Rather, we think that there are many additional cases where restrictive clauses do not appear to serve either of these goals but are present in licensing agreements nonetheless.

The next section reviews some economics literature relevant to this paper. Section "Model Setup" presents a simple model of two types of component providers producing both inside projects on a platform and outside projects that may involve diseconomies of scope. In sect. "Equilibrium" we analyze equilibrium with and without exclusivity clauses, and compare the social welfare. Section "Extensions and Conclusion" presents some discussion and conclusions.

Literature

This paper is part of the literature on the *quality* of members of platforms and the policies platforms adopt to enhance that quality. Hagiu (2009) studies the exclusion of some low-quality types in order to

---

[1] One piece of evidence for quality commons in shopping malls is that rental contracts commonly include "cotenancy" clauses that result in reduced rent if other tenants leave the mall. See, e.g., Fung (2018).

increase average quality of one side of the market. Boudreau (2010) challenges the conventional wisdom that more components are always better for a platform. He makes a useful distinction between "platforms," which are surrounded by thousands, even hundred thousands, of components and "alliances" where the number of components is smaller and the contractual relations can be more individual. While the model of this paper applies to alliance situations in some cases, it is primarily designed for the case of platforms where the sheer number of firms requires a broad-brush approach to contracting. In the closest paper to this one, Stennek (2014) discusses exclusive contracting in pay-TV as a way to increase the incentives for program providers to invest in quality, but he focuses on asset specificity rather than diseconomies of scope. A recent paper on platforms setting minimum requirements for complementors is Hagiu and Wright (2018).

Boudreau and Hagiu (2009) introduce the idea that the platform is similar to a regulator, and may try to regulate quality. They discuss the demise of the Atari video game platform of the early 1980s, when a flood of low-quality components was blamed for wrecking the market in a sort of Gresham's Law of video games. Recent, more successful platforms have tried to avoid such problems with tighter management of component quality. Three examples from Boudreau and Hagiu are particularly relevant to the present model. First, the TopCoder platform runs a series of programming contests to match clients with programmers. To maintain the coders' focus on the contests, "TopCoder imposes its control over all interactions with customers; software developers do not interact with final customers" (p. 176). Thus, TopCoder excludes outside contact which might be in the narrow self-interest of an individual programmer but would reduce the value of the platform as a whole. Second, the Roppongi Hills real estate development in Tokyo adopted an "only one" policy whereby stores that locate there are required to produce unique features *not* available anywhere else. Third, Boudreau and Hagiu note that faculty members of business schools are usually subject to maximal thresholds on the time they may spend on consulting work. Again, the motivation is to maintain the quality of their input into the platform (here, the school).

Other closely related work concerns technology licensing and exclusive territories. Schuett (2012) presents a model of field-of-use restrictions (FOUR) in technology licensing. His review of the empirical literature suggests that 30–50% of all technology licenses include field-of-use restrictions. Like us, he focuses on cost heterogeneity among the licensee firms as an efficiency motivation for restrictive licenses. He analyzes the potential for restrictions to include noncontractible, producer-surplus-reducing overlap, and he shows that licensors can use royalties to mitigate this contracting problem. The literature on exclusive territories strikes a similar theme, but there is an essential difference. With exclusive territories, there is another dealer, in a contractual relationship with the producer, who would otherwise be engaged in commerce within the excluded territory. Thus the producer is excluding a firm from producing the *same* product as it is protecting, while our interest is in producers excluding a *different,* noncompeting, product. That said, the end goal is the same: to increase the quality of the protected product.

The timing of this model has the platform first announcing its contractual offerings to component providers and its pricing to consumers. Then the component providers choose whether or not to join the platform, and after observing the number of component providers, consumers decide to join the platform. This is the sequential game structure pioneered in Katz and Shapiro (1985) and Church and Gandal (1992); both those papers showed that multiple equilibria are possible: there is always a zero equilibrium where the platform receives no components. Hagiu (2006) analyzed this timing in the context of two-sided pricing, paying particular attention to the question of whether the platform precommits to consumer pricing at Stage 1 or chooses it *ex post* at Stage 3. He finds that precommitment is optimal as long as sellers coordinate on the positive rather than zero equilibrium, an assumption that we follow here. The sequential game structure is contrasted with simultaneous games where buyers and sellers join the platform at the same time. This structure was used in the original two-sided markets literature pioneered by Rochet and Tirole (2003) and Armstrong (2006).

Model Setup

We model a single platform that connects consumers with components. Component providers are potentially multiproduct firms. For each component provider i, the "inside product," product 1, is specific to the platform and has quality level q_{1i}. The "outside product," product 2, can be sold directly to consumers without use of the platform; its quality is q_{2i}. To isolate the effects that are internal to the platform, we let these products be unrelated in demand (cross elasticity of 0). Thus, we rule out any foreclosure incentive whereby the platform tries to shut down production of product 2 in order to increase its own demand directly.

Timing

The game is in three stages. First, the platform chooses a fixed access price P_1 that consumers will pay, and it also chooses whether components will be restricted from working on the outside project if they join the platform.

In the second stage, a population of n component providers choose whether or not to accept the contract. Component providers have cost function

$$C(q_{1i}, q_{2i}) = q_{1i}^2 + \gamma q_{1i} q_{2i} + q_{2i}^2$$

where the term γ determines the component's type, namely whether it is a substitute type or not. Let $\gamma = 0$ with probability ρ and $\gamma = 1$ with probability $(1 - \rho)$, these are the neutral and substitute types, respectively. Let $n_1 \leq n$ denote the number of component providers who actually join the platform.

Component providers receive revenue

$$R(q_{1i}, q_{2i}) = \beta D + p_2 q_{2i}$$

where β is a parameter, D is the number of consumers who end up signing up for the platform, and p_2 is the price per unit of quality of the outside product. Note that a component provider's revenue from its inside product does not depend directly on q_{1i}, but only indirectly through the number of consumers who sign up for the platform. We can think of the component providers being engaged in pure team production with their inside products.

In the third stage, consumers choose whether or not to subscribe to the platform, and they also buy the outside product in a separate decision. For goods on the platform, consumers only care about *average* quality,

$$\overline{q_1} = \sum_{i=1}^{n_1} q_{1i}$$

A consumer of type θ who pays to access the platform receives utility

$$U_1 = v n_1 \bar{q}_1 - P_1 - \theta$$

Note that consumers care about both the number of components and the average quality of components on the platform. Parameter v indexes the value of quality to the consumer, while consumer type parameter θ is distributed uniformly on $[0,1]$.

Consumers also may purchase the outside products of the components. Denote the number of components that offer outside projects by $n_2 \leq n$. The utility a consumer receives from these projects is

$$U_2 = (w - p_2) n_2 \bar{q}_2$$

where w is a parameter indexing reservation utility for quality of the outside projects and $\bar{q}_2$ is the average quality of the outside projects. Here the motivation for using the average is purely for easy comparability with U_1.

Note that the two utilities are additively separable, so they do not interact in the consumer's decision of whether or not to buy the platform. We will only need the utility U_2 for welfare analysis, not for finding the equilibrium of the game. For simplicity, we assume that p_2 is fixed.

From here on, we will drop the i subscript from the quality choices because it is the average quality that matters, so every component will behave identically.

Third Stage: Consumers

We now solve backward to find a subgame perfect equilibrium. Given $\bar{q}_1$, and n_1, all consumers with $U_1 > 0$ will subscribe to the platform, so

$$D(P_1, n_1, \bar{q}_1) = v n_1 \bar{q}_1 - P_1$$

provided that $D \leq 1$.

Second Stage: Components

In Stage 2, the exclusivity provision has already been set. The components must decide whether or not to enter, and if so what quality to choose.

One possible case is that a component (of either type) chooses not to join the platform and to specialize on the outside product. In this case,

$q_1 = 0$, and the cost functions of the neutral and substitute types are identical. An outside specialized component solves

$$\max_{q_2} \Pi_2 = p_2 q_2 - q_2^2$$

This yields an optimal $q_2 = \frac{1}{2}p_2$ and outside payoff

$$\Pi_2^* = \frac{1}{2}p_2^2 - \left(\frac{1}{2}p_2\right)^2 = \left(\frac{1}{2}p_2\right)^2$$

The opposite case is that a component specializes on the inside product only. Either type might be required to do this by contractual obligation, and a substitute type might do it voluntarily to avoid diseconomies of scope. An inside-specialized component sets $q_2 = 0$, which again makes the cost functions of the two types identical. The component then solves

$$\max_{q_1} \Pi_1 = \beta D(P_1, n_1, \overline{q_1}) - q_1^2$$

Denote the value of this problem by Π_1^*.

Finally, there is the case of a component that pursues both projects at the same time. An unspecialized neutral component solves

$$\max_{q_1, q_2} \Pi_{12}^N = \beta D(P_1, n_1, \overline{q}_1) + p_2 q_2 - q_1^2 - q_2^2$$

Denote the maximum by Π_{12}^{N*}.

If the substitute type pursues both projects, it solves

$$\max_{q_1, q_2} \Pi_{12}^S = \beta D(P_1, n_1, \overline{q_1}) + p_2 q_2 - q_1^2 - q_1 q_2 - q_2^2$$

Denote this maximum by Π_{12}^{S*}.

Each type of component will select between the above three candidate optima, selecting the one with the highest payoff.

First Stage: Platform

In Stage 1, the platform maximizes its own revenue $P_1 D(P_1, n_1, \overline{q}_1)$. The demand function takes the form $P_1 D(P_1, n_1, \overline{q}_1) = v n_1 \overline{q}_1 - P_1$ so the first-order condition for access price P_1 is

$$\frac{\partial P_1 D(P_1, n_1, \overline{q}_1)}{\partial P_1} = (v n_1 \overline{q}_1 - P_1) - P_1 = 0$$

For now, we leave aside participation constraints for the components, but we will return to them in the next section. Assuming participation is not an issue, the platform optimally sets

$$P_1 = \frac{vn_1 \bar{q}_1}{2}$$

That is, the platform sets the consumer access price at one-half the demand intercept, i.e. the point where price elasticity of demand is unitary. This is the standard result for any monopolist with zero marginal cost and a linear demand curve.

Equilibrium

We analyze the platform's maximization problem for two different types of contracts: (i) an unmanaged platform which restricts neither q_1 nor q_2, and (ii) a restrictive contract that specifies $q_2 = 0$ but leaves q_1 unmanaged.

In all of what follows, we focus on the cases where both component types join the platform ($n_1 = n$) and check that the relevant participation constraints are met. Clearly, it would not be in the platform's interest to promote the case where zero components join the platform. The case of only neutral components joining is somewhat more interesting, but the platform would only prefer this case if the indirect network effect v and/or the proportion of substitute types $(1 - \rho)$ were low. If these were low enough, there would be little opportunity cost of excluding substitute types. However, the point of this paper is to study platform strategies with respect to substitute types, so we are most interested in the case where the substitute types are too numerous and/or too valuable to ignore.

An Unmanaged Platform

If the platform does not place any constraints on the behavior of the firms, then the neutral types will never specialize. Their profit maximization problems are:

$$\max_{q_1, q_2} \Pi_{12}^N = \beta D(P_1, n_1, \bar{q}_1) + p_2 q_2 - q_1^2 - q_2^2$$

Since this function is additively separable in q_1 and q_2, the optima are determined independently of one another. Note that any one component's inside quality decision affects the average only a little: $\frac{\partial \bar{q}_1}{\partial q_1} = \frac{1}{n}$. This means the first-order condition for a neutral type's inside quality optimum is:

$$\frac{\partial \Pi_{12}^N}{\partial q_1} = \frac{\beta n v}{n} - 2q_1 = 0 \Rightarrow q_1 = \frac{\beta v}{2}$$

Since this does not take account of the spillover to the other $n-1$ components, there is a quality commons problem.

The first-order condition for a neutral type's outside quality is

$$\frac{\partial \Pi_{12}^N}{\partial q_2} = p_2 - 2q_2 = 0 \Rightarrow q_2 = \frac{p_2}{2}$$

Substituting the optimal solutions into the objective function gives the optimal profits:

$$\Pi_{12}^{N*} = \beta \left(v n_1 \bar{q}_1 - P_1 \right) + \frac{1}{4} p_2^2 - \frac{1}{4} (\beta v)^2$$

The problem of a substitute type is more complicated because its cost function includes diseconomies of scope between the two quality levels. If it chooses to pursue both projects, it solves

$$\max_{q_1, q_2} \Pi_{12}^S = \beta D(P_1, n_1, \bar{q}_1) + p_2 q_2 - q_1^2 - q_1 q_2 - q_2^2$$

Assuming the interior solution exists, the first-order conditions are:

$$\frac{\partial \Pi_{12}^S}{\partial q_1} = \frac{\beta n v}{n} - 2q_1 - q_2 = 0$$

$$\frac{\partial \Pi_{12}^S}{\partial q_2} = p_2 - q_1 - 2q_2 = 0$$

Solving simultaneously gives:

$$q_1 = \frac{2}{3} \beta v - \frac{1}{3} p_2 \text{ and } q_2 = \frac{2}{3} p_2 - \frac{1}{3} \beta v$$

Substituting the optimal solutions into the objective function and simplifying gives the optimal profits:

$$\Pi_{12}^{S*} = \beta(vn\bar{q}_1 - P_1) + \frac{1}{3}p_2^2 - \frac{1}{3}(\beta v)^2$$

This leads immediately to a result:

Lemma 1. Nonspecialized substitute types have higher payoffs than neutral types due to free riding: $\Pi_{12}^{S*} > \Pi_{12}^{N*}$

The result in Lemma 1 is typical of the payoffs to free riders (or really "reduced effort riders") in pure team production settings. The result would not be so stark if we further supposed that neutral types tend to be larger firms, with a higher multiple β in their revenue functions and no corresponding multiple in their cost functions due to economies of scale in *quantity* as opposed to the constant returns to scale in *quality* that we have focused on here. Since this paper focuses on managing free-riding behavior, it makes sense to emphasize that behavior. Even if we added a size multiple, it would still be true that substitute types have higher revenue per unit of quality that they contribute to the platform, which is the essence of the problem from the platform's point of view.

Alternatively, the substitute type could specialize in the inside project only. In that case, it would solve:

$$\max_{q_1} \Pi_1 = \beta D(P_1, n_1, \bar{q}_1) - q_1^2$$

The solution is $q_1 = \frac{\beta v}{2}$ and the payoff is

$$\Pi_1^* = \beta(vn\bar{q}_1) - \left(\frac{\beta v}{2}\right)^2$$

Based on the above, we can see that if substitute types specialize, they behave like neutral types, and the average quality level is $\bar{q}_1^H = \frac{\beta v}{2}$. If substitute types pursue both projects, the average quality level is

$$\bar{q}_1^L = \rho \frac{\beta v}{2} + (1 + \rho)\left(\frac{2\beta v}{23} - \frac{1}{3}p_2\right)$$

which is lower than q_1^H. This lower quality q_1^L is decreasing in the price of the outside project p_2.

Lemma 2. Average component quality is inefficiently low in an unmanaged equilibrium.

Proof This follows from the quality commons formulation where average quality enters the payoff function. Is substitute components specialize and all components colluded on quality, they would maximize $\beta(vn\bar{q}_1) - \bar{q}_1^2$ by choosing $\bar{q}_1$ and set $\bar{q}_1 = \frac{1}{2}\beta vn$. If the substitute types remained unspecialized, the collusive quality would maximize, by choice of $\bar{q}_1$, a weighted average of Π_{12}^N and Π_{12}^S.

From here on, we will assume that on an unmanaged platform, the substitute type prefers to pursue both projects rather than specializing:

Outside Project Value (OPV) Assumption: The price of the outside project, w, is neither too low nor too high, so that a substitute-type component chooses an interior solution pursuing both projects whenever it chooses to participate in the platform:

$$\frac{\beta v}{2} \leq p_2 \leq 2\beta v \qquad \text{(OPV)}$$

Our justification is simple: the point of this paper is to discuss components that might problematically pursue outside projects. If such components independently give up on such projects in order to specialize on the platform, or if they find such projects so valuable that they do not join the platform, they are not of interest here.

A Platform with Restrictive Licenses

We have seen that the inside quality level is too low due to the commons problem when the platform is unmanaged. Suppose that the only tool available to the platform to manage this problem is to ban outside projects, thus forcing $q_2 = 0$. Clearly such exclusivity will have a high cost, since neither type of component will receive the profits from the outside product. But there is also a benefit because the inside quality produced by the substitute components will rise. Our question in this section is whether the quality increase can be enough to outweigh the cost and still meet the participation constraints.

From the previous section, it is clear that with the outside product removed, both types behave like the inside-specialized substitute type. We have seen that this leads to an optimal quality choice

$$q_1 = \frac{\beta v}{2}$$

and inside-specialized payoff

$$\Pi_1^* = \beta\left(vn\frac{\beta v}{2} - P_1\right) - \left(\frac{\beta v}{2}\right)^2$$

The first question is whether components would participate in a platform offering such a contract (and no other alternative). This depends on whether the profits from the inside product exceed the value of the outside product, namely whether

$$\Pi_1^* \geq \Pi_2^* \Rightarrow \beta\left(vn\frac{\beta v}{2} - P_1\right) - \left(\frac{\beta v}{2}\right)^2 \geq \frac{1}{4}p_2^2 \qquad \text{(PC)}$$

Predictably, the participation constraint is met as long as the price the components receive for outside quality is not too high.

The next question is whether the platform actually gains from such a contract, and the answer here is clear:

Lemma 3. As long as the participation constraint is met, the platform gains from restricting outside projects. The amount of this gain is

$$v^2 n^2 \left(\frac{\left(\bar{q}_1^H\right)^2 - \left(\bar{q}_1^L\right)^2}{4} \right)$$

Proof The platform's payoff is simply $P_1 D(P_1, n_1, \bar{q}_1)$. This is always increasing in $\bar{q}_1$, so any policy that increases average quality will increase the platform's payoff. As discussed above, the optimal interior value is $P_1 = \frac{vn\bar{q}_1}{2}$. Thus, the platform gain is $\frac{(vn\bar{q}_1^H)^2}{4} - \frac{(vn\bar{q}_1^L)^2}{4}$ which simplifies as above.

Now let us turn to welfare analysis. This comes in four parts: the change in platform producer surplus, the change in neutral-type producer surplus, the change in substitute-type producer surplus, and the change in consumer surplus. Lemma 3 already showed the gain in platform producer surplus.

Lemma 4. A neutral type's payoff is higher on an exclusive versus unmanaged platform if

$$\frac{\beta v n}{2}\left(\overline{q}_1^H - \overline{q}_1^L\right) - \frac{1}{4}p_2^2 > 0$$

Proof We are looking for the payoff difference

$$\Pi_1^*\left(\overline{q}_1^H\right) - \Pi_{12}^{N*}\left(\overline{q}_1^L\right)$$

This can be written out

$$\beta\left(vn\overline{q}_1^H - P_1\right) - \left(\frac{\beta v}{2}\right)^2 - \beta\left(vn\overline{q}_1^L - P_1\right) - \frac{p_2^2}{2} + \left(\frac{\beta v}{2}\right)^2 + \left(\frac{p_2}{2}\right)^2$$

If in both cases the platform sets P_1 equal to its optimal interior value, this can be simplified to the above.

This lemma is easily interpreted; the neutral types see a gain in producer surplus due to the higher quality of the platform but a loss equal to its profits from the outside project.

Under OPV, a similar producer surplus comparison applies to a substitute type. It will gain less because it loses the advantage of free riding on the neutral types' quality, but it does receive some benefit from avoiding diseconomies of scope.

Lemma 5. A substitute type's payoff is higher on a restrictive versus unmanaged platform if

$$\frac{\beta v n}{2}\left(\overline{q_1}^H - \overline{q_1}^L\right) + \frac{1}{12}(\beta v)^2 - \frac{1}{3}p_2^2 > 0$$

Proof We are looking for the payoff difference

$$\Pi_1^*\left(\overline{q}_1^H\right) - \Pi_{12}^{*S}\left(\overline{q}_1^L\right)$$

This can be written out

$$\beta\left(vn\bar{q}_1^H - P_1\right) - \frac{1}{4}(\beta v)^2 - \beta\left(vn\bar{q}_1^L - P_1\right) - \frac{1}{3}p_2^2 + \frac{1}{3}(\beta v)^2$$

Again assuming an interior optimum for P_1, we have the expression in the lemma.

An interesting implication here is that depending on the parameters, either the neutral type *or* the substitute type may gain more (or lose less) from the imposition of the restriction. By comparing the expressions in Lemmas 4 and 5, one can see that a neutral type's payoff increases by more than a substitute type's payoff if $\beta v < p_2$. This condition may or may not hold under OPV, so either situation is possible. Interestingly, when it does not hold, the substitute types, which are the source of the problem for the platform, are nonetheless more willing to accept the solution.

Finally we turn to the change in consumer surplus. This has two components, the loss of utility from the outside projects and the change in utility due to the increased quality of the inside projects.

Lemma 6. Consumers' payoffs are higher under an exclusive platform than an unmanaged one if

$$v^2 n^2 \left(\frac{(\bar{q}_1^H)^2 - (\bar{q}_1^L)^2}{4}\right) - (w - p_2)\bar{q}_2 nM > 0$$

Proof Each of the M consumers loses utility $U_2 = (w - p_2)\bar{q}_2 n$ from the outside products. Under OPV, the average quality of the outside products is

$$\bar{q}_2 = \rho\left(\frac{p_2}{2}\right) + (1 - \rho)\left(\frac{2}{3}p_2 - \frac{\beta v}{3}\right)$$

Consumer surplus from the platform is the net utility squared, since we have assumed a demand curve with slope -1. Thus, the change in consumer surplus from the platform is

$$\left(\frac{vn\bar{q}_1^H}{2}\right)^2 - \left(\frac{vn\bar{q}_1^L}{2}\right)^2 = v^2 n^2 \left(\frac{(\bar{q}_1^H)^2 - (\bar{q}_1^L)^2}{4}\right)$$

This is the same as the platform's profit gain, since the platform's optimal price takes one-half of the surplus from the consumers.

Combining the above Lemmas, we can see several main results.

Most obvious, if the consumer valuation of the outside projects, w, is high, then there can be a social welfare loss from exclusivity despite its efficiency advantages.

Second, the difference between $\bar{q}_1^H$ and $\bar{q}_1^L$ is key to any gains from exclusivity. This difference, in turn, is increasing in the percentage of substitute types and the price of the outside project.

Finally, the price of the outside project enters directly since it reflects lost producer surplus from moving to a restrictive regime. However, it also negatively affects consumer surplus from these outside projects.

EXTENSIONS AND CONCLUSION

In the above, we assumed that both types of components are willing to participate in the restrictive regime. And we showed that it is possible that one or both types actually gain from the restriction. Even if they do not gain, they may still be willing to participate as long as the platform can make a take-it-or-leave-it offer. But considering the obvious loss from abandoning the outside project, one might expect that in many cases at least one component type would *not* be willing to participate under the restriction. Since the platform always gains from the restriction, we now ask whether there are other ways to ensure participation even when the participation constraints of the model above are not satisfied.

Menu of Contracts. We have considered a single contract offered to two different types of agents. It seems natural in this situation for the platform to offer a menu of two contracts. Properly designed, these contracts would separate the two types, allowing the neutral types to pursue their "harmless" outside projects and giving the substitute types incentives to accept an exclusive contract which would improve everyone's welfare.

In actual fact, we rarely see such a menu offered. Several reasons are possible, but the one that seems most likely is that the incentive compatibility constraint would require that the neutral type pay a substantial fee for the privilege of an unrestricted contract. This would violate principles of fairness and put the platform in the position of expropriating considerable surplus from its most valuable components.

Relaxed Restriction

Another way to meet the neutral types' participation conditions would be to relax the restriction that $q_2=0$. Instead, a restriction that q_2 must be less than or equal to some positive amount could be stipulated. This would increase the profits of the neutral type, and to a lesser degree the substitute type, and make their participation more likely. It has two disadvantages, however. First, it would reduce the substitute types' optimal q_1, partially negating the value of the restrictive regime. Second, it is less easy to monitor than the simple prohibition on outside projects.

Decrease Consumer Price

A final, and intriguing, solution to a binding participation constraint is for the platform to lower its price to consumers below the level $P_1 = \frac{vn\bar{q}_1}{2}$ analyzed above. By doing so, the platform gives up some profit (at least relative to what it would receive if the participation constraint could be ignored) but it also expands the size of its customer base. This does not change the optimal choice of q_1, but it does change the level of the profit of a component firm

$$\beta(vn\bar{q}_2 - P_1 - \left(\frac{\beta v}{2}\right)^2$$

raising it by β for every dollar of price reduction. An interesting implication is that such a move would also improve social welfare, since the platform behaves as a monopolist with the usual associated dead-weight loss. Thus, even in the case where the restrictive regime would otherwise be socially costly, it might become socially beneficial if coupled with are reduced P_1 to ensure participation.

Other Extensions

The model could be elaborated with respect to platform ownership and decision-making. We can expect a platform to make different decisions regarding restrictions if it is owned privately, owned by some mixture of the component types, or owned by the consumers. Another extension is that the outside projects, rather than being independent, might be collected on a platform of their own. Then outside project restrictions would serve two goals—the efficiency goal we have described here and a demand-side goal of reducing the number of components on the competing platform.

This paper presents a model for thinking about the quality of components on a platform when those components produce both indirect network effects and a quality commons whereby average quality matters. The model includes some firms whose incentives are less aligned with those of the platform due to outside opportunities that are privately valuable but negatively impact the quality commons due to diseconomies of scope. Restricting outside projects can increase the average platform quality. This causes both gains and losses from a social welfare perspective, so it may or may not be a desirable behavior from a regulatory policymaker's point of view. Still, it is important for policymakers to consider these cost-side objectives and not jump to the conclusion of market power.

References

Armstrong, M. 2006. Competition in Two-Sided Markets. *RAND Journal of Economics* 37: 668–691.

Boudreau, K. 2010. Open Platform Strategies and Innovation: Granting Access vs. Devolving Control. *Management Science* 56 (10): 1849–1872.

Boudreau, K., and A. Hagiu. 2009. Platform Rules: Regulating the Ecosystem Around a Multi-Sided Platform. In *Platforms, Markets and Innovation*, ed. Annabelle Gawer. Edward Elgar.

Church, J., and N. Gandal. 1992. Network Effects, Software Provision, and Standardization. *The Journal of Industrial Economics* 40 (1): 85–103.

Fung, E. 2018. Sears Closing Stores Is a Blessing for Some Landlords, a Curse for Others. *Wall Street Journal*, October 16.

Gilbert, R.J., and M.L. Katz. 2011. Efficient Division of Profits from Complementary Innovations. *International Journal of Industrial Organization* 29 (4): 443–454.

Hagiu, A. 2006. Pricing and Commitment by Two-Sided Platforms. *RAND Journal of Economics* 37: 720–737.

Hagiu, A. 2009. Quantity vs. Quality and Exclusion by Two-Sided Platforms. papers.ssrn.com.

Hagiu, A., and J. Wright. 2018. Platform Minimum Requirements. Working Paper.

Hogendorn, C., and S. Yuen. 2009. Platform Competition with 'Must-Have' Components. *Journal of Industrial Economics* 57 (2): 294–318.

Katz, M., and C. Shapiro. 1985. Network Externalities, Competition, and Compatibility. *The American Economic Review* 75 (3): 424–440.

Rey, P., and T. Vergé. 2008. Economics of Vertical Restraints. In *Handbook of Antitrust Economics*, ed. Paolo Buccirossi, 353–390. MIT Press.

Rochet, J., and J. Tirole. 2003. Platform Competition in Two-Sided Markets. *Journal of the European Economic Association* 1 (4): 990–1029.

Schuett, F. 2012. Field-of-Use Restrictions in Licensing Agreements. *International Journal of Industrial Organization* 30 (5): 403–416.

Stennek, J. 2014. Exclusive Quality—Why Exclusive Distribution May Benefit the TV-Viewers. *Information Economics and Policy* 26: 42–57.

CHAPTER 11

QoS Investment, Vertical Integration and Regulation of the Internet

Edmond Baranes and Cuong Hung Vuong

Introduction

The development of businesses based on the internet, that is constituted by millions of interconnected networks worldwide, has increased demand for bandwidth and investment in broadband infrastructure. Various policies have been implemented to enhance facilities-based competition or encourage ISP to make sufficient investment. Furthermore, one specific technical feature of the internet is that a network is technically controlled by a monopolistic ISP that could prioritize, slow down or even block the circulation of different types of data packets in the lower layers of the network. This activity is generally not accepted by net neutrality regulation which forbids discrimination against any applications or content from its network. In addition to this, ISP often engages in different businesses rather than solely providing internet services. Therefore, it is always worth studying incentives of an ISP to invest in infrastructure accounting for complex relationships with different market players including content providers (CP) and end users.

E. Baranes · C. H. Vuong (✉)
MRE, University of Montpellier, Montpellier, France
e-mail: hung-cuong.vuong@umontpellier.fr

From end users' perspectives, because of perfect complementarity, the qualities of content and internet affect their consumption. In other words, perceived quality of one content is increasing in both individual qualities of content and internet services. However, one might conceive that a monopolistic ISP, that is vertically integrated with a CP, has incentive to not provide network access to a rival CP, and consequently reducing access to content by end users. Since internet is usually monitored by competition authorities, that behavior is considered as abuse of market power and certainly prohibited. In a less extreme context, an ISP might want to provide preferential treatment, favoring its own content, thanks to its structure capacity. The profit maximizing ISP, however, could welcome development of CPs who bring high value to end users and hence increase demand for internet. In short, the integrated ISP must consider few trade-offs when deciding on whether to provide relatively higher quality of service (QoS) to its affiliated content. Our model is therefore set up to investigate whether a vertically integrated ISP always has incentive to provide asymmetric internet quality in favor of its affiliated CP and detriment the rival CP, and that non-discriminatory quality regulation might be necessary in this context.

In precision, this paper provides a simple conceptual framework in which a monopolistic ISP determines QoS provided to two vertically differentiated CPs, and one of them is integrated with the ISP. The competing CP can access to the internet at free of charge. Both the ISP and CPs adopt usage-based pricing to end users. Furthermore, we consider that quality of internet service is costly or ISP's investment is increasing in QoS of the internet. Then we analyze optimal investment under an unregulated regime whereby the ISP is allowed to provide different qualities to the two CPs. We find that first, there exists a range of market parameters that the ISP provides positive levels of internet's quality to both contents. Second, if quality of the affiliated content is sufficiently low compared to the quality of the competing content, it is optimal for the integrated ISP to provide relatively higher internet QoS to the rival CP. This is because the competing content provides relatively high utility to consumers and hence raises demand for internet service. Furthermore, because perceived qualities of contents are highly disparate, the ISP can charge relatively higher for internet service in this case. When we can show that welfare might be improving if non-discriminatory quality regulation is imposed.

Related literature. Our article considers that the ISP, that has monopolistic positioning in offering internet service, is also present in content market. General economic analysis in this context can be found in Riordan (2008) and Rey and Tirole (2007). Additionally, due to complementarity between content and internet, we could find similar reasoning from economic studies in the software industry (see Carlton and Waldman [2002], Casadesus et al. [2010]). Our research is especially related to well-established economic studies on the impact of network neutrality on the investment of the network operators (Schuett [2010], Krämer et al. [2013] and Greenstein et al. [2016] and references cited therein for overviews). We build a model in which the ISP could invest to provide positive quality of internet service, otherwise it can only provide minimum quality of internet service. We, however, do not analyze issues related to congestion since our focus is on investment on quality of internet service to enhance average data transmission speed. Like our research, Njoroge et al. (2012) and Haucap and Klein (2012) also focus on the level of QoS of broadband network in the context of complementarity between internet and contents. Additionally, in our model, the monopolistic ISP can strategically assign different qualities of internet which is related to the classical monopoly problem as analyzed in Mussa and Rosen (1978).

There are few studies on impacts of vertical integration on competition, investment and diversity of content in the internet. Bandyopadhyay et al. (2010) consider the case of vertical integration between broadband service providers and content suppliers. In their model, the ISP's vertical integration decision is endogenized. They show that it is not always the case that the integrated ISP degrades the quality of access services to the competing content thus incentivizing provision of the network advantage to the independent content. Similarly, Brito et al. (2014) consider that the ISP can offer a fast lane to CPs and analyze the effects produced by net neutrality on the ISPs' incentives to invest and then extend their analysis to the case in which one of the two ISPs is integrated with content. Ultimately, they show that the discriminatory regime may not always produce incentives to the integrated ISP to favor its own content. While these papers provide interesting insights into the potential impact of vertical integration, the results are limited to how the broadband providers may discriminate between content providers under regulation. In the present paper, we address the issue of vertical integration between broadband and content in a more complete analysis, how vertical integration may affect ISP's optimal investment and subsequent impacts on competition in the market for content services. Also we can compare the price for the internet access services to end users and welfare in the regulated and unregulated regimes.

Indeed, Broos and Gautier (2017) is a closed paper to the present paper as they also focus on complementarity between internet service and content. In their model, services are both horizontally and vertically differentiated, and they can show monopoly ISP may want to exclude competitor if it is of inferior quality and the ISP cannot ask for a surcharge for its use. Furthermore, Dewenter and Rosch (2016) employ a two-sided model where CPs compete for advertisers and their content is free. They conclude that exclusion only takes place when the competitor offers perfect homogeneous content and the ISP has a local monopoly over its internet access customers or if network effects are strong. This is because of positive network effect that leads to a higher willingness to pay and therefore to larger markets and higher profits. Unlike these studies, our paper investigates ISP's incentive in a broader context than exclusion decision, that is asymmetric provision of internet QoS in favor the affiliated CP.

The remainder of the paper is structured as follows. In section two, we set up the model. We analyze ISP's optimal investment in an unregulated regime whereby the ISP is allowed to determine different qualities of internet service provided to each content in section three. In section four, we identify the ISP's optimal investment decision under non-discriminatory quality regulation and therefore we can evaluate the effectiveness of the regulation by comparing social welfare between two regimes. A brief conclusion is provided in section "Conclusion".

The Model

In the present model, the monopolistic ISP provides the internet platform to two vertically differentiated content services CPs (CP_A and CP_B) and end users, which can be represented as follows (see Fig. 11.1).

The ISP. The monopolistic ISP charges end users based on their usage for the broadband access services,[1] denoted as p (see Fig. 11.1). CP_B is integrated with the ISP while the competitor (CP_A) can access the

---

[1] In practice, Comcast's XFINITY plan, Cisco and Telus data plans, Google Project Fi prices can be considered as usage-based prices.

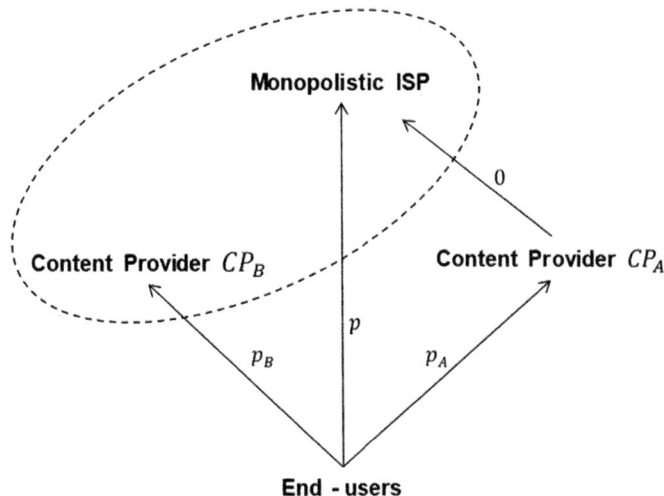

Fig. 11.1 The model's structure (*Source* Authors' design)

internet at free of charge. For simplicity, we assume that the ISP's fixed and marginal costs of production are zero. This would allow us to strictly focus on analyzing ISP's decision to invest in internet QoS. Specifically, without investment, the ISP provides minimum quality of network service, normalized to 0.

Since the ISP can fully control internet QoS or speeds of delivering data packages of each content to end users (and thereby affect perceived quality of that content), we assume that the ISP could technically provide different internet's qualities (denoted as μ_A and μ_B) to CP_A and CP_B, (respectively). Furthermore, we assume that the ISP's unit cost associated with the provision of quality μ_i is $c(\mu_i) = \frac{\mu_i^2}{2}$. This implies that the ISP could make separable investment to provide different qualities to each CP. Therefore, for simplicity, we assume that the ISP's investment is additive,[2] that is:

---

[2] In the context of the long-distance telephony provided by a vertically integrated monopolist, Economides (1999) also assumes for the separable additive cost function in provision of qualities of both long-distance lines as well as local lines. Furthermore, empirical evidence on additive cost in the US telecom industry is provided in Beresteanu (2005).

$$C(\mu_A, \mu_B) = c(\mu_A) + c(\mu_B) = \frac{\mu_A^2}{2} + \frac{\mu_B^2}{2}$$

This important assumption substantially reduces complications arising from interaction among the cost of investment in different levels of quality provided to both CPs, while it allows us to focus markedly on the role of vertical differentiation and substitution of contents. This assumption is appropriate in our model since we do not consider the cost of investment associated with deployment the whole network or migration from "*old*" to "*new*" technology, and therefore we can ignore indivisibilities or fixed cost of investment.[3] It can be easily checked that $C(\mu_A, \mu_B)$ is increasing and convex in both qualities μ_A and μ_B (i.e., $C'(\ldots) > 0$ and $C''(\ldots) > 0$).

CPs. Without loss of generality, we set quality of content B to A to *1* and γ, respectively, where $\gamma > 0$ represents for level of differentiation between contents. If $\gamma > 1$, we say content B has higher quality than content A and vice versa. Both the fixed and marginal costs of CPs are normalized to *0*. We also assume that CPs' pricing is linear, and hence we denote p_A and p_B as unit prices that an end user must pay for each unit of demand of content A and B, respectively.[4]

End users. There is a unit mass of end users accessing contents via the internet. From end users' perspectives, internet and content are perfect complementary, and hence we consider that an end user consumes one unit of content i, his/her perceived quality is the sum of both qualities (of content i and internet). The minimum quality of internet service is normalized to *0*, and hence perceived quality from consuming content i is simply the quality of content i in this case. We adopt the model by Singh and Vives (1984), and assume for a quadratic utility function from consuming q_A and q_B units of content A and content B, respectively, as follows:

$$U(q_A, q_B) = \alpha_A q_A + \alpha_B q_B - \left(\frac{1}{2}\right)\left(q_A^2 + 2\beta q_A q_B + q_B^2\right) \quad (11.1)$$

---

[3] Indeed this assumption is especially related to the classical situation in which a monopoly selling a broad range of qualities (Mussa and Rosen 1978).

[4] In practice, content services can be free, but we consider that consumers dislike most advertising displays (banner ads, pop-up, videos…) or sales' links when they access to content. Therefore our assumption can be interpreted as the cost for consumers to remove unwanted adverts appeared on free access content.

where α_A and α_B are perceived qualities of content A and content B, respectively, and β is the degree of substitutability of contents. Since end users are assumed to consume both contents, we strictly focus our analysis on the case that $0 < \beta < 0.73$, although a broader range could be analyzed in the same current framework.[5]

Thus consumer's maximization problem is given by:

$$\underset{q_A, q_B}{max} \{U(q_A, q_B) - (p_A + p)q_A - (p_B + p)q_B\} \tag{11.2}$$

The demand Q addressed to the ISP is simply equal to sum of demands for both contents, that is $Q = q_A + q_B$.

ISP's Optimal Qualities Under Unregulated Regime

Our interest is to analyze ISP's optimal provision of the internet's qualities provided to both CPs which can be considered as a two-stage game as follows:

Stage 1. The ISP determines the level of internet QoS provided to each CP.

Stage 2. Price competition takes place and end users determine the consumption of each content based on their perceived utility.

The game would be solved deriving Subgame Perfect Nash Equilibrium. Particularly, in stage 2, qualities are fixed, and hence we solve for the traditional maximization problem provided in (11.1) and (11.2), and obtain linear demand for content i depending on firms' prices:

$$q_A = \left(\frac{\mu_A - \beta\mu_B + \beta(p_B + p) - (p_A + p) + \gamma - \beta}{1 - \beta^2} \right) \tag{11.3}$$

$$q_B = \left(\frac{\mu_B - \beta\mu_A + \beta(p_A + p) - (p_B + p) + 1 - \gamma\beta}{1 - \beta^2} \right) \tag{11.4}$$

It can be seen that demand for content i is decreasing in its composite usage price, which is $(p_i + p)$, but increasing in competitor's composite

---

[5] This market configuration is also useful to reduce significantly technical issue in our analysis in the next section.

price, which is $(p_j + p)$. Furthermore, the demand for a content is proportional to the internet QoS, but disproportional to the level of the internet QoS provided to the competitor.

The demand for internet is derived from aggregating (11.3) and (11.4):

$$Q = \left(\frac{\mu_A + \mu_B - (p_A + p) - (p_B + p) + \gamma + 1}{1 + \beta} \right) \quad (11.5)$$

We can see from (11.5) that due to the perfect complementarity of internet and contents, demand of the internet is negatively affected by prices of both contents and internet. Also, the ISP can charge higher internet price to end users for higher levels of internet's qualities.

The profit function of the integrated ISP which is composed of revenues from both internet and content services as well as the cost of providing internet QoS as the following:

$$\pi_{ISP} = pQ + p_B q_B - \frac{\mu_A^2}{2} - \frac{\mu_B^2}{2} \quad (11.6)$$

The profit's function of competing firm is:

$$\pi_A = p_A q_A \quad (11.7)$$

Substituting demand from (11.3), (11.4) and (11.5) into (11.6) and (11.7), we have:

$$\pi_{ISP} = \frac{-2p_B^2 + (-\beta\mu_A - \beta\gamma + 2\beta p + \beta p_A + \mu_B - 2p + 1)p_B}{2(1-\beta^2)} \quad (11.8)$$

$$\pi_A = \frac{[(\beta\mu_B - \mu_A - \beta(p + p_B) + (p + p_A) + \beta - \gamma]p_A}{(\beta^2 - 1)} \quad (11.9)$$

The integrated ISP must set optimal internet price and price of content B taking the price of content A as given. Similarly, CP_A sets price of content A taking internet price and price of content B as given. Hence, we solve simultaneously the firm maximization problems provided in (11.8) and (11.9), we can derive equilibrium prices based on qualities as follows:

$$p = \frac{2\mu_A + \beta\mu_B + 2\gamma + \beta}{6} \quad (11.10)$$

In (11.10), due to the complementarity between internet and content, the internet's price is increasing in the internet qualities provided to both CPs.

$$p_A = \frac{\mu_A - \beta\mu_B + \gamma - \beta}{3} \tag{11.11}$$

$$p_B = \frac{(3-\beta)\mu_B - 2\mu_A - \beta - 2\gamma + 3}{6} \tag{11.12}$$

It can be seen from (11.11) and (11.12) that the price of content i is increasing but decreasing in internet qualities provided to CP_i and the competitor, respectively.

Substituting (11.10), (11.11) and (11.12) into (11.3) and (11.4), we can obtain equilibrium quantities based on the internet qualities as follows:

$$q_A = \frac{\mu_A - \beta\mu_B + \gamma - \beta}{3(1-\beta^2)} \tag{11.13}$$

$$q_B = \frac{(3-\beta^2)\mu_B - 2\beta\mu_A - \beta^2 - 2\beta\gamma + 3}{6(1-\beta^2)} \tag{11.14}$$

Subsequently, the demand for internet is:

$$Q = \frac{(3+\beta)\mu_B + 2\mu_A + \beta + 2\gamma + 3}{6(1+\beta)} \tag{11.15}$$

One can see that *ceteris paribus*, for a unit increase of quality provided to the affiliated CP would lead to a relatively higher demand for internet than that provided to the competing CP. This is because of the impact of vertical integration between the ISP and CP_B that internalizes the externalities arising from the prices of the complementary services under vertical separation.

We now replace equilibrium's prices and quantities (11.10), (11.11), (11.12), (11.13) and (11.14) into (11.8) to obtain the ISP's profit function based on the internet qualities:

$$\pi_{ISP} = \frac{(18\beta^2 - 14)\mu_A^2 + 8(\gamma - \beta - \beta\mu_B)\mu_A + (13\beta^2 - 9)\mu_B^2 + (-10\beta^2 - 8\beta\gamma + 18)\mu_B - 5\beta^2 - 8\beta\gamma + 4\gamma^2 + 9}{36(1-\beta^2)} \tag{11.16}$$

Taking the first-order conditions of the profit maximization problem in (11.16), we can obtain the ISP's optimal provision of internet qualities provided to the affiliated CP and the competing CP. We will subscript "U" for the equilibrium values in this section.

$$\mu_B^U = \frac{5\beta^2 + 4\gamma\beta - 7}{13\beta^2 - 7} \quad (11.17)$$

$$\mu_A^U = \frac{2(2\beta - \gamma)}{13\beta^2 - 7} \quad (11.18)$$

It is straightforward to obtain equilibrium demands for each content by substituting optimal qualities in (11.17) and (11.18) into (11.13) and (11.14), we have:

$$q_A^U = \frac{3(2\beta - \gamma)}{13\beta^2 - 7} \quad (11.19)$$

$$q_B^U = \frac{3\beta^2 + 5\beta\gamma - 7}{13\beta^2 - 7} \quad (11.20)$$

Since $0 < \beta < 0.73$, we have:

$$\begin{cases} q_A^U > 0 \text{ if } \gamma > \gamma^{(1)} = 2\beta \\ q_B^U > 0 \text{ if } \gamma < \gamma^{(2)} = \frac{4\beta}{7-5\beta^2} \end{cases}$$

Therefore, in the remainder of the paper, our analysis only focusses on the market configuration in which $\gamma^{(1)} < \gamma < \gamma^{(2)}$ and $0 < \beta < 0.73$. In this range, we can check that $\mu_A^U > 0$ and $\mu_B^U > 0$.

Interestingly, we have $\mu_A^U > \mu_B^U$ if $\gamma > \gamma^{(3)}$ where:

$$\gamma^{(3)} = \frac{7 - 5\beta^2 + 4\beta}{4\beta + 2} \quad (11.21)$$

We have thus established the following proposition.

Proposition 1 In the unregulated internet, if quality of the competing content is sufficiently high ($\gamma > \gamma^{(3)}$), then the profit-maximizing-integrated

ISP provides relatively higher quality of internet service to the competing CP than that provided to the affiliated CP.

Proposition 1 indicates that when the degree of vertical differentiation is sufficiently small, i.e., $\gamma < \gamma^{(3)}$, we have $\mu_A^U < \mu_B^U$. In other words, the integrated ISP has an incentive to provide a higher internet quality to its content. This is because first, it could gain higher market share as can be seen from (11.14). Second, it benefits from higher demand for internet since from (11.15), we can see that, *ceteris paribus*, an increase in internet quality provided to the affiliated content would lead to a relatively higher of consumption of internet service in comparison with that provided to the competing content.

However, we can see that $\mu_A^U > \mu_B^U$ when $\gamma > \gamma^{(3)}$. This is because the integrated ISP has two revenues: from content B and from internet service. From (11.10) and (11.15), both internet's price and demand are higher in the quality of the competing content. Therefore, for-profit maximizing, the integrated ISP would provide preferential QoS to the competitor when the competitor offers sufficiently high-quality content.

To this end, we can easily compute equilibrium's prices, firms' profits, consumer surplus and social welfare under the unregulated regime. In precision, substituting (11.17) and (11.18) into (11.10), (11.11) and (11.12), we have:

$$p^U = \frac{3\beta^3 - \beta + (5\beta^2 - 3)\gamma}{13\beta^2 - 7} \tag{11.22}$$

$$p_A^U = \frac{3(1 - \beta^2)(2\beta - \gamma)}{13\beta^2 - 7} \tag{11.23}$$

$$p_B^U = \frac{(1 - \beta)(3\beta^2 + 5\beta\gamma - 6\beta + 3\gamma - 7)}{13\beta^2 - 7} \tag{11.24}$$

We can check that $p^U > 0$ and $p_A^U > 0$. However, $p_B^U > 0$ iff $\gamma < \frac{7 - 3\beta^2 + 6\beta}{\gamma + 5\beta}$. So when the competing CP provides sufficiently high-quality content, the vertically integrated ISP must subsidy its content, i.e., $p_B^U < 0$.

The equilibrium firm profits are positive as follows:

$$\pi_{ISP}^{U} = \frac{5\beta^2 + 8\beta\gamma - 2\gamma^2 - 7}{2(13\beta^2 - 7)} > 0 \tag{11.25}$$

$$\pi_{A}^{U} = \frac{9(1-\beta^2)(2\beta - \gamma)^2}{(13\beta^2 - 7)^2} > 0 \tag{11.26}$$

From (11.1) and (11.2), we can compute the equilibrium consumer surplus:

$$CS^U = \frac{45\beta^4 + 72\beta^3\gamma - 5\beta^2\gamma^2 - 90\beta^2 - 64\beta\gamma + 9\gamma^2 + 49}{(13\beta^2 - 7)^2} > 0 \tag{11.27}$$

Finally, social welfare is simply the sum of firms' profits and consumer surplus from (11.25), (11.26) and (11.27), we have:

$$W^U = \frac{38\beta^4 + 248\beta^3\gamma - 49\beta^2\gamma^2 - 144\beta^2 - 192\beta\gamma + 41\gamma^2 + 98}{(13\beta^2 - 7)^2} > 0 \tag{11.28}$$

These equilibrium values are benchmark levels to assess the effectiveness of non-discrimination quality regulation to be analyzed in the next section.

Investment Under Non-Discriminatory Regulation

In this section, the integrated ISP is obliged to provide identical quality of internet service to both contents, denoted as μ. We are interested in analyzing how this regulatory restriction affects ISP's investment and subsequently the social welfare.

The profit function of the competing CP is still as shown in (11.7), while profit of the integrated ISP is different from (11.6) as follows:

$$\pi_{ISP} = pQ + p_B q_B - \mu^2 \tag{11.29}$$

The ISP's optimal internet quality is also analyzed in a two-stage game as in the previous section. Since we only put restriction of the identical internet's quality in stage 1, quantities and prices in stage 2 are similar as those provided (11.10), (11.11), (11.12), (11.13) and (11.14) where we replace μ_A and μ_B by μ.

Then the profit function of the integrated ISP given in (11.29) becomes:

$$\pi_{ISP} = \frac{(31\beta^2 - 8\beta + 9\gamma^2 - 23)\mu^2 + (18 - 10\beta^2 - 8\beta\gamma - 8\beta + 8\gamma)\mu + 9 - 5\beta^2 - 8\beta\gamma - 4\gamma^2}{36(1 - \beta^2)} \quad (11.30)$$

We compute the first-order conditions to obtain the ISP's optimal quality from stage 1. It is noted that the equilibrium values in this section are subscripted with "R".

$$\mu^R = \frac{5\beta + 4\gamma + 9}{31\beta + 23} > 0 \text{ for any } \beta, \gamma > 0 \quad (11.31)$$

The intuition in (11.31) is simple. Due to perfect complementarity between content and internet, and the fact that cost function is convex, the ISP could optimally provide a positive finite level of the internet's quality.[6]

Comparing with the internet's qualities under both regulatory regimes, we have:
$$\begin{cases} \mu_B^U < \mu^R < \mu_A^U \text{ if } \gamma > \gamma^{(3)} \\ \mu_B^U > \mu^R > \mu_A^U \text{ if } \gamma < \gamma^{(3)} \end{cases}$$ where the value of $\gamma^{(3)}$ is provided in (11.21).

To examine the effect of non-discriminatory quality regulation on the social welfare, we first identify the equilibrium's prices under this regulatory regime:

$$p^R = \frac{6\beta^2 + 11\beta\gamma + 7\beta + 9\gamma + 3}{31\beta + 23} \quad (11.32)$$

$$p_A^R = \frac{3(\beta + 1)(1 - 4\beta + 3\gamma)}{31\beta + 23} \quad (11.33)$$

---

[6] This result is principally consistent with our finding in the companion paper examining the case in which the ISP and CPs are vertically separated (see Baranes and Vuong [2020]).

$$p_B^R = -\frac{6\beta^2 + 11\beta\gamma - 11\beta + 7\gamma - 13}{31\beta + 23} \tag{11.34}$$

Then we can easily compute the corresponding firms' profits, consumer surplus and subsequently the social welfare. In precision, the profit of the vertically integrated ISP is:

$$\pi_{ISP}^R = \frac{5\beta^2 + 8\beta\gamma - 3\gamma^2 - 2\gamma - 8}{(\beta - 1)(31\beta + 23)} \tag{11.35}$$

Similarly, profit of the competing CP is given by:

$$\pi_A^R = \frac{9(1 + \beta)(4\beta - 3\gamma - 1)^2}{(1 - \beta)(31\beta + 23)^2} \tag{11.36}$$

Consumer surplus (11.1) and (11.2) becomes:

$$CS^R = \frac{180\beta^3 + (288\gamma + 180)\beta^2 - (77\gamma^2 - 154\gamma + 257)\beta - 85\gamma^2 - 118\gamma - 265}{2(\beta - 1)(31\beta + 23)^2} \tag{11.37}$$

Lastly, social welfare in the non-discriminatory quality regulation is total sum of (11.35), (11.36) and (11.37), we have:

$$W^R = \frac{202\beta^3 + (1216\gamma + 266)\beta^2 - (425\gamma^2 - 722\gamma + 627)\beta - 385\gamma^2 - 318\gamma - 651}{2(\beta - 1)(31\beta + 23)^2} \tag{11.38}$$

It is of interest to compare social welfare with and without regulatory restriction provided in (11.28) and (11.38). Let $DW = W^R - W^U$, then we have:

$$DW = \frac{\begin{bmatrix} (5\beta^2 + 4\beta\gamma - 4\beta + 2\gamma - 7) \\ 476\beta^5 + (6184\gamma - 5076)\beta^4 + (7478\gamma - 9343)\beta^3 \\ -(2238\gamma + 307)\beta^2 - (5152\gamma - 6541)\beta - 1412\gamma + 2849 \end{bmatrix}}{2(13\beta^2 - 7)^2(\beta - 1)(31\beta + 23)^2} \tag{11.39}$$

It can be seen from (11.39) that, the denominator of the R.H.S. is negative, so $DW > 0$ if the nominator is negative, that is $\gamma^{(4)} < \gamma < \gamma^{(3)}$ where:

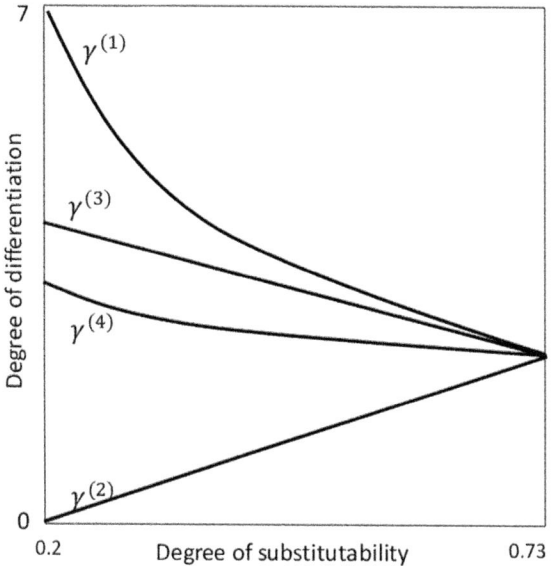

Fig. 11.2 The ISP's optimal qualities and social welfare under both regimes (*Source* Authors' computation)

$$\gamma^{(4)} = \frac{476\beta^5 - 5076\beta^4 - 9343\beta^3 - 307\beta^2 + 6541\beta + 2849}{-2(3092\beta^4 + 3739\beta^3 - 1119\beta^2 - 2576\beta - 706)} \quad (11.40)$$

We can now state the regulatory impact on social welfare in the following proposition.

Proposition 2 Non-discriminatory quality regulation, requiring the integrated ISP to provide identical internet QoS to both CPs, could enhance social welfare when $\gamma^{(4)} < \gamma < \gamma^{(3)}$.

Proposition 2 provides clear conditions that non-discriminatory quality regulation could be welfare improving. We can illustrate graphically the ISP's optimal qualities and compare the impact of investment on social welfare between both regimes in the following figure.

For illustrative purpose, in Fig. 11.2 we only present the range of substitutability that are not close to independent,[7] precisely $0.2 < \beta < 0.73$. There are positive demands for both contents when $\gamma^{(2)} < \gamma < \gamma^{(1)}$. Furthermore, in this range, the ISP provides positive internet qualities to the two CPs under both unregulated and regulated regimes.

Regarding the levels of internet qualities, the ISP can provide preferential treatment if left unregulated, and hence in the upper range of the curve $\gamma^{(3)}$, we have $\mu_B^U < \mu_A^U$ and in the lower range of the curve $\gamma^{(3)}$, we have $\mu_B^U > \mu_A^U$. As explained, if the competitor provides sufficiently high-quality content, the profit-maximizing ISP would provide it with a relatively higher quality of internet service than that provided to the affiliated CP under unregulated environment.

It is worth noting that under non-discriminatory quality regulation, social welfare is improved (i.e., $DW > 0$) when $\gamma^{(4)} < \gamma < \gamma^{(3)}$. In this range, the ISP, if left unregulated, provides a relatively high internet quality to the affiliated content, but low internet quality to the competitor in comparison with the optimal quality provided under regulation (that is, $\mu_B^U > \mu^R > \mu_A^U$). Furthermore, we compare equilibrium content prices between two regimes, and find while the price of the affiliated content is higher, the price of the competing content is lower under the unregulated environment, that is, $p_B^U > p_B^R$ and $p_A^R > p_A^U$ in this range. This result suggests an important role of non-discriminatory quality regulation of the internet in shaping social welfare under certain market configuration.

Conclusion

The present paper contributes to the existing literature of regulation in the digital economy in the context that the monopolistic ISP is integrated with one CP. We only limit our analysis in the range of market parameters that the ISP provides positive levels of internet QoS to both CPs if left unregulated. Our paper shows that the integrated ISP does not always have incentive to provide relatively higher quality of internet

---

[7] When β is near 0, then $\gamma^{(1)}$ will be infinite.

service to the affiliated content. In other words, if the competing CP offers sufficiently high-quality content to end users, that could bring great value for the network, the integrated ISP, if left unregulated, could offer preferential treatment to the competing CP due to perfect complementary of the internet and content.

Furthermore, in this paper, we can show that non-discriminatory quality regulation might enhance social welfare in a specific range of market parameters, suggesting a need for careful implementation of the regulation in practice. Finally, further research might incorporate other elements including an option for a CP to enhance transmission speed using content delivery network (CDN) technology and competition between ISPs and CDN providers.

References

Bandyopadhyay, S., H.K. Cheng, and H. Guo. 2010. Net Neutrality and Vertical Integration of Content and Broadband Services. *Journal of Management Information Systems* 27: 243–276.

Baranes, E., and C.H. Vuong. 2020. Investment in Quality Upgrade and Regulation of the Internet. CESifo Working Paper No. 8074.

Beresteanu, A. 2005. Nonparametric Analysis of Cost Complementarities in the Telecommunications Industry. *RAND Journal of Economics* 36: 870–889.

Brito, D., P. Pereiraz, and J. Vareda. 2014. On the Incentives of an Integrated ISP to Favor Its Own Content. 20th ITS Biennial Conference, Rio de Janeiro, Brazil, 30 Nov.–03 Dec. 2014: The Net and the Internet—Emerging Markets and Policies, International Telecommunications Society (ITS).

Broos S., and A. Gautier. 2017. The Exclusion of Competing One-Way Essential Complements: Implications for Net Neutrality. *International Journal of Industrial Organization*. Elsevier 52 (C): 358–392.

Carlton, D., and M. Waldman. 2002. The Strategic Use of Tying to Preserve and Create Market Power in Evolving Industries. *RAND Journal of Economics* 2002:194–220.

Casadesus-Masanell, R., B.J. Nalebuff, and D. Yoffie. 2010. Competing Complements. Working paper.

Dewenter, R., and J. Rosch. 2016. Net Neutrality and the Incentives (Not) to Exclude Competitors. *Review of Economics*, De Gruyter 67 (2): 209–229.

Economides, N. 1999. Quality Choice and Vertical Integration. *International Journal of Industrial Organization* 17: 903–914.

Greenstein, S., M. Peitz, and T. Valletti. 2016. Net Neutrality: A Fast Lane to Understanding the Trade-Offs. *Journal of Economic Perspectives* 30: 127–150.

Haucap, J., and G.J. Klein. 2012. How Regulation Affects Network and Service Quality in Related Markets. *Economics Letters* 117: 521–524.

Krämer J., L. Wiewiorra, and C. Weinhardt. 2013. Net Neutrality: A Progress Report. *Telecommunications Policy* 37 (9): 794–813.

Mussa, M., and S. Rosen. 1978. Monopoly and Product Quality. *Journal of Economic Theory* 18: 301–317.

Njoroge, P., A. Ozdaglar, N.E. Stier-Moses, and G. Weintraub. 2012. Investment in Two Sided Markets and the Net Neutrality Debate. *Review of Network Economics* 12: 355–402.

Rey, P., and J. Tirole. 2007. A Primer on Foreclosure. In *Handbook of Industrial Organization*, vol. 3, ed. M. Armstrong and R. Porter. Amsterdam: North-Holland.

Riordan H. M. 2008. Competitive Effects of Vertical Integration. In *Handbook of Antitrust Economics*, ed. Paolo Buccirossi. Cambridge: MIT Press.

Schuett, F. 2010. Network Neutrality: A Survey of the Economic Literature. *Review of Network Economics* 9 (2): 1–15.

Singh, N., and Vives, X. 1984. Price and Quantity Competition in a Differentiated Duopoly. *Rand Journal of Economics* 15: 546–554.

CHAPTER 12

Operations of Internet Platform Intermediaries

Rob Frieden

INTRODUCTION

Several segments of the Information, Communications and Entertainment ("ICE") marketplace have dominant intermediaries that operate a platform needed by both upstream sources of content and downstream consumers. These ventures can achieve market dominance in a "winner-take-all" (Malik 2015) competition by creating the dominant platform (Schumpeter 2017) used by consumers. In the markets for broadband carriage and many internet service market segments, such as social networking, winning ventures quickly can accrue scale and efficiency advantages as more and more consumers join the bandwagon and subscribe (Gal and Elkin-Koren 2017).

Successful insertion of an intermediary platform has generated both positive and negative impacts on consumer welfare, competition, the rate of innovation, employment and other key factors. On the positive side, intermediaries can promote efficiency and positive network externalities (Katz and Shapiro 1985; Moffatt 2016) where the overall value of a network and its ability to generate consumer benefits grow as more users participate. On the negative side, intermediaries, operating without

R. Frieden (✉)
Penn State University, University Park, PA, USA
e-mail: rmf5@psu.edu

significant competition, can extract high prices from both upstream and downstream participants, erect strong barriers to market entry, acquire competitors and use comparative advantages to dominate in both core and related markets such as the collection, processing and sale of "Big Data" (Helveston 2016) about subscriber behavior. Additionally, invisible, bad actors, operating upstream from the intermediary, have largely undetected and unchecked opportunities to distort markets and elections, interfere with broadband users' reasonable privacy expectations and threaten trust in such essential institutions as the news media.

Economists use the term two-sided (or multisided) markets to identify platform functions where transactions occur both upstream and downstream from the intermediary (Rochet and Tirole 2003, 2006; Armstrong 2006; Filstrucchi et al. 2014). The business models used by intermediaries rely on a strategic calibration of prices, often appearing to provide "free," or subsidized services to users on one side of the platform, typically downstream consumers. Consumers can access valuable services with zero financial payments, but they do have to offer something of great value: permitting intermediaries to compile information about their wants, needs, desires, internet uses, searches, locations and other behavior that can be processed and marketed to advertisers for better targeting of their commercial pitches. Privacy intrusions (Pasquale 2013) and the commodification of consumer behavior, so-called surveillance capitalism (Zuboff 2019), generate significant value that a platform operator can capture often without subscribers fully understanding and quantifying the potential for reduced benefits.

This paper identifies defects in the ways most governments currently respond to allegations of harm to consumers and competition. Governments can refrain from regulating access and tolerate market concentration as the proper reward for ventures offering desirable content and carriage services. Alternatively, they can impose *ex ante* safeguards to remedy anticipated harms to competition and consumers such as market concentration, price discrimination, reduced consumer welfare and captured consumer surpluses. Between these options, governments can rely on *ex post* review of complaints by courts or expert regulatory agencies to offer calibrated remedies.

The paper analyzes a U.S. Supreme Court case that assesses both downstream and upstream impacts in a two-sided market. It recommends that courts and government agencies address marketplace distortions by recalibrating existing tools to examine the competitive and consumer impacts on both sides of an intermediary's platform.

Consumer Benefits from Two-Sided Markets

Intermediaries have operated in many marketplaces for centuries (Cohen 2017). Examples include travel agents, broadcast networks, newspapers and credit card issuers. Emerging broadband, digital platforms have enhanced the power and impact of such ventures resulting in vast changes to "the traditional equilibria of supply and demand, blurring the lines between owners and users, producers and consumers, workers and contractors, and transcending the spatial divides of personal and professional, business and home, market and leisure, friend and client, acquaintances and stranger, public and private" (Lobel 2016, p. 90) Digital broadband platform operators can accrue substantial consumer benefits even as they increase market shares. A "win-win" scenario combines ample benefits for platform operators and consumers by enhancing the value proposition in commercial transactions.

Digital broadband platform operators can quickly acquire scale economies (Perritt 2017) and efficiency gains by spreading costs over a growing global population of users. The incremental cost to add an additional participant approaches zero, because many internet-mediated markets have high initial, investment costs, but very low incremental costs incurred when adding users. Additionally, these platforms can accrue positive networking externalities (Shapiro and Varian 1997; Lemley and McGowan 1998; Newman 2012) as subscribership grows. When intermediaries reach a critical mass of popularity, nonusers see the advantages in joining the bandwagon which further enhances the comparative attractiveness of a single platform operator *vis-à-vis* other competitors and options.

Platform intermediaries must deliver a compelling value proposition to generate consumer use, particularly when alternatives exist, with low entry barriers and switching costs. The combination of competitive necessity and more efficient operations can readily translate into the offering of lower priced products and services to consumers, particularly because two-sided platform operators can calibrate how much to charge each side:

> [P]rofit-maximizing prices may require charging one side less than the marginal cost of serving that side. Empirical surveys of industries based on ... [two-sided platforms] find many examples of prices that are low, or even negative, so that customers on one side are incentivized to participate in the platform. (Evans and Noel 2005, p. 668)

Consumer Costs from Two-Sided Markets

Immediate and longer term costs offset readily identifiable benefits from two-sided platforms. In the short term, ventures like Amazon enhance consumer welfare by offering a growing inventory of products and services at lower prices, the product of operational efficiencies and the willingness to eschew profits in exchange for increasing market share and scope. However, in the longer term, consumers may suffer from the loss of competition from "bricks and mortar," local vendors as well as from the consequences of ever more accurate assessment of their price sensitivity and increasingly invasive collection of subscribers' consumption behavior and the brokering of such data by largely unregulated ventures (Khan 2017; Kuempel 2016). At some point, online platform operators may consider their market position sufficiently impenetrable so that they can refrain from aggressive price cutting and forgoing near term profitability (Khan 2019).

Additionally, these operators may have so developed data analytics that they can quite accurately set and frequently modify prices with an eye toward maximizing profits (Fleming 2015). Dynamic pricing refers to the ability of product and service vendors to change prices quickly by collecting and analyzing data about current consumer demand (Calo 2014; Adame 2016). Rather than set a fixed price, only occasionally raised or lowered, vendors can make frequent pricing changes based on current marketplace conditions or a specific buyer's profile.

While such dynamic pricing arguably represents an efficiency enhancing, fine-tuning of price setting, consumers may consider it unfairly discriminatory. With the ability to track current demand for a product or service, intermediaries can use so-called surge pricing that imposes an algorithm calculated price. For example, on-demand ride-sharing services vary their quoted rates based on a current assessment of available cars and demand. While rates may fall below that charged by a tariffed taxi cab service, peak demand can trigger massively higher rates offered by new platform ventures such as Uber and Lyft. Amazon's algorithm temporarily quoted a $2630.52 price for a used paperback book usually offered for $1.99 (Streitfeld 2018).

Subscriber Data Value and Lock-In Cost Missing in the Cost/Benefit Analysis

One can readily assess the benefits of access to intermediary platforms, but the costs are not as readily determined. Consumers may wrongly assume that they have free access, because in most instances access to content triggers no upfront subscription payment. The free access conclusion fails to consider three hidden and not easily quantifiable costs: (1) the increase in the price of advertised goods and services, possibly better calibrated through data mining resulting in surge prices; (2) the monetary value accruing to intermediaries when they acquire, collate, analyze and sell consumer data, as well as auction advertising placements on their web sites (Bodie et al. 2017; Woodcock 2017; Hacker and Petkova 2017); and (3) the elimination of competition.

Broadband intermediaries have achieved remarkable success in developing techniques to monitor, surveil, collect, analyze, collate and sell subscriber data. This reduces the value proposition of what the intermediary offers, because the ability to "mine" subscriber data has a value that can provide a substantial, new revenue stream from freely collected consumer data.

Deficiencies in Existing Government Oversight Models

Most governments have failed to revise existing legal, regulatory and jurisprudential frameworks for application to issues raised by the onset of digital broadband intermediary platforms. This section addresses how traditional governmental strategies ignore fundamental differences between bricks and mortar and internet-mediated transactions.

As a threshold matter, governments decide whether and how to intervene in a specific industry sector. They may opt to rely entirely on marketplace forces, confident that competition will force stakeholders to operate in ways that deliver a compelling value to consumers without anticompetitive practices. Other governments may pursue the opposite: an interventionist approach, imposing *ex ante* rules and regulations, such as network neutrality (Frieden 2015) and common carrier regulation. Between these polar opposites, two alternative, possibly complementary, *ex post* strategies exist: (1) apply antitrust, consumer protection and prohibitions on unfair trade practices to remedy proven harms; and (2) use dispute resolution through litigation and complaint filing procedures to fashion remedies that typically impose monetary fines and compulsory modification of business practices.

Each of the legacy models fails to achieve an ideal balance between governmental regulatory forbearance and intervention, primarily because the assumptions, strategies and tactics applied do not make essential adjustments reflecting the difference between digital, broadband networking and preexisting channels of commerce. Without modification of market definition and impact assessment, governments risk false positives, which trigger unnecessary marketplace intervention, or by reaching false negatives, which fail to trigger important safeguards based on an incorrect determination that no harm to consumers or competition has, or will occur.

A Realistic Assessment of Platform Costs and Benefits

Consumers and governments may not fully understand the trade-offs when digital, broadband intermediaries dominate many ICE market segments. One can readily appreciate the upside consumer benefits in having access to advertiser-supported content and internet markets subsidized by ventures willing to forego short-term profits for longer term market share and product diversification. A more difficult undertaking calculates what direct and indirect costs consumers incur, presently and in the future, for the opportunity to participate in "winner-take-all" two-sided markets.

Prevailing economic doctrine, widely embraced by government legislators, judges and regulators, favors an inclination not to intervene in the marketplace, when identifiable, near term cost savings and other welfare enhancements flow to consumers. Much revered, so-called Chicago School marketplace assumptions (Bork 1978; Posner 1979; Crane 2014) and antitrust prescriptions may not make sense for digital, platform markets including the view that rational commercial actors (such as Amazon) never would pursue below market pricing given the unlikely opportunity to recoup current losses in the future. Likewise, a laser focus on efficiency and consumer welfare, as espoused by Robert Bork, may require a longer time span that considers whether immediate and easily measured, short-term consumer welfare enhancements are partially or completely offset over time. Such analysis requires scrutiny of both downstream and upstream market effects. Lina Khan (2017) makes a compelling case supporting this approach.

It has become increasingly clear that consumers must contribute more value, than what they might infer from widespread promotion of "free" and subsidized access. Even in the short run, the value proposition from participating in two-sided markets may decline as consumers begin to understand the monetary value of the network usage data they generate and consent to having platform operators use for dynamic pricing of their goods and services and as a marketable commodity for sale to upstream advertisers.

In the longer term, the commodification of consumer data may accrue the greatest strategic and financial advantages for ventures that already have successfully exploited positive network externalities and already have acquired large market shares. This advantage stifles innovation and competition if consumers cannot freely change their platform subscription and take their business and data to another platform. In the internet ecosystem, consumers often lack complete information about what they must pay and what they lose in exchange for the opportunity to become a subscriber. Few consumers may have the disposition and wherewithal to undertake regular cost/benefit analyses as well as a determination whether to stick with the status quo, or to seek better terms and conditions. Such inertia enhances the ability of incumbent firms to maintain their market dominance.

Digital broadband consumers may likely suffer more significant, but not readily quantifiable harms, as digital, broadband intermediaries find new and more precise ways to maximize profits from both upstream and downstream sources. Real or perceived lock-in by incumbent firms helps maintain their market dominance.

Government agencies with jurisdiction to monitor such actions appear ill-equipped to provide effective oversight based on their adherence to now questionable economic and antitrust theory, the inability or unwillingness to consider costs and benefits on both sides of the two-sided market and their emphasis on short-term consumer benefits that may not seem as generous as initially estimated.

The Way Forward

Regulatory agencies with jurisdiction to safeguard consumers and reviewing courts should better calibrate the tools they use to investigate the potentially harmful effects of platform intermediaries on competition and consumers, with emphasis on the potential for privacy intrusions, unfair trade practices, market concentration and anticompetitive practices. The goals for recalibration should focus on acquiring a better understanding of platform operator practices and their impacts rather than serve as a justification for more intrusive government oversight. Such a holistic approach can better assess the costs and benefits generated by platform intermediaries.

Assess Impacts on Both Sides of a Platform

To achieve greater clarity on the potential for beneficial and harmful impact, courts and government agencies should examine platform operations on both upstream and downstream market sides. Using a cost–benefit analysis, they may determine that harmful impacts on one side are offset by benefits on the other side. In other instances, they may identify greater harms or benefits.

By examining both sides of a digital, broadband platform market, courts and regulatory agencies can enhance the accuracy of their assessment of competition and whether consumers benefit or suffer from doing business with intermediaries having substantial market share. In turn, they can better calibrate a remedy, or reach an empirically supported conclusion that no market intervention is necessary.

Insights from Ohio v. American Express

A U.S. Supreme Court decision involving a credit card issuer, provides insight on current disputes about the proper scope of market definition and analysis of platform intermediaries (Ohio v. American Express Co. 2018). The case addressed how courts should define the relevant market (Katz and Sallet 2018) so that they do not make the mistake of finding anticompetitive harms where little or none exists, a false positive, and perhaps also to avoid decision-making that ignores actually occurring consumer harms, a false negative.

In *Ohio v. American Express*, the conservative majority of the Supreme Court, endorsing recent economic doctrine championed by academics (Filistrucchi et al. 2014; Evans and Schmalensee 2008; Evans and Noel 2005), upheld a decision by the Second Circuit Court of Appeals to reject a lower court's relevant market determination in an antitrust review of an alleged vertical restraint of trade. The Court endorsed the finding that the lower court should have assessed consumer impacts on both sides of markets served by the credit card issuing company: the downstream users of cards and the upstream vendors accepting cards for payment. The alleged vertical restraint involved so-called anti-steering contractual language that prohibited vendors, agreeing to accept American Express credit cards, from trying to persuade customers to use a different card that imposed lower "swipe fee" processing costs on the vendor.

While suggesting that its relevant market and impact analysis required consideration of how anti-steering provisions affected both merchants and consumers—ostensibly a complete two-sided market assessment—the District Court focused solely on how the anti-steering contractual language helped maintain higher swipe fees that harmed both credit card issuer competition and consumers with apparently no offsetting benefits (United States v. American Express Co. 2015). This court also determined that American Express had market power, because it imposed 20% swipe fee increases over a 5-year period without losing market share in terms of the number of vendors accepting its cards and its market share of credit card transactions. The court determined that in the absence of the anti-steering provisions, merchant swipe fees would have been lower as would consumer costs. The court also considered corroborating evidence the decision by the Discover credit card company to abandon its business model of offering comparatively lower fees as an inducement for more vendors to accept it for payment and in turn to acquire greater market share of credit card usage. As the company having the smallest market share, Discover sought to differentiate its card with merchants, but could not acquire more market share, because vendors could not encourage customers to use it.

Both the Second Circuit Court of Appeals and the Supreme Court opted to examine impacts on both sides of the credit card platform marketplace. They concluded that to assess the complete impact of a credit card company's anti-steering contractual language, courts should identify and consider the consequences of any positive or negative impact.

The lower court was deemed to have failed to consider how anti-steering rules could have positive impacts that could offset the negative impacts the lower court identified, as well as impact the relationships and interactions between both market segments.

The appellate courts undertook a comparison of costs and benefits affecting both vendors and credit card users. While anti-steering rules mandated by credit card issuers can constitute an illegal vertical restraint on trade, by reducing competition among credit card companies, the courts considered the potential for offsetting, positive financial impact on credit card users through more generous and diversified benefits, e.g., financial rebates and enhanced travel services.

The Second Circuit Court of Appeals explained the need to examine both sides of the credit card market, in light of variance in costs incurred by both vendors and credit card users having impact on both sides of the platform operated by a credit card issuer. Considering the interdependency of product and service vendors and consumers using credit cards, the court identified two joint market effects not considered by the lower court: (1) impact of anti-steering rules on the level of card issuer market competition; and (2) the impact of credit card issuer anti-steering rules on their incentives to offer usage inducements to consumers. While the credit card marketplace is concentrated with only four companies and evidences substantial barriers to market entry, the court noted the ease with which consumers can shift card allegiance based on many factors including the costs incurred by using a specific card as well as the financial inducements offered by credit card issuers to encourage consumer loyalty.

The Supreme Court conservative majority affirmed the Second Circuit Court of Appeal's analysis and conclusion that the lower court should have assessed the consumer impact of transactions occurring on both sides of the credit card issuer's platform. The Court determined that both sides of an intermediary platform require examination, because two inter-related transactions take place, each of which affect both upstream and downstream participants. By examining the marketplace impact on both sides of the American Express platform, the Court identified consumer and competitive benefits that offset the harm to consumers identified by the District Court. Identifying this benefit would not occur if a court solely examined impacts on just one side of the intermediary's transactions, when identifying what constitutes the relevant market for credit card services. A reviewing court could avoid a false positive finding of anticompetitive harm to consumers from credit card anti-steering contractual terms, by compiling a complete evidentiary record including an assessment about the potentially favorable impact of such contractual language on both vendors and consumers.

A dissenting opinion, written by Justice Breyer and joined by the three other liberal Justices, disputed the lawfulness of the majority opinion's two-sided market examination. He strongly asserted that the lower court had no reason to expand its market impact analysis, noting that no antitrust case precedent supports doing so. Additionally, he noted case precedent does not favor judicial netting or balancing of competitive benefits and harms occurring in different markets.

The sole focus on the immediate impact of higher swipe fees on downstream credit card users ignores other factors that might reduce or eliminate a finding of anticompetitive harm such as higher prices to consumers. A more nuanced, calibrated and granular analysis considers the credit card ecosystem as both two-sided and segmented by card issuer marketing strategies. Swipe fee pricing strategy constitutes a key differentiator for which credit cards vendors would accept, but other factors come into play, particularly on the consumer side. Some consumers might want a credit card that offers generous financial rebates and other subsidies, such as airline miles. Others might want one that offers a low short-term interest rate on balance due transfers. Others might willingly pay for the privilege of tapping a benefit-rich inventory, including airline lounge access, "free" baggage allowances on specific airlines, concierge-provided travel assistance and early opportunities to buy broadway and concert tickets.

In this more segmented marketplace, a credit card company might execute a strategy of demanding comparatively higher swipe fees of vendors to generate more generous and desirable credit card user rewards. Another company might use lower swipe fees as an incentive for more vendors to accept purchases using the card. Arguably, such differentiation promotes a competitive marketplace both in terms of what inducements credit card companies must offer consumers and which card a consumer will use for each transaction.

The *American Express* case emphasizes the need for courts and by extension, regulatory agencies, to consider the interdependent relationship between upstream and downstream market participants in terms of their impact on each other and in terms of their relationship with the platform intermediary. In the credit card ecosystem, the availability of alternative credit cards and the ease with which consumers can change allegiances evidence a competitive credit card platform marketplace with significant consumer sensitivity to comparative costs and benefits accruing from the use of specific cards. Some credit card users attempt to maximize downstream subsidies and rebates by acquiring many different cards and strategically using them, e.g., Card A for gasoline, Card B for airline tickets and Card C for restaurants based on which card offers the best rebates and other benefits (see Katz 2019 for an elaboration of these issues).

The division of the Supreme Court, on a liberal vs. conservative fulcrum in this case, may identify what constituencies and economic doctrine each faction favors. The majority persuasively demonstrated that in the credit card ecosystem, two complementary and inter-related transactions take place. Justice Breyer, in dissent, suggested that the complementary relationship between the products has no applicability to the purpose of assessing whether a litigating party has satisfied its evidentiary burden of proof and for defining the market segments for credit card transactions. The Court majority considered card user and card accepting vendors as jointly participating in transactions that affect each other and thereby bind them and their markets together.

Consider Whether and How Lock-In Exists

Courts and regulatory agencies should consider the service options available to digital, broadband subscribers. In some instances, they have ample choices that prevent lock-in and evidence a competitive marketplace. However, in other instances, lock-in occurs, because consumers have few alternatives, or they incur costs, inconvenience or reduced benefits if they leave the dominant platform.

Lock-in can occur even when alternative options exist. For example, an electronic commerce site, like eBay, may steer subscribers to a former affiliated electronic funds transfer platform operated by PayPal, even though cheaper, alternative payment systems exist. Consumers have incentives to use PayPal, because the eBay site appears to favor and expedite such transactions and most vendors prefer to receive payment via PayPal. The preference for PayPal and the greater ease consumers have in using the preferred payment system generate substantial motivations to take the promoted and preferred path of least resistance.

Courts and regulatory agencies should consider the potential for lock-in beyond simply assessing whether a specific market segment has multiple platform operators. The existence of alternatives, by itself, does not evidence a competitive marketplace which can self-regulate. In the absence of viable service alternatives, courts and regulators should consider downstream consumers' quality of experience to ensure that the apparent preference for a single platform option promotes convenience and enhances consumer welfare.

Assess Market Impacts, Rather Than Simply Calculate Market Share

As noted, courts and regulators generally refrain from reaching conclusions about market competitiveness based solely on calculations showing a concentrated market, or one dominated by a single venture. Large firms having high market share may evidence a firm's superior business skill, or the need for ventures to accrue economies of scale to thrive in a specific market segment.

On the other hand, market dominance may have significant and potentially adverse impacts on consumers and the potential for competition. Significant harm may arise because a firm can leverage dominance in one market to dominate other market segments. For example, Google dominates the market for internet search and advertising, despite ample alternatives. Regulatory or judicial intervention is not warranted simply because Google has acquired substantial market share in internet search. However, the company's success in dominating the search market also translates into substantial market share in the auctioning of advertising opportunities to search consumers (Newman 2014), making it possible for the company to impose anticompetitive terms and conditions.

Courts and regulators may need to consider the inter-relationship between a venture's successes in two or more markets, because dominance in combined, or interdependent markets, may trigger new or greater risks for consumers. Just as platform intermediary operation affects both downstream and upstream users, so too can market success in one market generate uncontested opportunities to extend market power elsewhere. Such leverage may have adverse impacts on the potential for new competition, even from innovative ventures.

Conclusion

Digital broadband technologies and markets have reached a critical mass of market penetration and efficiency enhancements highlighted by embedded platforms. The internet ecosystem has many market segments predominated by single ventures that have acquired dominance in "winner-take-all" competition that rewards ventures best able to exploit positive network externalities. Intermediaries have offered significant, identifiable benefits to consumers, who also incur offsetting costs, not all of which can be easily quantified or measured. Intermediary platforms operators can calibrate cost recoupment from both upstream and downstream users. In many instances, downstream consumers have benefitted from subsidies and pricing strategies that reduce, or eliminate direct, out of pocket costs. However, subsidy payers, such as advertisers, eventually recover their costs through higher charges for goods and services.

In light of enhancements in the acquisition, analysis and marketing of consumer behavior data, both vendors and platform intermediaries now have more diversified and extensive ways to generate higher revenues and profits. Such data mining can impose new costs on consumers who must tolerate ever more extensive privacy intrusions in exchange for access to so-called free services. Enhanced consumer surveillance can impose lower or higher costs as exemplified by dynamic pricing that frequently changes rates through algorithmic analysis of overall demand, as well as a prediction of a prospective customer's intensity of preference for a particular good, or service.

In light of the mixed impacts of embedded intermediaries on competition and consumers, legislatures, courts and regulators should apply new tools for assessing current and prospective impacts. Unfortunately, the speed of innovation and the convergence of technologies and markets have exceeded the ability of governments to stay current. The tools used to assess market impacts have become ill-suited and poorly calibrated to meet new challenges (Brandenburger et al. 2017). Conventional competition policy and economic theories lack an emphasis on identifying both short-term and longer term consequences of platform operations. While immediate consumer welfare enhancement supports regulatory forbearance, governments need to consider whether and how longer term impacts will remain benign, or favorable.

In too many instances, governments have overstated consumer benefits and the absence of competitive harm. Most courts and regulatory agencies have not considered an intermediary's impact on both upstream and downstream markets, failed to consider fully whether and how subscriber lock-in has occurred and generated rationales excusing substantial market concentration based on short-term consumer benefits that may not be as generous in light of offsetting privacy intrusions.

Going forward, governments should appreciate that platform intermediaries seek to maximize profits and that the conferral of benefits to consumers may be offset by negative impacts on both consumers and competition, even in the short term. A more holistic examination of impacts, without placing a premium on short-term consumer benefits, would achieve a more accurate assessment of the likely mixed impacts generated by platform intermediaries.

References

Adame, V. 2016. Consumers' Obsession Becoming Retailers' Possession: The Way That Retailers Are Benefiting from Consumers' Presence on Social Media. *San Diego Law Review* 53: 653–700.

Armstrong, M. 2006. Competition in Two-Sided Markets. *The RAND Journal of Economics* 37 (3): 668–691.

Bodie, M.T., M.A. Cherry, M.L. McCormick, and J. Tang. 2017. The Law and Policy of People Analytics. *University of Colorado Law Review* 88: 962–1042.

Bork, Robert H. 1978. *The Antitrust Paradox: A Policy at War with Itself.* New York: Basic Books.

Brandenburger, R., L. Breed, and F. Schoning. 2017. Merger Control Revisited: Are Antitrust Authorities Investigating the Right Deals? *Antitrust* 31: 28–34.

Calo, R. 2014. Digital Market Manipulation. *George Washington Law Review* 82: 995–1051.

Cohen, J.E. 2017. Law for the Platform Economy. *University of California Davis Law Review* 51: 133–204.

Crane, D.A. 2014. The Tempting of Antitrust: Robert Bork and the Goals of Antitrust Policy. *Antitrust Law Journal* 79: 835–853.

Evans, D.S., and M. Noel. 2005. Defining Antitrust Markets When Firms Operate Two-Sided Platforms. *Columbia Business Law Review* 2005 (3): 667–702.

Evans, D.S., and R. Schmalensee. 2008. Markets with Two-Sided Platforms. *Issues in Competition Law and Policy* 1: 667–693.

Filstrucchi, L., D. Geradin, E. van Damme, and P. Affeldt. 2014. Market Definition in Two-Sided Markets: Theory and Practice. *Journal of Competition Law and Economics* 10 (2): 293–339.

Fleming, L. 2015. How Much Does J. Crew Really Know About You? The Harsh Reality of a Mega-Retailer's Privacy Policy. *Syracuse Journal of Science and Technology* 31: 1–38.

Frieden, R. 2015. Ex Ante Versus Ex Post Approaches to Network Neutrality: A Comparative Assessment. *Berkeley Technology Law Journal* 30: 1562–1612.

Gal, M.S., and N. Elkin-Koren. 2017. Algorithmic Consumers. *Harvard Journal of Law and Technology* 30: 309–353.

Hacker, P., and B. Petkova. 2017. Reining in the Big Promise of Big Data: Transparency, Inequality, and New Regulatory Frontiers. *Northwestern Journal of Technology and Intellectual Property* 15: 1–42.

Helveston, M.H. 2016. Consumer Protection in the Age of Big Data. *Washington University Law Review* 93: 859–917.

Katz, Michael. 2019. Multisided Platforms, Big Data, and a Little Antitrust Policy. *Review of Industry Policy* 54 (4): 695–716. https://doi.org/10.1007/s11151-019-09683-9, https://link.springer.com/article/10.1007%2Fs11151-019-09683-9.

Katz, Michael, and J. Sallet. 2018. Multisided Platforms and Antitrust Enforcement. *Yale Law Review* 127: 2142–2175.

Katz, M.L., and C. Shapiro. 1985. Network Externalities, Competition and Compatibility. *American Economic Review* 75: 424–440.

Khan, Lina M. 2017. Amazon's Antitrust Paradox. *Yale Law Journal* 126 (3): 564–907. https://www.yalelawjournal.org/note/amazons-antitrust-paradox.

Khan, Lina M. 2019. The Separation of Platforms and Commerce. *Columbia Law Review* 119: 973–1098. https://columbialawreview.org/content/the-separation-of-platforms-and-commerce/.

Kuempel, A. 2016. The Invisible Middlemen: A Critique and Call for Reform of the Data Broker Industry. *Northwestern Journal of International Law and Business* 36: 207–234.

Lemley, M.A., and D. McGowan. 1998. Legal Implications of Network Economic Effects. *California Law Review* 86: 479–611.

Lobel, O. 2016. The Law of the Platform. *Minnesota Law Review* 101: 87–166.

Malik, O. 2015. In Silicon Valley Now, It's Almost Always Winner Takes All. *The New Yorker*. Retrieved from http://www.newyorker.com/tech/elements/in-silicon-valley-now-its-almostalways-winner-takes-all.

Moffatt, M. 2016. Introduction to Network Externalities. Retrieved from https://www.thoughtco.com/introduction-to-network-externalities-1146145.

Newman, J.M. 2012. Anticompetitive Product Design in the New Economy. *Florida State University Law Review* 39: 681–733.

Newman, N. 2014. Search, Antitrust, and the Economics of the Control of User Data. *Yale Journal on Regulation* 31: 401–454.

Ohio v. American Express Company. U.S. Supreme Court, No. 16–1454, 585 U.S. ___, 138 S. Ct. 2274 (2018). Retrieved from https://www.supremecourt.gov/opinions/17pdf/16-1454_5h26.pdf.

Pasquale, F. 2013. Privacy, Antitrust and Power. *George Mason Law Review* 20: 1009–1024.

Perritt, H.H., Jr. 2017. Keeping the Internet Invisible: Television Takes Over. *Journal of Technology Law and Policy* 21: 121–196.

Posner, R.A. 1979. The Chicago School of Antitrust Analysis. *University of Pennsylvania Law Review* 127: 925–948.

Rochet, J.-C., and J. Tirole. 2003. Platform Competition in Two-Sided Markets. *Journal of the European Economic Association* 1: 990–1029.

Rochet, J.-C., and J. Tirole. 2006. Two-Sided Markets: A Progress Report. *The RAND Journal of Economics* 37: 645–667.

Schumpeter. 2017. The University of Chicago Worries About a Lack of Competition. *The Economist*. Retrieved from http://www.economist.com/news/business/21720657-itseconomists-used-champion-big-firms-mood-has-shifted-university-chicago.

Shapiro, Carl, and Hal Varian. 1997. *Information Rules: A Strategic Guide to the Network Economy*. Cambridge, MA: Harvard Business School Press.

Streitfeld, D. 2018. Amazon's Curious Case of the $2,630.52 Used Paperback. *The New York Times*, July 16. Retrieved from https://www.nytimes.com/2018/07/15/technology/amazon-used-paperback-book-pricing.html.

United States v. American Express Company, 88 F. Supp. 3d 143 (E.D.N.Y. 2015); 838 F.3d 179 (2d Cir. 2016) *cert. granted*, Ohio v. Am. Express Co., No. 16–1454 2017 WL 2444673 (Oct. 16, 2017).

Woodcock, R.A. 2017. Big Data, Price Discrimination, and Antitrust. *Hastings Law Journal* 68: 1371–1420.

Zuboff, S. 2019. *The Age of Surveillance Capitalism: The Fight for a Human Future at the New Frontier of Power*. New York: PublicAffairs.

CHAPTER 13

Escalating Instability of Network Neutrality Policy in the U.S.

Barbara A. Cherry

INTRODUCTION

Since the early 2000s, telecommunications policies in the U.S. have been unstable, particularly for policies related to providers of broadband internet access services (BIAS), such as network neutrality. Since the federal presidential and Congressional elections in 2016, the instability of federal network neutrality policy continues through the Federal Communications Commission's (FCC's) adoption of the *Restoring Internet Freedom Order* in 2017, reversing the FCC's *Open Internet Order* of 2015 that had established *Open Internet Rules*. Moreover, the *2017 Order* creates a new phase of federal–state policy instability through a sweeping declaration of federal preemption, escalating tensions between the federal government and the states' regulatory authority.

What are the reasons for this policy instability and escalation? This paper is part of a larger body of research to answer this question (Cherry 2010, 2015). Inquiry flows from the observation that this policy instability arises during a period of pronounced political dysfunction in the

B. A. Cherry (✉)
Indiana University, Bloomington, IN, USA
e-mail: cherryb@indiana.edu

U.S., resulting from severe party polarization and divided government, that also coincides with the era of deregulatory policymaking. The analysis examines the emergence of this period of political dysfunction and its impact on deregulatory policies from the perspective of coevolving political and economic systems. There are long temporal arcs of evolution within and among political and economic systems. These include evolution within and among the system of political governance under the U.S. Constitution and the economic system of capitalism, both fueled by technological innovations.

Dating from the nineteenth century, this coevolution in the U.S. has led to increased reliance on the corporate form—an artificial, legal entity—as a tool or mechanism for collective action by humans to conduct activities within the economic system. Over time, in numerous and varying ways, corporations have leveraged their economic power to effectuate changes in both the economic and political systems for their own benefit. In the late twentieth and early twenty-first centuries, corporations' deregulatory strategies have also further evolved. To pursue their deregulatory policy objectives, these strategies include increased reliance on judicial litigation and administrative decisions, novel interpretations or even mischaracterizations of the law, and judicial retrenchment of individual rights and legal remedies. Thus, a further question is what developments within and among the political and economic systems in the U.S. Have contributed to the evolution of these deregulatory strategies?

By examining these questions as a general matter (not confined to the telecommunications industry), one finds that the instability of network neutrality policy is symptomatic of more fundamental systemic developments under U.S. political governance. For this reason, instability of U.S. network neutrality policy is likely to continue for an extended period of time. Moreover, the new phase of disruption in U.S. network neutrality policy, by exacerbating policy instability through federal preemption of states' regulatory authority, further complicates the achievement of political and legal resolution.

INSTABILITY OF U.S. NETWORK NEUTRALITY POLICY

For the telecommunications policy debate on network neutrality, the resulting political dynamic has been a legal battle over the classification of BIAS—as a telecommunications service (i.e. common carriage service) or as an information service under federal law. This is because under the federal statutory framework, the legal classification of the service determines the scope of FCC jurisdiction and applicable regulatory framework. In turn, the legal classification of the service has changed, depending upon which political party has the majority of FCC commissioners.

Beginning in 2002, a Republican majority of FCC commissioners classified BIAS provided by cable companies and later telephony companies as an information service in 2002 and 2005, respectively. After two D.C. Circuit Court of Appeals decisions in *Comcast v. FCC* (2010) and *Verizon v. FCC* (2014), identifying legal flaws in asserted FCC jurisdiction under analysis grounded in classification of BIAS as an information service,[1] a Democratic majority of FCC commissioners reclassified all BIAS service as a telecommunications service and accordingly established open internet rules in the FCC's *Open Internet Order* (2015), known as the network neutrality order, which was upheld upon judicial review in June 2016 by the D.C. Circuit Court of Appeals in *US Telecom Association v. FCC*.

As a result of the November 2016 federal elections—under which the Republican Party controls the Presidency, a majority in both the House and Senate of Congress, and a majority of the FCC commissioners—the regulation of BIAS service is again in flux. In May 2017, the FCC opened a Notice of Proposed Rulemaking (NPRM), titled "In the Matter of Restoring Internet Freedom." Pursuant to this NPRM, the FCC adopted an order in December 2017, reclassifying BIAS service as an information service, reversing the FCC's 2015 *Open Internet Order*, and repealing most of the 2015 *Open Internet Rules*.

The 2017 *Restoring Internet Freedom Order* disrupts, again, the federal regulation regime for BIAS service. Numerous lawsuits have already been filed challenging the FCC order, which has been consolidated in an appeal pending before the D.C. Circuit Court of Appeals, *Mozilla v. FCC*.[2] Thus far, the U.S. Congress is divided as to how to

---

[1] In a 2010 order, while the majority of FCC commissioners were under Democratic control, the FCC attempted to adopt open internet rules while preserving classification of BIAS as an information service. However, in *Verizon v. FCC*, the Court found the FCC lacked jurisdiction to impose some of these rules while maintaining legal classification of BIAS as an information service.

[2] An update regarding *Mozilla v. FCC* is provided in section "Epilogue".

resolve the matter legislatively. On April 10, 2019, the U.S. House of Representatives under a Democratic Party majority passed the Save the Internet Act, which would reclassify BIAS as a telecommunications service and impose requirements that were adopted by the FCC in the 2015 *Open Internet Rules*. However, on the same day, the Republican Senate majority leader, Mitch McConnell, declared this legislation to be "dead on arrival" in the U.S. Senate.

Moreover, under the *2017 Order*, this disruption is more acute than simply a change in federal jurisdictional authority. This is because the FCC preempts any State or local government measures that would effectively impose rules or requirements that the FCC preempted or decided to refrain from imposing under this order. (The *2015 Order* contained no comparable provision of preemption.) As a result, this preemption *also* significantly disrupts states' regulatory authority under the longstanding, historical dual jurisdictional (joint federal-state) regulatory framework applied to telecommunications services in the U.S.

The states are responding to this preemption in various ways. Twenty-two state attorneys general filed appeals challenging the FCC's 2017 *Restoring Internet Freedom Order*, which have been consolidated with all other parties' appeals in the case now pending before the federal D.C. Circuit Court of Appeals. In addition, some states have introduced and/or passed legislation to impose their own network neutrality requirements, essentially inviting judicial litigation to test the validity of the FCC's preemption of state jurisdictional authority. The National Association of Regulatory Utility Commissioners (NARUC) is also considering how state regulatory commissions should respond to the FCC 2017 Order.

Emergence of Political Dysfunction in the U.S.

Examination of the political dysfunction that has arisen in the U.S. reveals that the instability of network neutrality policy is a reflection of the general, underlying dynamic of political dysfunction in the U.S. After describing fundamental aspects of the U.S. governance structure, this section briefly describes important changes in political party dynamics and their effects on policymaking. It also examines how corporations fomented political movements to take advantage of these changes in the context of extending the earlier deregulatory movement. The collective effects of these developments are heightened political dysfunction and policy instability. For network neutrality, these effects are manifested in policy outcomes that flip back and forth with shifts in the political party holding the majority of commissioners at the FCC before and after intermittent judicial rulings.

U.S. Governance Structure of Federalism and the Distinctive Role of the Judiciary

Political institutions are fragmented by design under the U.S. Constitution. Under a system of federalism, sovereignty is shared between the federal government and the states. The federal government is divided into three branches (legislative, executive, judicial) under the doctrine of separation of powers, with checks and balances among the three branches. Each state also has its own constitution, which likewise provides separation of powers among three branches of government with checks and balances among the branches. As to those matters for which the federal government is sovereign, federal law governs; and as to those matters for which states are sovereign, state law governs.

Under U.S. governance the judiciary has a distinctive role. In this respect, the judiciary has two important functions. One of the key functions of the judiciary is the common law judicial process, which the states retained and inherited from England. The federal judiciary established under the U.S. Constitution also subsequently utilized a common-law process. This process consists of judge-made law, on a case-by-case basis through the principle of case precedent. Thus, the common-law system is a lawmaking function that relies on the private enforcement mechanism of litigation.

Another key function is judicial review. Under judicial review, a court may invalidate laws and decisions by the legislative and executive branches—including administrative agencies—that are incompatible with a higher authority. In this regard, U.S. constitutional law is the highest authority, and statutes have higher authority than administrative law or common law. The federal judiciary has the definitive power under federal law, and the state judiciaries have the analogous power with regard to interpretation of their respective state law. As courts may not initiate cases, but rather cases must be brought before them, the judicial review function also relies on the private enforcement mechanism of litigation.

It is pursuant to the process of judicial review that federal courts are the final authority as to the constitutionality of federal law under the U.S. Constitution, and similarly the state courts are the final authority as to the constitutionality of state law under State constitutions. Given this primacy of judicial review, the framework of the U.S. Constitution and judicial doctrines of its interpretation that have evolved thereunder over time provide legal principles for determining when the federal government may or may not preempt state governments.

Adversarial Legalism Under Divided Government and Party Polarization

Under the governance structure briefly described in this section, the U.S. has developed policymaking through a process referred to as adversarial legalism.

"[A]dversarial legalism"—mean[s] policymaking, policy implementation, and dispute resolution by means of lawyer-dominated litigation. Adversarial legalism can be distinguished from other methods of governance and dispute resolution that rely instead on bureaucratic administration, or on discretionary judgment by experts or political authorities, or on the judge-dominated style of litigation common in most other countries. While the U.S. often employs these other methods too, it relies on adversarial legalism far more than other economically advanced democracies (Kagan 2001, p. 3).

Given its pervasiveness and deep roots in American political institutions and values, Kagan asserts that adversarial legalism "is best viewed not merely as a method of solving legal disputes but as a mode of governance" (2001, p. 5).

In the 1960s, with the rise of party polarization and divided government,[3] adversarial legalism began to increase dramatically through the creation of new federal and state statutory regimes reliant on private litigation for their enforcement (Kagan 2001, pp. 38–39; Farhang 2010, pp. 221–222). "The conjuncture of divided government and party polarization, primarily with Democratic legislators facing Republic presidents, over roughly the past four decades has been an important cause of

---

[3] Divided government describes a situation in which one political party controls the executive branch while another party controls one or both houses of the legislative branch.

the explosion of federal statutory litigation beginning at the end of the 1960s" (Farhang 2010, p. 222).

Between the mid-1960s and mid-1970s, federal and state governments enacted an unprecedented wave of regulatory statutes concerning pollution control, land use, consumer protection, and nondiscrimination in employment and education. They also mandated a more legalistic and punitive approach to regulatory enforcement, not only by regulatory officials but also by advocacy groups acting as "private attorneys general" (Kagan 2001, pp. 38–39, citations omitted).

As a result, both federal and state policymaking increasingly relied on litigation, within a governance structure that by design was already heavily dependent on private enforcement mechanisms.

Phase of Hyperpartisanship

Brownstein (2008, pp. 17–18) shows that relations between the Republican and Democratic parties in the U.S. have moved through four distinct phases during the history of modern U.S. politics: high partisanship (1896–1938); bipartisan negotiation (1938–mid-1960s); period of transition (mid-1960s–mid-1990s); hyperpartisanship (mid-1990s–to date). The rise of adversarial legalism described in this section on hyperpartisanship occurred during the third phase of transition.

Central to the current dynamic of political dysfunction is the fourth phase of hyperpartisanship, dating from the mid-1990s. "The defining characteristics of this age of hyperpartisanship are greater unity within the parties and more intense conflicts between them," arising largely from Republican backlash to post-World War II liberal policies (Brownstein 2008, p. 15). In addition, the parties are now divided by ideology rather than along arbitrary lines of religion, race, or economic status; and "[i]deologically, culturally, and geographically, the electoral coalitions of the two major parties have dispersed to the point where they now represent almost mirror images of each other" (Brownstein 2008, p. 16).

Moreover, the persistence of hyperpartisanship is enabled by gerrymandering, that is, the political manipulation of redistricting, creating a feedback loop that increases political tribalism through the primary election system that increases the ideological purity of winning party nominees (Friedman and Mandelbaum 2012, Chapter 12). Although gerrymandering virtually guarantees reelection of incumbents and thereby inhibits contestable interparty elections (Neuborne 2015, pp. 82–86), its coupling with hyperpartisanship creates intraparty competition of candidates within the primary election system and the resulting feedback loop of ideological purity that Friedman and Mandelbaum describe.

This hyperpartisanship is coupled with Americans also being closely divided, in that "[t]he ideological differences between the parties are as great as at any time in the past century. But the country is split almost exactly in half between the two sides" (Brownstein 2008, p. 19). This "is an unprecedented and explosive combination. Voters for the losing side always feel unrepresented when the other party wins unified control over government" (Brownstein 2008, p. 19). Yet, under this dynamic, "the *intensity* of belief among average voters isn't the principal problem" (p. 23, emphasis in original). "American politics ... is breaking down because too few political leaders resist the rising pressures inside the parties for ideological and partisan conformity that make it more difficult to bridge our disagreements. ... [T]he problem isn't too many ideologues but too few conciliators willing to challenge the ideologues, and partisan warriors on each side demanding a polarized politics" (p. 25). In other words, "What's unusual now is that the *political system* is more polarized than the country" (p. 25, emphasis in original).

With regard to the evolution of telecommunications policies, the Telecommunications Act of 1996 (TA96) was enacted toward the end of the third phase of transition when bipartisan consensus was still possible on some issues. However, as an implementation of TA96 continues in the 2000s, the combination of hyperpartisanship and closely divided Americans has created a political dynamic whereby FCC policy decisions regarding network neutrality shift back and forth—successive rulings reversing prior ones as to factual determinations and conclusions of law—as the political party holding the majority of FCC commissioners changes over time.

Rise of Market Populism as a Political Mythology

During the third phase of political party relations (mid-1960s–mid-1990s) described by Brownstein as a transition, market populism arose as a political mythology. Under market populism, "American leaders in the nineties came to believe that markets were a popular system, a far more democratic form of organization than (democratically elected) governments" (Frank 2001, p. xiv). Frank describes market populism as pro-corporate populism, which rode atop a 30-year backlash against the social and cultural changes of the 1960s. It arose from a corporate movement, abetted by "revolutionary" management theory, for the purpose of extending the deregulatory movement of the Reagan administration into the 1990s by politically and socially legitimizing corporations (Frank 2001, pp. 170–219).

To speak of pro-corporate populism is to raise one of the great political enigmas of the last thirty years. Historically populism was a rebellion *against* the corporate order, a political tongue reserved by definition for the non-rich and the non-powerful.... Populism was the American language of social class.

But beginning in 1968 this primal set piece of American democracy seemed to change its stripes. The war between classes had somehow reversed polarity. It was now a conflict in which the patriotic, blue-collar "silent majority" (along with their employers) faced off against a new elite, the "liberal establishment" and its spoiled, flag-burning children. This new ruling class—earned the people's wrath not by exploiting workers or ripping off family farmers, but by showing contemptuous disregard for the wisdom and the values of average Americans. The backlash erected an entire new social hierarchy according to which the "normal Americans" were at the bottom as usual—but the people at the top weren't the millionaires or the owners, they were those sneering kids who dodged the draft, along with their liberal parents and the various minorities and criminals those parents seemed so determined to pamper (Frank 2001, pp. 25–26, emphasis in original).

Frank opens his book with the observation that the passage of TA96 is an artifact of market populism.

In February 1996, Congress passed the Telecommunications Act, a typical economic artifact of the Age of Clinton [during which] [d]eregulation was one of the central tenets of the free-market faith of the nineties, something that both President Clinton and the Republican Congress agreed was in the best interests of all. For them the Telecommunications Act was just another great push in the triumphant rollback of "big government" that had already been underway for fifteen years. As such, it was passed by a huge majority acting in the finest spirit of bipartisanship (Frank 2001, p. ix).

Rise of the Tea Party Movement and the Mythology of Its Origins

Political dysfunction during the fourth phase of hyperpartisanship (mid-1990s–to date) intensified with the rise of the Tea Party movement within the Republican Party, which generally opposes excessive taxation and government intervention in the private sector. "Nearly every contemporary story or media account of the Tea Party movement in America describes it as a largely spontaneous, uncoordinated, populist movement that emerged in 2009 after Obama took office. ... It's a nice story. It's also a carefully crafted myth" (Nesbit 2016, pp. 24–25).

Contrary to the myth, Nesbit's book "reveals the Tea Party not as a sudden emergence due to reactionary movements against big government but rather the behind-the-scenes, secret strategy of wealthy corporations and individuals that began in the early 1990s to control the GOP. It uncovers the hidden alliances made to further that purpose" (2016, pp. 3–4). Rather, "[t]he truth is that the funding mechanisms, central messages, and key pillars of the Tea Party movement were many years in the making. It is not a true grassroots movement but was strategically built to deflect anger away from corporations and redirect it at the government. A network had already been developed, funded by Koch and tobacco industry money, and only awaited a suitable opponent, which they found in Barack Obama" (p. 25).[4]

As Nesbit recounts in detail, the basic infrastructure of the network that would become known as the Tea Party movement can be traced to the Koch Industries' founding in 1984 of Citizens for a Sound Economy, a hybrid nonprofit organization engaged in educational and political activities, to create grassroots support for deregulation, corporate tax cuts, and other right wing, corporate causes (2016, pp. 11–17).

While the Kochs clearly didn't much like party politics, they chose the Republican Party as the likeliest target of opportunity for the Libertarian brand. ... In their political utopia, there would be no national government—or one that was a small fraction of its size and scope today. At a minimum, the Kochs hoped to create a movement within the Republican Party that reflected their own philosophy—that the best government was a very limited government (Nesbit 2016, p. 11).

In the 1990s, two of the largest tobacco companies, Philip Morris and RJ Reynolds, joined forces with the Kochs "to create antitax front

---

[4]Koch refers to the Koch brothers, Charles and David, who own Koch Industries, the largest private oil company in the world.

groups in a handful of states that would battle any tax that moved" (Nesbit 2016, p. 13). However, the tobacco companies' involvement did not become public for almost twenty years.

What didn't become public until nearly twenty years later was that these themes of a Tea Party anti-tax, anti-regulation, and anti-government revolt were then developed almost simultaneously by two of the largest tobacco companies—Philip Morris and R.J. Reynolds—under the guise of political and business coalitions to fight excise taxes of all sorts, including cigarette taxes. In successive phases in the 1990s, with the Kochs' CSE as its core mobilization network partner, Philip Morris and RJR helped create state-based anti-tax and anti-regulation propaganda campaigns such as Get Government Off Our Back, Enough is Enough, and the Coalition Against Regressive Taxation (Nesbit 2016, p. 14).

It is during the fourth phase of hyperpartisanship that FCC decisions related to network neutrality began; and as hyperpartisanship intensified with the rise of the Tea Party movement, so did the instability of FCC policies as the majority of FCC commissioners shifted back and forth between the Republican and Democratic parties.

Conclusion: Underlying U.S. Political Dysfunction Drives Instability of Network Neutrality Policy

Upon reexamining the trajectory of U.S. network neutrality policymaking described in section "Instability of U.S. Network Neutrality Policy" within the context of the more general trajectory of changing political dynamics and dysfunction discussed in section "Emergence of Political Dysfunction in the U.S.", it becomes clear that the former is a microcosm of the latter.

The Telecommunications Act of 1996 (TA96) was enacted by Congressional legislation during the late stage of the third phase of political party relations, referred to as the phase of transition between the second phase of bipartisanship and the fourth phase of hyperpartisanship. Bipartisanship agreement on deregulatory telecommunications policy in TA96 was facilitated by the rise of pro-corporate, market populism as a political mythology during the 1990s. TA96 modified the preexisting regulatory framework based on joint federal–state jurisdiction, reflective of the U.S. governance structure of federalism, with delegation of regulatory authority to the FCC and state commissions.

Enforcement of and further policymaking within TA96 heavily relies on adversarial legalism, which is a mode of governance that has increased dramatically during the rise of party polarization and divided government. Moreover, implementation of TA96 under heightened adversarial legalism has and continues to occur during the fourth phase of hyperpartisanship between the political parties. As for network neutrality policy in particular, the partisan legal battle at the FCC over the classification of BIAS as a telecommunications or information service began in the early 2000s during this phase of hyperpartisanship. Furthermore, it is this conjuncture of adversarial legalism and hyperpartisanship that has created the politically volatile conditions underlying the trajectory of inconsistent FCC rulings related to service classification and imposition of open internet rules. As the political party holding the majority of FCC commissioners shifts between the Republican and Democratic parties, so too do the rulings, where the opportunity or necessity for further rulings is often triggered by federal appellate court decisions upholding or reversing prior FCC rulings. In addition, the two most recent FCC rulings in 2015 and 2017 occurred after the emergence of the Tea Party Movement. The rapidity with which the FCC's *2015 Order* under a Democratic majority of commissioners was reversed by the *2017 Order* under a Republican majority reflects the increased intensity of hyperpartisanship under Tea Party influence, particularly given that in 2016, in *US Telecom Association v. FCC*, the D.C. Circuit Court of Appeals had just *upheld* the FCC's *2015 Order* in its entirety.[5]

The *2017 Order* is significant not only in the rapidity of its reversal of the *2015 Order* but also in its substance. First, the *2017 Order* reversed FCC policy by reclassifying BIAS as information services, thereby reducing the scope of FCC (federal) jurisdictional authority, and vacating most of the 2015 Open Internet rules (other than those related to transparency). Adversarial legalism continues as the legal validity of the *2017 Order* has already been challenged and an appeal is pending before the D.C. Circuit Court of Appeals. Furthermore, reflective of the current phase of hyperpartisanship in Congress, a Democratic-controlled House

---

[5] By contrast, FCC rulings in 2010 and 2015 under a Democratic majority of commissioners were triggered by D.C. Circuit Court of Appeals decisions in 2010 and 2014 identifying legal flaws in asserted FCC jurisdiction under the analysis grounded in classification of BIAS as an information service.

has passed the Save the Internet Act to overrule the *2017 Order*, whereas the Republican Majority Leader of the Senate has declared this legislation dead on arrival.

Second, the *2017* Order went further by asserting a broad scope of federal preemption, prohibiting any state or local government measures that would effectively impose rules or requirements that the FCC has chosen to preempt or decided to refrain from imposing under its order. In so doing, the *2017 Order* disrupts longstanding state regulatory authority rooted in the U.S. Constitution framework of federalism. The scope of this federal preemption is perceived as being particularly, politically provocative. As a result, this federal preemption is fomenting a backlash of further partisan policymaking and adversarial legalism within the states to challenge the federal encroachment on their sovereign powers. As of mid-2019, five states have enacted state network neutrality laws (California, New Jersey, Oregon, Vermont, Washington), and approximately thirty states have or are considering some form of legislative action. Many, but not all, of the states that have chosen not to act are Republican controlled (Rogers 2018). In this way, a new phase of federal–state policy instability has ensued, escalating tensions between the federal and state governments.

Thus, reflective of the current phase of hyperpartisanship in the U.S., the instability of network neutrality is ongoing. Until the underlying political dynamics change, the policy instability will likely continue. Moreover, given the lack of any apparent noticeable shift in such dynamics in the near future, the instability could persist for an extended period of time.

Epilogue

After this paper was written, the D.C. Circuit Court of Appeals released its decision in *Mozilla v. FCC*, the appeal of the 2017 *Restoring Internet Freedom Order*. Although it is beyond the scope of this paper to examine *Mozilla v. FCC* in detail—given its length (of 186 pages) and complexity of factual and legal analyses—some observations are provided here as to how the decision is unusual, reflective of the underlying political instability of the network neutrality debate.

As to its form, the court's decision in *Mozilla v. FCC* was released *per curiam*. This means it is unsigned, and reflects the will of the court collectively rather than that of any particular judge. The *per curiam* portion of the opinion is 146 pages, followed by a significant concurring opinion by Judge Millett, a one-paragraph concurring opinion by Judge Wilkins, and a substantive dissent by Judge Williams.

A *per curiam* opinion under the circumstances of this case is unusual, and may be considered even controversial. As Robbins (2012) explains, "the author of a *per curiam* opinion is meant to be institutional rather than individual, attributable to the court as an entity rather than to a single judge… [and for] *de facto* courts of last resort such as the federal courts of appeals, should be limited to a very narrow class of opinions in which the use of formulaic, boilerplate language has already extinguished any sense of individuality" (p. 1199). However, "when separate opinions follow a *per curiam* decision …, the existence of separate opinions reveals that the matter is not routine and well-settled, but rather that some important aspect of the case is subject to conflicting interpretations. When the separate opinion is a concurrence, this suggests that the court could have reached the result in a different manner. A dissent, on the other hand, calls into question both the court's reasoning and the result…[T]he result is not automatic and reasoning is not formulaic" (Robbins 2012, pp. 1240–1241, footnotes omitted). The controversiality of a *per curiam* opinion in such circumstances is illustrated by the U.S. Supreme Court's *per curiam* opinion in *Bush v. Gore* (2000).

[T]he Supreme Court in *Bush v. Gore* issued a *per curiam* to defuse charges of political bias in deciding the winner of the 2000 presidential election. Despite the appearance of unity provided by the *per curiam* label, there was publically acknowledged discord among the Justices as evidenced by the accompanying five signed opinions. Critics charged that the use of the *per curiam* in this case was improper because it "failed to package the majority opinion as a restrained solution, based on a substantial consensus" (Robbins 2012, pp. 1204–1205).

> Review of the substance of the *per curiam* opinion in *Mozilla v. FCC*, in juxtaposition with the analyses in the concurring and dissenting opinions, indeed reveals not only lack of unanimity among the judges but also the deep reservations by the majority of judges in the opinion. Furthermore, the reservations and disagreements among the judges, briefly described below, also align with the political party affiliations of their appointments to the D.C. Circuit Court of Appeals: the judges asserting substantial reservation with the *per curiam* opinion, Judge Millett and Judge Wilkins, were appointed by President Obama, a Democrat; and the

judge dissenting in part with the *per curiam* opinion, Judge Williams, was appointed by President Reagan, a Republican. By contrast, the opinions in the earlier network neutrality cases – *Comcast Corp. v. FCC* and *Verizon v. FCC* – were signed (not *per curiam*). In comparison, the *per curiam* opinion in *Mozilla v. FCC* can arguably be seen as an attempt to mask the hyperpartisanship underlying the political party appointments of the judges.

With regard to the substance of the *per curiam* opinion in *Mozilla v. FCC*, first, the Court upholds the FCC's classification of BIAS as an information service; but two of the three judges—that is, the majority—did so with substantial reservation. Judge Millett opens her concurring opinion (of 16 pages) with the following paragraph.

I join the Court's opinion in full, but not without substantial reservation. The Supreme Court's decision in *National Cable & Telecommunications Ass'n v. Brand X Internet Services*, 545 U.S. 967 (2005), compels us to affirm as a reasonable option the agency's reclassification of broadband as an information service based on its provision of Domain Name System ("DNS") and caching. But I am deeply concerned that the result is unhinged from the realities of modern broadband service (J. Millett, concurring opinion, p. 1).

Judge Wilkins, in his own concurring opinion (comprising solely of the following paragraph), agrees with Judge Millett.

I join the Court's opinion in full. As Judge Millett's concurring opinion persuasively explains, we are bound by the [U.S.] Supreme Court's decision in *National Cable & Telecommunications Ass'n v. Brand X Internet Services*, 545 U.S. 967 (2005), even though critical aspects of broadband internet technology and marketing underpinning the Court's decision have drastically changed since 2005. But revisiting *Brand X* is a task for the Court—in its wisdom—not us (J. Wilkins, concurring opinion, p. 1).

Thus, the majority of the judges clearly state that they consider themselves bound by previous U.S. Supreme Court precedent, notwithstanding the apparent conflict between the legal result and critical aspects of current broadband internet technology and marketing. Yet, their further assertion, that the task lies with the U.S. Supreme Court to address their reservation, is highly significant. Given that the U.S. Supreme Court's acceptance of such cases upon appeal is purely discretion, and given the stature of the D.C. Circuit Court of Appeals in federal regulatory matters, such statements can be readily interpreted within the legal community as signals both for parties to consider appeal to the U.S. Supreme Court as well as for the U.S. Supreme Court to accept such as appeal.

Second, although the *per curiam* opinion in *Mozilla v. FCC* upholds the FCC's classification of BIAS as an information service, it also vacates the FCC's preemption of state and local measures to regulate.

[T]he Court concludes that the Commission has not shown legal authority to issue its Preemption Directive, which would have barred states from imposing any rule or requirement that the Commission "repealed or decided to refrain from imposing" in the Order or that is "more stringent" than the Order (*Per curiam*, slip opinion, p. 13).

Thus, the states may impose rules or requirements that the FCC has declined to impose. However, there is disagreement among the judges as to this holding. Judge Williams, who does not have reservations with regard to upholding classification of BIAS as an information service, dissents from this portion of the *per curiam* opinion. Judge Williams would have affirmed the FCC's preemption directive given that the FCC's classification is being upheld.

Taken as a whole, in upholding the FCC's service classification but vacating its preemption directive, the *per curiam* opinion in *Mozilla v. FCC* effectively "splits the baby" among the litigants. In so doing, the *per curiam* opinion as to its form belies a tenuous compromise in substance among the judges. Moreover, *Mozilla v. FCC* does not provide closure but merely prolongs the present instability of U.S. network neutrality policy. There are multiple potential forums for further developments: future revisions or actions by the FCC; Congressional legislation to clarify ambiguities in TA96; states passing their own network neutrality laws; and the possibility of an *en banc* rehearing by the whole D.C. Circuit Court of Appeals or an appeal to the U.S. Supreme Court. Yet, given the prevailing hyperpartisanship in the U.S., the likelihood of a politically stable resolution to network neutrality policy arising from one or more of these forums appears remote.

REFERENCES

Brownstein, Ronald. 2008. *The Second Civil War: How Extreme Partisanship Has Paralyzed Washington and Polarized America*. New York, NY: Penguin Books.

Cherry, Barbara A. 2010. Consumer Sovereignty: New Boundaries for Telecommunications and Broadband Access. *Telecommunications Policy* 34: 11–22.

Cherry, Barbara A. 2015. Technology Transitions Within Telecommunications Networks: Lessons from U.S. v. Canadian Policy Experimentation Under Federalism. *Telecommunications Policy* 39: 463–485.
Comcast Corp. v. FCC, 600 F.3d 642 (D.C. Cir. 2010).
Declaratory Ruling, Report and Order, and Order. 2017. In the Matter of Restoring Internet Freedom, WC Docket No. 17-108, FCC 17-166, Federal Communications Commission.
Farhang, Sean. 2010. *The Litigation State: Public Regulation and Private Lawsuits in the U.S.* Princeton, NJ: Princeton University Press.
Frank, Thomas. 2001. *One Market Under God: Extreme Capitalism, Market Populism, and the End of Economic Democracy.* New York, NY: Anchor Books.
Friedman, Thomas L., and Mandelbaum, Michael. 2012. *That Used to Be Us.* Publisher: Picador, Reprint edition.
Kagan, Robert A. 2001. *Adversarial Legalism: The American Way of Law.* Cambridge, MA: Harvard University Press.
Mozilla v. FCC, Case No. 18-1051, __ F.3d __ (D.C. Cir. 2019).
Nesbit, Jeff. 2016. *Poison Tea: How Big Oil and Big Tobacco Invented the Tea Party and Captured the GOP.* New York, NY: Thomas Dunne Books.
Neuborne, Burt. 2015. *Madison's Music On Reading the First Amendment*, 1st ed. New York, NY: The New Press.
Report and Order on Remand, Declaratory Ruling, and Order. 2015. In the Matter of Protecting and Promoting the Open Internet, GN Docket No. 14-28, FCC 15-24, Federal Communications Commission.
Robbins, Ira P. 2012. Hiding Behind the Cloak of Invisibility: The Supreme Court and *Per curiam* Opinions. *Tulane Law Review* 86: 1197–1242.
Rogers, Kaleigh. 2018. Which States Have Net Neutrality Laws? *Motherboard, vice.com*, April 9. Motherboard. Retrieved from https://motherboard.vice.com/en_us/article/ywx5pw/which-states-have-net-neutrality-laws?utm_campaign=Newsletters&utm_medium=email&utm_source=sendgrid.
US Telecom Association v. FCC, 825 F.3d 674 (D.C. Cir. 2016).
Verizon v. FCC, 740 F.3d 623 (D.C. Cir. 2014).

CHAPTER 14

Multisided Markets and Platform Dominance

James Alleman, Edmond Baranes, and Paul N. Rappoport

INTRODUCTION

There can be no doubt that the GAFAM companies[1]—Facebook, Amazon, Apple, Microsoft, and Google, as well as Twitter—have transformed society since their emergence; the changes wrought by their services have had ripple effects that are both positive and negative. On the positive side, soaring consumer access to information, news, has been stimulated by the ever-more ubiquitous and falling prices of broadband

---

[1] Facebook, Amazon, Apple, Netflix, and Google are referred to by the acronym FAANGs. In Europe, Microsoft is added to the list and Netflix is eliminated. They are referred to as GAFAMs.

J. Alleman (✉)
University of Colorado, Boulder, CO, USA
e-mail: james.alleman@colorado.edu

E. Baranes
University of Montpellier, Montpellier, France

P. N. Rappoport
Temple University, Philadelphia, PA, USA

© The Author(s) 2020
J. Alleman et al. (eds.), *Applied Economics in the Digital Era*,
https://doi.org/10.1007/978-3-030-40601-1_14

fixed and mobile bandwidth. E-government has transformed the delivery of public services.

However, negative effects have likewise been stark. Certainly, there have been huge disruptions caused by e-commerce. Retail industries, industrial supply chains, banking, and publishing are just a few obvious examples. Brick and mortar stores are closing. State tax collectors are fighting the loss of sales tax collections.

The internet giants give rise to three issues: privacy, threats to democracy, economic dominance. Privacy is not simply the collection of data for commercial use to target ads to users, but has been used in more malevolent ways: political manipulation, collecting data from children, and government surveillance. The Cambridge Analytica was a wakeup call on the malevolent use of users' data. The use of social media on Facebook and Twitter, in particular, to spread misinformation and facilitate fraud has raised legitimate concerns about their responsibility for undermining democratic institutions, instigating cyber-bullying, enabling identity theft, and distorting public opinion.

On the other hand, social media has also facilitated the Arab Spring, the Orange Revolution, and March for Our Lives. Policy makers and regulators are caught between conflicting values of free speech and expression on the one hand and the desire to mitigate these and other injuries to social and government institutions on the other. As important as these issues are, we do not address them in this paper except to note that their significant market power (SMP) and dominance can magnify, aid, and abet these malevolent actions. Here, we are concerned with the economic consequences of the SMP of the internet giants. What is the cost to the consumers and society of this market power? It is significant—approximately $54 billion in the case of Alphabet (Google) and $33 billion in the case of Facebook in 2018, for a total of $87 billion.

Because Facebook and Google and other internet giants are two-sided markets their economic rents are "hidden" from the public. On the user-side of the market, prices are zero—"free." On the other side of the market, Facebook's and Google's revenues are derived from advertising which appears when the users go to various web sites. Facebook and Google can extract exorbitant prices for ads, since they are virtually the only source that can target ads directly to potential customers. This is where the economic rents are not so obvious. Since 1980, the antitrust authorities have focused, with few exceptions, on the users'-side of the market, which in the case of internet giants, is free. Where is the price gouging? On the other side of the market!

The Supreme Court's decision in *American Express* addressed a two-sided transactional market, not the case of a "media" platform which is non-transactional such as Facebook and Google and many other internet platforms (*Ohio et al. v. American Express Co. et al.* 2017; Katz 2018). Each internet platform has a different business plan e.g. Amazon is much different from Google and Facebook, the focus of this paper.

The European Commission has alleged that "...Google treats and has treated more favorably, in its general search results pages, Google's own comparison-shopping service 'Google Shopping' and its predecessor service 'Google Product Search' compared to rival comparison-shopping services" (European Commission 2015). More recently it was fined $1.5 billion Euros for antitrust violations in the advertising practices. Its third fine since 2017 (Satariano 2019) in 2018 was $5 billion (SEC Alphabet 2018).

The United States has not been as aggressive, in large part due to the deregulation trend in the eighties and Bork's *Antitrust Paradox* (1978) book which argued that unless consumer welfare and competition were adversely impacted, there was no need for antitrust intervention. This paper shows that both these conditions exist in the internet space, despite the hands-off approach of the US regulatory authorities and acceptance in many judicial decisions.[2]

The paper is divided into six sections beginning with this section. The next section briefly describes the internet platform markets: Why and how Facebook and Google are an economic threat and why economic incentives promote anticompetitive behavior. The third section develops the theory of two-sided market for the platforms. The fourth section estimates the economic cost of these monopoly platforms; this paper focuses on Facebook and Google, but the methodology could be applied to other dominant platforms. The fifth section examines potential remedies and solutions, and why remedies require more than the internal controls that have been proposed by the firms. The last section summarizes and makes tentative conclusions.

---

[2] Bork's book has many critiques—among both economists and lawyers. See Khan (2017).

THE ECONOMIC THREAT OF DOMINANT INTERNET PLATFORMS

Network Externalities

Facebook, Google, and other internet giants gain their position from the existence of positive externalities, i.e. network effects, such that the service becomes more valuable to users, the more users there are.[3] Adding users makes the service more valuable, which attracts more users, which makes it more valuable, etc. The expansion of network effects is underpinned by a business model that provides their services free to end-users and relies on advertising revenues for profits.[4]

The Markets[5]

The percentage of the population that uses social media has grown dramatically over the last decade. In the United States, as of 2015, data indicate that over eighty percent (81%) of the adult population use social media as shown in Fig. 14.1 (Statista 2018). When one examines what sites are visited, Facebook dominates, with twice as many visits as any other site (Fig. 14.2).[6]

Facebook also dominates the users of social media in the United States. Nearly eighty percent (79%) of users belong to Facebook while less than one-third of the users have accounts on the other social media sites. In these terms, Facebook dominates the field, and this is reflected in its revenue of nearly 40 billion dollars (Fig. 14.3). Its latest results show a growth of revenue of 40%. Facebook is so dominated that Google+, Google's social media platform exited the market even though it ranked fourth in visits in 2015 (Welch 2019).

With respect to search, Google dominates with over three-quarters (75.8%) of the search advertising revenue in the United States (see Fig. 14.4). Some estimates suggest it is even higher, approximately ninety percent. On the other hand (being good economists), Khan

---

[3] That is not to say that these firms do not exhibit economies of scale and scope, but this is not their main competitive advantage.

[4] Other platforms have different business strategies, for example, Amazon (Khan 2017).

[5] For more details on the size, distribution, and other characteristics of the Internet Platform giants, see Alleman and Taschdjian (2019).

[6] Recall, Facebook owns Instagram, which ranks fifth in visits.

14 MULTISIDED MARKETS AND PLATFORM DOMINANCE 307

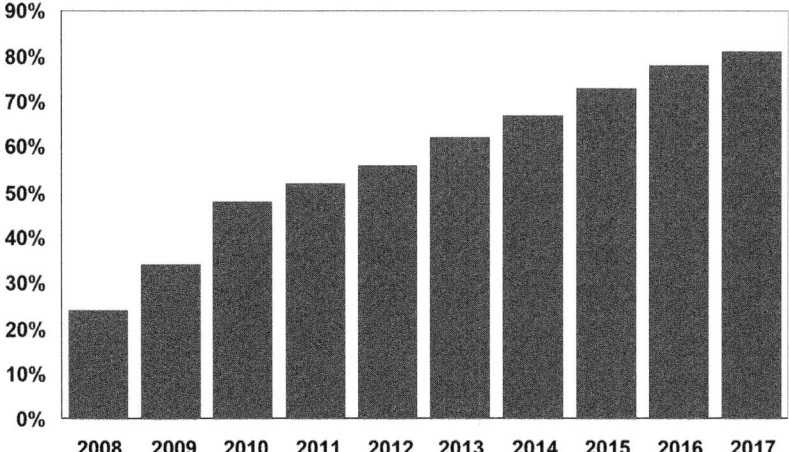

Fig. 14.1 Population who use social media in the U.S. (*Source* Social media usage in the United States [https://www-statista-com.colorado.idm.oclc.org/study/40227/social-social-media-usage-in-the-united-states-statista-dossier/])

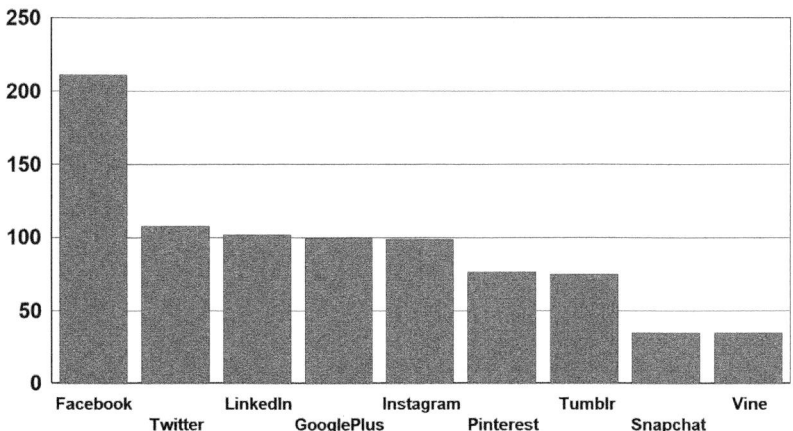

Fig. 14.2 Visitors to social network site in the U.S. (*Source* Social media usage in the United States [https://www-statista-com.colorado.idm.oclc.org/study/40227/social-social-media-usage-in-the-united-states-statista-dossier/])

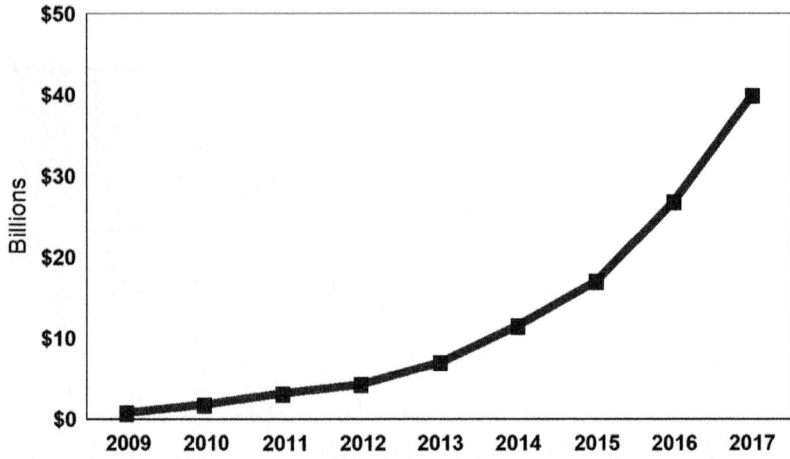

Fig. 14.3 Facebook's advertising revenue (*Source* Social media usage in the United States [https://www-statista-com.colorado.idm.oclc.org/study/40227/social-social-media-usage-in-the-united-states-statista-dossier/])

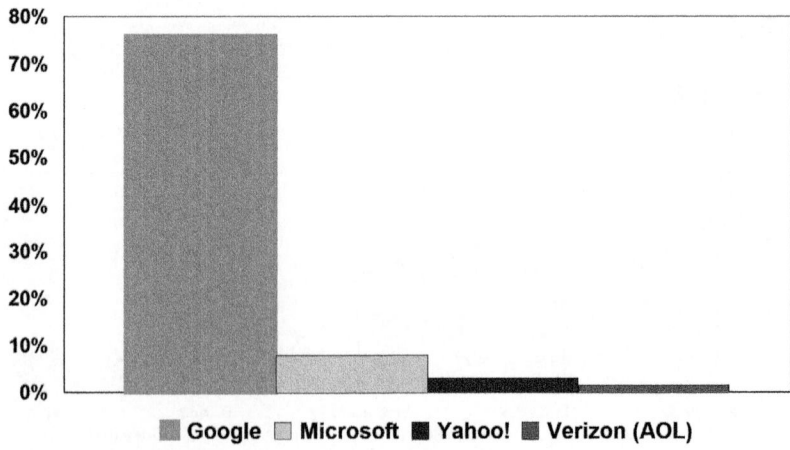

Fig. 14.4 Search market revenue share, U.S. (*Source* Social media usage in the United States [https://www-statista-com.colorado.idm.oclc.org/study/40227/social-social-media-usage-in-the-united-states-statista-dossier/])

(2018) reports that Amazon is Googles' biggest rival in terms of search. Forty percent of product search begins with Amazon (Khan 2017).

The percentage of digital advertising has been growing, primarily, at the expense of TV and newspapers. Digital advertising has doubled between 2010 and 2015—growing to one-third of the US market as of 2015. At that growth rate it could reach 50% of the United States' total advertising revenue (Statista 2018).

Two-Sided/Multisided Markets

Under the Chicago School doctrine as expressed by Bork (1978, 1993), the Department of Justice (DOJ) and Federal Trade Commission (FTC) do not see any consumer harm in the dominance of internet platforms, since the consumers' price is zero with Facebook and Google (Khan 2017; Bork 1993).[7] They are not, apparently, aware of the research over the last decades on two-sided and multisided markets (Parker *et al.* 2005).

A two-sided market is one in which the firm sells to the consumer on one side and the advertiser, for example, on the other side.[8] Newspapers are among many examples. In some cities, newspapers are free to the consumer, but "paid for" by the advertisers. In other cases both sides of the market have a positive price. The former, free to the consumer, represents the Facebook–Google model: The latter represents major newspapers or the online *Wall Street Journal* model.[9] What the DOJ and FTC policy have failed to consider is the second side of the market. The policy does not consider the harm done to the consumers by the high price of the advertising as shown below. These prices get reflected to consumers in the higher prices they must pay for the goods and services advertised.

With network effects, the rationalization for "free-to-the-user" is amplified. Since the user is, presumably, more price sensitive, lower

---

[7] The 2019 $5 billion settlement with Facebook was based on privacy violations, not economic issues (Kang 2019).

[8] For simplicity, the analysis is developed as a two-sided market, but more correctly they are multisided markets (MSM). See Evans and Noel (2008), Evans and Schmalensee (2012), Katz (2018), Katz and Sallet (2018), OECD (2018b).

[9] Bosteon (2019) surveys the economic and legal aspects of "zero-priced" markets.

or "zero-priced" on this side of the market will increase the number of users, and the value to the advertiser—the second side of the market. Google and Facebook do not have to be concerned with advertisers switching to other platforms since they are virtually the only game in town.

Google and Facebook dominate worldwide digital advertising revenues. In the United States, Google/YouTube and Facebook/Instagram together garnered 63.1% of net digital ad revenues of $86.4 billion in 2017 (eMarketer 2017).

The small-but-significant-and-non-transitory-increase-in-price (SSNIP) test, which is normally applied to test for market power as well as to define relevant markets, is not helpful in the case of retail services that have zero prices.[10] But it is applicable to advertising. If Facebook or Google were to permanently increase the price for advertising, would advertisers shift their purchases, either to competing venues or by reducing advertising overall? Since the platforms' ads are targeted with laser-like focus, the advertisers are not likely to leave since they do not have a good substitute for the platform.

Analysis of Harm

Monopoly Prices

In a two-sided market, an internet platform like Google serves as a digital advertising intermediary between sellers and potential customers through its control over ad spending. Because of its unique focus on customers' demographics, Google is able to exercise monopoly power through an auction of prime ad space to sellers who want "front page" access to potential customers.[11] In so doing, the internet platform obtains billions of dollars at the expense of sellers and indirectly from the consumers of the products advertised.[12]

---

[10] In contrast to Amazon's announced increase in the price on Amazon Prime by nearly twenty percent (Stewart 2018).

[11] The "front page" is the first screen the viewer sees.

[12] To understand how the AdWords auction works see https://www.wordstream.com/articles/what-is-google-adwords or https://blog.tryadhawk.com/google-adwords/how-google-adwords-works/. From the source, see https://ads.google.com/home/.

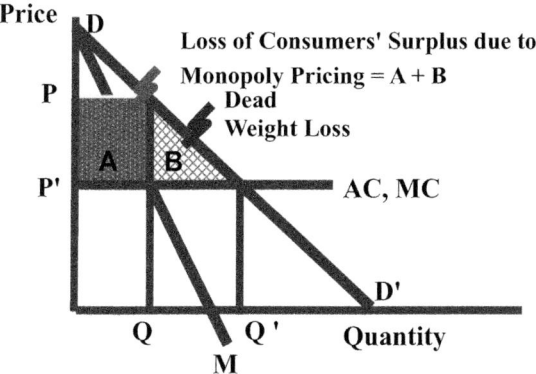

Fig. 14.5 Platform effect: demand for internet advertising (*Source* Authors' construction)

To illustrate the economic harm that a two-sided monopoly can do, we simplify the model for advertising prominence on a digital platform like Google. On the consumer-side of the market, the access for search is "free," except for ads, pop up, and other click-bait distractions. On the sellers' side of the market, there is access to millions of customers and laser-focused placement of advertising and top billing for the sellers from customers' searches. This comes at a direct cost to sellers and indirect cost to customers. In this advertising market, a firm like Google charges dynamic-monopoly prices.[13] Assume the marginal cost is constant, not an unrealistic assumption when the firm has tremendous economies of scale coupled with network effects (Shapiro and Varian 1997).[14] Using traditional microeconomics techniques, the maximum economic profit is determined, as shown in Fig. 14.5 i.e. at the point that marginal cost (MC) equals marginal revenue (MR), profit is maximized by setting

---

[13] Google, for example, holds auctions continually, to set the price paid for advertising (Varian 2019). Thus, it can extract the greatest revenue out of this side of the market on an ongoing basis. It is practicing first-degree price discrimination on a continuous basis. This is the best of all possible worlds for a monopolist.

[14] Positive network effect shifts the derived demand for advertisers to the right, that is as more subscribers are added to the platform, the more valuable the platform is to the advertisers.

the price at P and selling Q. The area A represents the economic profit. In contrast, the social welfare maximizing price would be P' and the quantity sold would be Q'.[15]

Increase in Product Prices

In terms of welfare, the loss of consumers' surplus is the area under the demand curve above the marginal cost curve plus the over-charging by the internet platform in Fig. 14.5—the areas A plus B. Thus, the total economic loss of the monopoly pricing are the areas A plus B. Although indirect, this harms the consumers (as will be shown below), although this harm seems to go unrecognized by the antitrust authorities.

Okay, so you charge the advertisers a lot, what is the harm? Who pays for the advertising? In the long run, the consumers pay in the price of the products advertised (not rocket science). The price of these products is increased by the cost of the advertising. Simplify, once again, the additional cost that is added to the price of the products which were advertised, as shown in Fig. 14.6. (Note this is the seller's market, a different market from Fig. 14.5.) The price has to be increased for P' to P to cover the cost of the advertising. The increased cost is a social loss of A plus the consumers' surplus loss of B".

Via this method the monopoly rents are borne by the consumers.

By identifying both sides of the market the analyses identifies consumers' harm and follows the Chicago school of antitrust, although the Chicago school usually only examines one side of the market. The analysis does not negate the Brandeisian Movement which is concerned with the harm caused by undue market power including fewer innovations, stifling of entrepreneurs, squeezing suppliers, and crushing labor not simply a focus on efficiency concerns (Khan 2018). Brandeisian concerns only amplify the problems with market power in the digital space discussed above.

---

[15] In perfect competition (and other conditions), the pricing at marginal cost gives the best possible outcome for the society (if perfect competition ever existed except in the minds of the economists); hence the "welfare maximizing" price. The issue is more complicated when economics of scale and scope exit. See Alleman and Rappoport (2006).

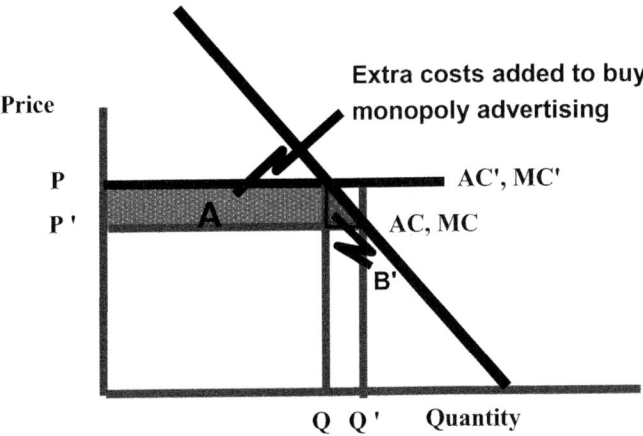

Fig. 14.6 Product effect: extra cost of product due to monopoly (*Source* Authors' construction)

Elimination of Competition

But this is not the only harm done by this monopoly market structure. Potential entrants will have difficultly entering the product market, since they must pay the cost of advertising to the monopoly. While only a sample of one, over 30% of the total operating cost for one start-up is the price it pays for Google AdWords.[16] Others have had similar complaints (Nicas and Weise 2019). This is a significant amount for any company, but in the case of Google, a click does not necessarily mean a conversion to revenue for the start-up, only for Google. Every click costs the start-up, but does not mean the "clicker" uses the service or buys the product. Potential start-ups may not have sufficient funds to cover this additional cost and are driven out of business or find it too expensive to even enter the market.

Since competition is reduced through the elimination of less capitalized firms, the remaining product market firms gain significant market

---

[16] Discussion with John Korbel, a StoreMe director, 9 April 2019. StoreMe is an on-demand luggage storage start-up serving the major cites on the east coast. See https://getstoreme.com.

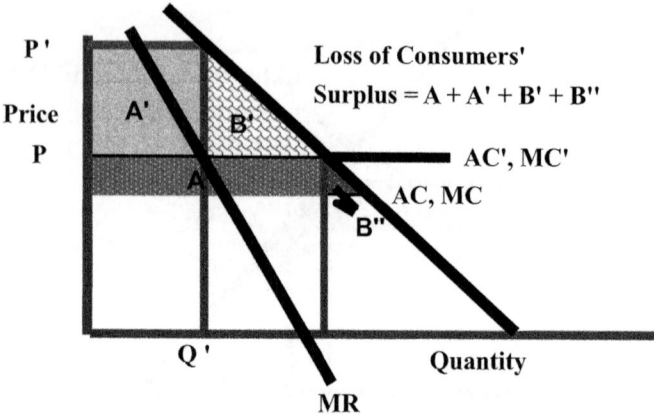

Fig. 14.7 Extra cost to consumers due to the elimination of competition (*Source* Authors' construction)

power (leveraged by the platform's monopoly prices) and, in the long run, will pass on these advertising costs to consumers through their own monopoly pricing. The resulting profits in the product seller's market as illustrated in Fig. 14.7 are A' (which reproduces Fig. 14.6 and adds the marginal revenue curve to determine maximum profit for the seller). The total loss in consumer surplus is the area $A'+B'+A+B''$ had the markets been competitive. Free is not free, it is an illusion.

This is not the only loss. How many innovations were not developed or lost due to the suppression or usurping of potential competitors?[17] What new businesses did not succeed or did not even get started due to the dominance of the internet giants? What are the GAFAM's impact on economic growth? Serious issues, but difficult to quantify.

To summarize, at a minimum, the social cost of the platform monopoly is:

- The economic rent (profit) extracted by the internet platform from advertisers and the loss of consumers' surplus;
- The increased cost to consumers of the products advertised on the internet platform, and to the extent it thwarts competition;

---

[17] See Kwoka (2014) for how past mergers retarded innovation and other deleterious non-price effects.

- The increased market power of the sellers because of the reduction of competition, thus resulting in higher prices to consumers and an additional loss of consumers' surplus. This is continuing, ongoing additional loss to consumers.

Estimation of Welfare Loss

A first approximation of the welfare loss is the (economic) profit of the internet platform, since this is derived from its monopoly position. This is large, $35.8 and $22.1 billion for Google and Facebook, respectively (SEC Alphabet 2018; SEC Facebook 2018).[18] This is represented by the area A in Fig. 14.5. But this is only part of the story; the total loss must consider the "dead-weight" loss, half of the profits under the assumption of a linear demand curve. The consumers' surplus is the profit plus the dead-weight loss, estimated below.

The loss of consumers' surplus was calculated based on SEC data for Google and Facebook. With the assumptions of a linear demand curve and a constant marginal cost (MC), this is enough to make an estimate of consumers' loss.

The revenues, costs and profits from Google and Facebook were obtained for 2018 from companies' Securities and Exchange Commission's 10-K filing (SEC Alphabet 2018; SEC Facebook 2018). Profits come directly from the companies' 10-Ks. It is represented as area A in Fig. 14.5. A constant marginal cost was assumed. This is a realistic assumption, given the strong economies of scale and network externalities. Since the marginal revenue (MR) exactly bisects the marginal cost (MC), the dead-weight loss is exactly half of the profits.

With the linear demand curve, the difference between the price (P) and marginal cost (MC) gives the height of the dead-weight loss calculation, and the linear demand curve means the width of the dead-weight loss will be equal to the quantity (Q). Then one has to simply calculate the area of a triangle to find the dead-weight loss, area B in Fig. 14.5. Add this to the total profits to determine the total loss in consumers' surplus, areas A+B. The results were significant—approximately $53.7 billion in the case of Google and $33.2 billion in the case of Facebook

---

[18] Alphabet's net income was $30.7 billion, but it had an EU fine of $5.1 in 2018. This was added to the income, since, presumably, it was unanticipated.

in 2018.[19] Recall, this is an annual estimation, so the loss continues year after year. This is the loss to consumers due to the monopoly of these two internet giants. To put this in perspective, it represents over 11% of online purchases for the United States in the same year (Internet Retailer 2019).

Remedies and Solutions

As noted in the introductions, the internet giants need some form of social control not only for the economic issues: monopoly profits, predatory pricing, elimination of competition, etc. but also for their threats to democratic and privacy (Alleman 2018; Wu 2016). The controls of the latter are important issues and can be handled, *inter alia*, by such tools as the "fairness doctrine," identifying sources of advertisements, or fake news. While the issues on privacy and threat to democracy are important and should be addressed, this paper only suggests solutions to the economic issues highlighted in this paper. By reducing the monopoly power of the internet giants, the solutions will buttress the tools to control these other areas.

Internal Tools: Promises Not Fulfilled

While Facebook and Google have called for regulation of internet platforms, this is self-serving in order to frame the regulation in their favor—it gets the "solutions" they want rather than what is best for the society. Facebook, Google, and others could internally fix their problems, if they desired to, but there are no incentives for them to do so. Indeed, they "break things" and ask for forgiveness later. They are always quick to apologize when they get caught doing something inappropriate, questionable, or illegal and promise to take corrective measures. However, the promises are not fulfilled in many cases. Indeed, Facebook, Google,

---

[19] The authors recognize that this is a crude estimate, but a low estimate. Since profits are "hidden" by a variety of tax avoidance schemes, these are under-estimates of the loss. Google's dynamic pricing algorithm may also lead to an under-estimation of the calculation. On the other hand, because Google only makes 96% of its revenue from advertisements this may overestimate the loss. Nor does this estimate account for a "normal" return on assets, which would lower the estimate. Nevertheless, it provides a rough order of magnitude of the problem and suggests further investigation.

and others have not and do not conform with the current rules.[20] How can we expect them to correct their behavior without external measures? We cannot, so we turn to examine external remedies.

Antitrust

Antitrust action is an obvious method to reduce the power of the giant internet platforms by breaking them up. Unfortunately, based on the lack of action on the various acquisitions of GAFAMs, this may not be a realistic strategy. Amazon's behavior illustrates how antitrust policy has been eroded. In May of 2018 Amazon raised the price of Amazon Prime by 20%, a clear sign of monopoly power (Rubin 2018). It also practices predatory pricing, thwarts competition, creates barriers to entry; but the current view of antitrust law, the neoclassical one, does not consider these practices deleterious (Khan 2017). For Facebook and Google, the DOJ and FTC, apparently, only view one side of the market—the consumers' side—not the advertisers' side, who pay excessive prices because of the unique market position of the platforms. But if these regulatory units were to look more closely at the old tools of antitrust—structure, conduct, and performance, as well as the advertising side of the market—they might have a different view (see Khan 2017).

Antitrust analysis as posited by the Chicago School is a poor tool. The precedent of imposing behavioral obligations on operators found to possess Significant Market Power (SMP) was established in the telecom industry for purposes of establishing interconnection between incumbents and new entrants. Such an approach should be examined for GAFAM companies.

Facebook and Google have significant market power as measured by their Herfindahl–Hirschman indices (HHI) by several different definitions of markets. They range from 8476 for Google in the market for search to 2024 for Facebook in the market for social media—representing "highly concentrated" to "moderately concentrated" markets (Alleman and Taschdjian 2019).[21] They have acquired

---

[20] See Singer (2018) for the several violations of regulatory rules by Facebook and Google.

[21] HHI is the sum of the squares of the market share of the largest firms in the market, namely, $HHI = \Sigma s_i^2$ where s is the market share of the ith firm. The United States Department of Justice would consider this to be "moderately concentrated" (HHI 1500 to 2000). The threshold for "highly concentrated" is 2500 in the United States (DOJ 2010).

many different firms with little or no antitrust scrutiny, many of which have become major parts of their business. Alphabet (Google) has acquired over 200 companies (Wikipedia 2019a). Facebook has acquired Instagram, WhatsApp, Oculus VR, and some sixty other companies, many of which could be spun off (Wikipedia 2019b). The breakup of these companies would be complex but feasible. A set of smaller firms would ameliorate some of the issues.

Regulation: Treat as Utility

An obvious method of regulating internet giants is treating them as a public utility (Feld 2019; Abernathy et al. 2019). They have become a vital infrastructure, critical to the economy, but under virtually no control. These internet platforms, particularly Facebook and Google, have suggested that they should be regulated, but they would set a self-serving agenda as noted above. Moreover, the firms have yet to control fake political posts from outside the country, "hate-speech," false news, and a variety of malevolent posts. Their discipline and response have not been adequate. As important as these issues are, we will only address the economic controls in this paper. As a utility, the regulatory authority could use a variety of tools which are reviewed below.

Data Portability

Internet giants such as Facebook and Google have collected a variety of data from their users[22] and supplemented with third-party data in order to pursue their business strategy of focused advertisements aimed at their users. It would be extremely difficult for any company to duplicate this data set. By making the data portable, it would loosen the power of network externalities which allows the internet giants to dominate the market. A requirement that this data be "portable" among internet platforms would go a long way to allowing competition into these markets. If correctly structured, this would allow more transparency. Moreover, users could select what data they wish to keep private and what can be

---

[22] User is used here to denote the suppliers of *their data* to the GAFAMs. The United States Congress is beginning to recognize the value of the data (Lohr 2019).

used publicly, or at least shared among the platforms. It would also allow the user-side of the market the ability to amend incorrect data, just as consumers can correct false financial information held by credit rating agencies. A secondary effect might be for the providers of the data—in this case, the household consumer—to recognize the value of the asset (financial details, addresses, personal and family details, marital status, hobbies, contact points, etc.) that they are giving away.

Rate of Return Regulation/Profit Constraints/Price-Caps

In the utility industries: power, water, telecommunications, etc. have large capital requirements which gives the utility economies of scale, and hence, monopoly control of the market. The regulatory solution to this market failure has been to control the rate-of-return (RoR) on assets in order to emulate the competitive outcome.[23] But in the case of internet giants, it is the network externalities which give the GAFAMs their monopoly power; hence, the RoR does not offer a solution. Network externalities have to be confronted.

In response to the issues associated with the RoR, many regulators adopted price-caps controls. This requires that the regulated firm could only increase its vector of prices by an index of inflation less a factor for efficiency. After its introduction in the United Kingdom, it has been extensively used in the telecommunications industry by countries around the world. Price has the valuable function of allocating valuable resources, thus in order for price-caps to be applied to the internet sector, prices would first have to be set at "costs" and then the price-caps could be implemented. Finding the correct "costs" would be an issue, but this could be overcome.

Summary/Tentative Conclusions

Traditional antitrust and regulatory tools rely heavily on static economic analysis to determine relevant markets and assess market power. Careless application of the tools to control monopoly power can lead to policy

---

[23] A long and extensive literature exists on the problems with this and price-caps methods of regulation.

errors that undermine the dynamism of technological change and reduce consumer welfare. In a dynamic environment, antitrust and regulatory oversight is necessary.

GAFAMs' pricing behavior has at least three consequences: It raises the prices for the goods advertised, since these costs must be covered by the selling party to stay in business. This is at least $35.8 and $22.2 billion for Google and Facebook, respectively. This harms consumers, obviously. An addition harm is the dead-weight loss generated by the monopoly pricing. The total welfare loss is at least $53.7 and $33.2 billion for Google and Facebook, respectively. Collectively, the two internet giants create a welfare loss of nearly $87 billion dollars! And, increasing continually going forward.

Second, it thwarts potential competition on the consumers' side of the market. Only those with deep financial pockets can afford the cost of advertising; the others will be excluded from entering the market. This harms consumers by eliminating competition in the market for the product being advertised, and, once again, ensuring a higher price for consumers. This higher price is in addition to the increase in price due to the cost of advertising. It amplifies the harm to consumers of the price increase due to advertising costs.

The irony is that the platform is eliminating competitors not from its market—it already has a monopoly—but for (from) its clients' market, the buyers of space/advertising on its platform. This has a secondary effect of increasing the market power of the seller of the products. To the extent that these firms have significant market power they can charge higher prices for the products they are advertising, causing additional consumers' harm.

GAFAM's platforms are eliminating competition and harming consumers, which are grounds for some form of social control—the combined effect makes an even stronger case for treating internet platform giants as a utility and regulating them and/or vigorous antitrust actions for injurious conduct and performance. The "market" approach will not correct the problems. Proactive action is needed.

References

Abernathy, Nell, Darrick Hamilton, and Julie Margetta Morgan. 2019. New Rules for the 21st Century: Corporate Power, Public Power, and the Future of the American Economy. A Roosevelt Institute Report, April.

Alleman, J., and P. Rappoport. 2006. Optimal Pricing with Sunk Cost and Uncertainty. In *The Economics of Online Markets and ICT Networks*, ed. R. Cooper, G. Madden, A. Lloyd, and M. Schipp. *Contributions to Economics*. Physica-Verlag HD. https://link.springer.com/chapter/10.1007/3-7908-1707-4_10 (24.04.2019).

Alleman, J. 2018. The Threat of Internet Platforms: Facebook, Goggle, etc. *Proceedings of ITS European Conference 2018*, Aug 1–3, Trento, Italy. https://ideas.repec.org/p/zbw/itse18/184926.html (24.04.2019).

Alleman, James and Martin Taschdjian. 2019. Antitrust Failure: Internet Giants. Working Paper.

Bork, Robert. 1978. *The Antitrust Paradox*. New York, NY: The Free Press.

Bork, Robert. 1993. *The Antitrust Paradox*, 2nd ed. New York, NY: The Free Press.

Bostoen, Friso. 2019. Online Platforms and Pricing: Adapting Abuse of Dominance Assessments to the Economic Reality of Free Products. *Computer Law & Security Review* 35: 263–280. https://www.itsworld.org/wp-content/uploads/2019/05/Friso-Bostoen.pdf (3.06.2019).

eMarketer. 2017. Google and Facebook Tighten Grip on US Digital Ad Market, September 21. https://www.emarketer.com/Article/Google-Facebook-Tighten-Grip-on-US-Digital-Ad-Market/1016494 (22.05.2019).

European Commission. 2015. Antitrust: Commission Sends Statement of Objections to Google on Comparison Shopping Service; Opens Separate Formal Investigation on Android. Brussels, April 15. europa.eu/rapid/press-release_IP-15-4780_en.htm (22.05.2019).

Evans, D.S., and M.D. Noel. 2008. The Analysis of Mergers That Involve Multisided Platform Businesses. *Journal of Competition Law and Economics* 4 (3): 663–695.

Evans, D.S., and R. Schmalensee. 2012. *The Antitrust Analysis of Multi-Sided Platform Businesses*. Coase-Sandor Institute for Law and Economics Working Paper No. 623.

Feld, Harold. 2019. The Case for the Digital Platform Act: Market Structure and Regulation of Digital Platforms. Roosevelt Institute. Public Knowledge, May. http://bit.ly/2QEexkR (17.05.2019).

Financial Times. 2018. WPP Squeezed by Advertisers and Disruption. March, 3–4, p. 12. Grow Your Business with Google Ads. https://ads.google.com/home/ (24.04.2019).

Internet Retailer. 2019. Ecommerce Represented 14.3% of Total Retail Sales in 2018..., March 13. https://www.digitalcommerce360.com/article/us-ecommerce-sales/https://www.digitalcommerce360.com/article/us-ecommerce-sales/ (29.05.2019).

Kang, Cecilia. 2019. F.T.C. Approves Facebook Fine of About $5 Billion. *New York Times*, June 12. https://nyti.ms/2XHcVc4 (18.07.2019).

Khan, Lina M. 2017. Amazon's Antitrust Paradox. *The Yale Law Journal*, January. https://www.yalelawjournal.org/pdf/e.710.Khan.805_zuvfyyeh.pdf (6.11.2017).

Khan, Lina M. 2018. The New Brandeis Movement: America's Antimonopoly Debate. *Journal of European Competition Law & Practice* 9 (3): 131–132. https://doi.org/10.1093/jeclap/lpy020 (18.07.2019).

Kwoka, John. 2014. *Mergers, Merger Control, and Remedies: A Retrospective Analysis of U.S. Policy*. Cambridge, MA: MIT Press.

Lohr, Steve. 2019. Calls Mount to Ease Big Tech's Grip on Your Data. *New Your Times*, July 25. https://nyti.ms/2YwENUU.

Nicas, Jack and Karen Weise. 2019. Anger at Big Tech Unites Noodle Pullers and Code Writers. Washington Is All Ears. *New York Times*, June 10. https://nyti.ms/2WpldVr.

OECD. 2018a. *Tax Challenges Arising from Digitalisation—Interim Report 2018: Inclusive Framework on BEPS*, OECD/G20 Base Erosion and Profit Shifting Project. Paris: OECD Publishing. https://doi.org/10.1787/9789264293083-en (10.06.2019).

OECD. 2018b. Rethinking Antitrust Tools for Multi-Sided Platforms. www.oecd.org/competition/rethinking-antitrust-tools-for-multi-sided-platforms.htm (16.05.2019).

Ohio et al. v. American Express Co. et al. 2017. https://www.supremecourt.gov/opinions/17pdf/16-1454_5h26.pdf (7.05.2019).

Park, Jon. 2018. How Google AdWords Works [Infographic]: Paying Per Click (PPC), Impression (CPM), and Choosing How Much to Bid, July 16. https://blog.tryadhawk.com/google-adwords/how-google-adwords-works/ (24.04.2019).

Parker, Geoffrey and Marshall W. Van Alstyne. 2005. Two-Sided Network Effects: A Theory of Information Product Design. *Management Science* 51 (10): 1494–1504. https://papers.ssrn.com/sol3/papers.cfm?abstract_id=1177443 (9.06.2018).

Rubin, Ben Fox. 2018. Amazon Prime's Price Went Up, But You Probably Won't Leave, Amazon Increased the Fee for Annual US Prime members, Taking It From $99 to $119. Though Some Folks Are Squawking, They'll Probably Pony Up Regardless, April 27. https://www.cnet.com/news/amazon-prime-price-went-up-but-you-probably-wont-leave-anyways/ (31.05.2019).

Satariano, Adam. 2019. Google Fined $1.7 Billion by E.U. for Unfair Advertising Rules. *New York Times*, March 20. https://nyti.ms/2UNmkxU (7.05.2019).

Securities and Exchange Commission (SEC), Alphabet. 2018. https://www.sec.gov/Archives/edgar/data/1652044/000165204419000004/googl0-kq42018.htm (28.05.2019).

Securities and Exchange Commission (SEC), Facebook. 2018. https://www.sec.gov/Archives/edgar/data/1326801/000132680119000009/fb-12312018x10k.htm#s3714AF39C3CC5682ADA657DA7DE15627 (28.05.2019).

Shapiro, Carl and Hal Varian. 1997. *Information Rules: A Strategic Guide to the Network Economy.* Boston: Harvard Business School Press.

Singer, N. 2018. Timeline: Facebook and Google Under Regulators' Glare *New York Times*, March 24. https://nyti.ms/2G6quK0.

Statista. 2018. Social Media in the United States. https://www-statista-com.colorado.idm.oclc.org/study/40227/social-social-media-usage-in-the-united-states-statista-dossier/ (6.05.2018).

Stewart, James B. 2018. Amazon, the Elephant in the Antitrust Room. *New York Times*, May 3. https://nyti.ms/2FFVlMy (6.05.2018).

U.S. Department of Justice (DOJ). 2010. *Horizontal Merger Guidelines (08/19/2010).* August 19. https://www.justice.gov/atr/horizontal-merger-guidelines-08192010#5c.

Varian, Hal. 2019. Presentation "Internet Platforms' Rising Dominance, Evolving Governance", February 10–11. Silicon Flatirons, Boulder, CO.

Welch, Chris. 2019. Google Begins Shutting Down Its Failed Google+ Social Network. April 2. https://www.theverge.com/2019/4/2/18290637/google-plus-shutdown-consumer-personal-account-delete (22.05.2019).

Wikipedia. 2019a. https://en.wikipedia.org/wiki/List_of_mergers_and_acquisitions_by_Alphabet (1.06.2019).

Wikipedia. 2019b. https://en.wikipedia.org/wiki/List_of_mergers_and_acquisitions_by_Facebook (1.06.2019).

WordStream. https://www.wordstream.com/articles/what-is-google-adwords (24.04.2019).

Wu, Tim. 2016. *The Attention Merchants: The Epic Scramble to Get Inside Our Heads.* New York: Alfred A. Knopf.

Wu, Tim. 2020. *The Master Switch: The Rise and Fall of Information Empires.* New York: Alfred A. Knopf.

Conclusion

Dear Readers, as you come to the end of this Festschrift, we are sure that you will agree with us that the untimely passing of Gary Madden is not only a loss to his family and friends, but to the academic community. As demonstrated by this small sampling of his interests: The spectrum of his research, his insights, innovative techniques, and rigorous scholarship; he will be sorely missed.

Appendix: Gary Madden's Publications

Books

International Handbook of Telecommunications Markets, Volume II, Emerging Telecommunication Networks (2003), ed. Gary Madden, Edward Elgar Publishing, Cheltenham, U.K.

International Handbook of Telecommunications Markets, Volume I, Traditional Telecommunications Markets (2003), ed. Gary Madden, Edward Elgar Publishing, Cheltenham, U.K.

International Handbook of Telecommunications Markets, Volume III, World Telecommunications Markets (2003), ed. Gary Madden, Edward Elgar Publishing, Cheltenham, U.K.

Frontiers of Broadband, Electronic and Mobile Commerce (2004), ed. Russel Cooper & Gary Madden, Physica-Verlag, Heidelberg, Germany.

The Economics of Online Markets and ICT Networks (2006), ed. Russel Cooper, Gary Madden, Ashley Lloyd & Michael Schipp, Physica-Verlag, Heidelberg, Germany.

The Economics of Digital Markets (2009), ed. Gary Madden & Russel Cooper, Edward Elgar Publishing, Cheltenham, U.K.

Regulation and the Performance of Communication and Information Networks (2012), ed. Gerald Faulhaber, Gary Madden & Jeffrey Petchey, Edward Elgar Publishing, Cheltenham, U.K.

Articles

Gary Madden (1988). "An econometric model of Australian passenger automobile demand: A segmented markets approach", *Economic Analysis and Policy*, 18(1).

Gary Madden, Harry Bloch & David Hensher (1963). "Australian telephone network subscription and calling demands: Evidence from a stated-preference experiment", *Information Economics and Policy*, 5(3).

Gary Madden & Paula Haslehurst (1995). "Causal analysis of Australian economic growth and military expenditure: A note", *Defense and Peace Economics*, 6(2).

Gary Madden, Ian Kerr, Paula Haslehurst & Garry Macdonald (1995). "An empirical analysis of Australian defense expenditure: 1970–1990", *Australian Economic Papers*, 34(65).

Gary Madden (1995). "Experimentation in economics: An overview of the stated preference experimental design method", *Australian Economic Papers*, 36(64).

Gary Madden & Harry Bloch (1995). "Productivity growth in Australian manufacturing: A vintage capital model", *International Journal of Manpower*, 16(1).

Gary Madden, Scott Savage & Michael Simpson (1996). "Information inequality and broadband network access: An analysis of Australian household data", *Industrial and Corporate Change*, 5(4).

Gary Madden, Michael Simpson & Peter Dawkins (1997). "Casual employment in Australia: Incidence and determinants", *Australian Economic Papers*, 36(69).

Gary Madden, Scott Savage & Steven Kemp (1997). "Measuring public sector efficiency: A study of economics departments at Australian Universities", *Education Economics*, 5(2).

Gary Madden, Scott Savage & Michael Simpson (1997). "Broadband delivery of educational services: A study of subscription intentions in Australian provincial centers", *Journal of Media Economics*, 10(1).

Gary Madden & Michael Simpson (1997). "A probit model of household broadband service subscription intentions: A regional analysis", *Information Economics and Policy*, 8(3).

Gary Madden & Michael Simpson (1997). "Residential broadband subscription demand: An econometric analysis of Australian choice experiment data", *Applied Economics*, 29(11936).

Gary Madden, Scott Savage & I. Bell (1997). "The direct and indirect links between trade and growth: A study of APEC members", *Singapore Economic Review*, 42(1).

Gary Madden, Steven Kemp & Michael Simpson (1998). "Emerging Australian education markets: A discrete choice model of Taiwanese and Indonesian student intended study destination", *Education Economics*, 6(2).

Gary Madden & Scott Savage (1998). "Interconnection in international telecommunications", *Telecommunications Policy*, 22(11).

Gary Madden & Scott Savage (1998). "CEE telecommunications investment and economic growth", *Information Economics and Policy*, 10(2).

Gary Madden & Scott Savage (1998). "Sources of Australian labor productivity change 1950–1994", *Economic Record*, 74(227).

Gary Madden, Scott Savage & Michael Simpson (1998). "Asia-Pacific telecommunications USO's: Current practice and future options", *Prometheus*, 16(4).

Gary Madden, Scott Savage & Su Yin Thong (1999). "Technology, investment and trade: Empirical evidence for five Asia-Pacific countries", *Applied Economic Letters*, 6(6).

Gary Madden, Scott Savage & Grant Coble-Neal (1999). "Subscriber churn in the Australian ISP market", *Information Economics and Policy*, 11(2).

Gary Madden, Scott Savage & Su Yin Thong (1999). "Technology, investment and trade: Empirical evidence for five Asia-Pacific countries", *Applied Economic Letters*, 6(6).

Gary Madden & Scott Savage (1999). "Telecommunications productivity, catch-up and innovation", *Telecommunications Policy*, 23(1).

Gary Madden, Scott Savage & James Alleman (2000). "Trade imbalance in international message telephone service", *Applied Economics*, 32(10).

Gary Madden, Scott Savage & Grant Coble-Neal (2000). "Internet adoption and use", *Prometheus*, 18(2).

Gary Madden & Scott Savage (2000). "Some economic and social aspects of residential internet use in Australia", *Journal of Media Economics*, 13(3).

Gary Madden & Scott Savage (2000). "Telecommunications and economic growth", *International Journal of Social Economics*, 27(7).

Gary Madden & Scott Savage (2000). "Market structure, competition and pricing in United States international telephone service markets", *Review of Economics and Statistics*, 82(2).

Gary Madden, Scott Savage & Andrew McDonald (2000). "Assessing the economic preconditions for a Yen bloc", *Australian Economic Papers*, 39(1).

Gary Madden, Scott Savage & Paul Bloxham (2001). "Asian and OECD international R & D spillovers", *Applied Economics Letters*, 8(7).

Gary Madden & Scott Savage (2001). "Productivity growth and market structure in telecommunications", *Economics of Innovation and New Technologies*, 10(6).

Gary Madden & Scott Savage (2001). "Regulation and international telecommunications pricing behavior", *Industrial and Corporate Change*, 10(1).

Gary Madden & Harry Bloch (2001). "The cost structure of Australian telecommunications", *Economic Record*, 77(2).

Gary Madden, Scott Savage & Craig Tipping (2001). "Understanding European Union international message telephone service demand", *Information Economics and Policy*, 13(2).

Gary Madden, Harry Bloch & Scott Savage (2001). "Economies of scale and scope in Australian telecommunications", *Review of Industrial Organization*, 18(2).

Gary Madden, Scott Savage & Michel Simpson (2002). "Broadband delivered entertainment services: Forecasting Australian subscription intentions", *Economic Record*, 78(2).

Gary Madden, Harry Bloch & Grant Coble-Neal (2002). "Labor and capital saving technical change in telecommunications", *Applied Economics*, 34(14).

Gary Madden, Scott Savage & James Alleman (2003). "Dominant carrier market power in US international telephone markets", *Applied Economics*, 35(6).

Gary Madden & Grant Coble-Neal (2003). "Internet use in rural and remote Western Australia", *Telecommunications Policy*, 27(2).

Gary Madden, Scott Savage & Jason Ng (2003). "Asia-Pacific telecommunications liberalization and productivity performance", *Australian Economic Papers*, 42(1).

Gary Madden & Grant Coble-Neal (2004). "Internet traffic dynamics", *Telerktronikk*, 4.

Gary Madden & Grant Coble-Neal (2004). "A cost of living index incorporating a network effect", *Applied Economics*, 36(19).

Gary Madden, Grant Coble-Neal & Scott Savage (2004). "United States internet penetration", *Applied Economic Letters*, 11(9).

Gary Madden & Grant Coble Neal (2004). "Economic determinants of global mobile telephony growth", *Information Economics and Policy*, 16(4).

Gary Madden & Grant Coble-Neal (2004). "A dynamic model of mobile telephony subscription incorporating a network effect", *Telecommunications Policy*, 28(2).

Gary Madden & Grant Coble-Neal (2005). "Australian residential telecommunications consumption and substitution patterns", *Review of Industrial Organization*, 26(3).

Gary Madden & Grant Coble-Neal (2005). "Forecasting international bandwidth capacity", *Journal of Forecasting*, 24(4).

Gary Madden (2005). "Regional development and business prospects for ICT and broadband networks", *Telecommunications Policy*, 29(2).

Gary Madden, Robert Fildes & Joachim Tan (2007). "Optimal forecasting model selection and data characteristics", *Applied Financial Economics*, 17(15).

Gary Madden & Joachim Tan (2007). "Forecasting telecommunications data with linear models", *Telecommunications Policy*, 31(1).

Gary Madden, James Alleman & Hak Kim (2008). "Real options methodology applied to the ICT sector: A survey", *Communications and Strategies*, 70(2).

Gary Madden & Joachim Tan (2008). "Forecasting international bandwidth capacity using linear and ANN methods", *Applied Economics*, 40(14).

Gary Madden (2010). "Economic welfare and universal service", *Telecommunications Policy*, 34(1).

Gary Madden & Russel Cooper (2010). "Estimating components of ICT expenditure: A model-based approach with applicability to short time-series", *Applied Economics*, 42(1).

Gary Madden & Erik Bohlin (2010). "Balancing competition and regulation", *Telecommunications Policy*, 34(1).

Gary Madden, Erik Bohlin & Aaron Morey (2010). "An econometric analysis of 3G auction spectrum valuations", RSCAS Working Papers, European University Institute.

Gary Madden, Pratompong & Erik Bohlin (2011). "The determinants of mobile subscriber retention in Sweden", *Applied Economic Letters*, 19(5).

Gary Madden & Aaron Morey (2011). "The impact of privatization and competition on telecommunication network growth and productivity", *International Journal of Management and Network Economics*, 2(1).

Gary Madden & Aaron Morey (2012). "Regulatory flexibility and the administrative allocation of licensing of 3G spectrum", *Applied Economics*, 45(13).

Gary Madden, Aaron Morey & Bill Theoharakis (2012). "On the statistical identification of technology disruption", *Telecommunications Policy*, 36(1).

Gary Madden, Walter Mayer, Chen Wu & Thien Tan (2013). "The forecasting accuracy of models of post-award network deployment: An application of maximum score tests", *International Journal of Forecasting*, 31(4).

Gary Madden, Erik Bohlin, Hajime Oniki & Thien Tan (2013). "Potential demand for m-government services in Japan", *Applied Economic Letters*, 20(8).

Gary Madden, Erik Bohlin & Hasnat Ahmad (2013). "On the interim pricing of unbundled local loop", *Telecommunications Policy*, 38(8).

Gary Madden, Erik Bohlin & Thien Tran (2013). "Spectrum licensing and flexible beauty contest designs", *Annals of Public and Cooperative Economics*, 84(3).

Gary Madden, Md. Shah Azam & T. Randolph Beard (2013). "Small firm performance in online markets", *Economics of Innovation and New Technologies*, 22(1).

Gary Madden, Erik Bohlin, Hajirne Oniki & Thien Tran (2013). "Potential demand for m-government services in Japan", *Applied Economic Letters*, 20(8).

Gary Madden & Hasnat Ahmad (2013). "3G spectrum auction after-market network deployment", *Applied Economic Letters*, 20(3).

Gary Madden & Thien Tan (2013). "Do regulators consider welfare when assigning spectrum via comparative selections?" *Applied Economic Letters*, 20(9).

Gary Madden, Johannes Bauer & Aaron Morey (2014). "Effects of economic conditions and policy interventions on OECD broadband adoption", *Applied Economics*, 46(12).

Gary Madden, Erik Bohlin, Paitoon Kraipornsak & Thien Tan (2014). "The determinants of prices in the FCC's 700 MHz spectrum auction", *Applied Economics*, 46(17).

Gary Madden, Erik Bohlin, Thien Tran & Aaron Morey (2014). "Spectrum licensing, policy instruments and market entry", *Review of Industrial Organization*, 44(3).

Gary Madden, Walter Mayer, Zhong Jin & Thien Tan (2014). "Modeling OECD broadband subscriptions in disequilibrium", *Technological Forecasting and Social Change*, 90.

Gary Madden, Paitoon Kraipornsak & Inayat Hussain (2014). "The determinants of digital terrestrial radio aftermarket", *Applied Economic Letters*, 22(8).

Gary Madden, Krishna Jayakar, Chun Liu & Eun-A Park (2015). "Promoting broadband and ICT access for disabled persons: Comparative analysis of initiatives in Asia-Pacific region", *The Information Society*, 31(4).

Gary Madden, Suphat Suphachalasai & Thanet Makjamroen (2015). "Residential demand for bundled fixed-line and wireless mobile broadband services", *Applied Economics*, 47(47).

Gary Madden, Ismail Saglam & Inayat Hussain (2015). "Spectrum auction designs and revenue variations", *Applied Economics*, 47(17).

Gary Madden, Paitoon Kraipornsak & Inayat Hussain (2015). "The determinants of digital terrestrial radio aftermarket coverage", *Applied Economic Letters*, 22(6).

Gary Madden, Aniruddha Banerjee, Paul N. Rappoport & Hiroaki Suenaga (2016). "E-commerce transactions, the installed base of credit cards, and the potential mobile E-commerce adoption", *Applied Economics*, 49(1).

Gary Madden, Christian Dippon & Hiroaki Suenaga (2016). "Do economic, institutional or political variables explain regulated wholesale unbundled local loop rate setting?" *Applied Economics*, 48(39).

Gary Madden, Maria Rosalia Vicente, Paul Rappoport & Andy Banerjee (2017). "A contribution on the nature and treatment of missing data in large market surveys", *Applied Economics*, 49(22).

Gary Madden & Hiroaki Suenaga (2017). "The determinants of price in 3G spectrum auctions", *Applied Economics*, 49(32).

Gary Madden & Maria Rosalia Vicente (2017). "Assessing eHealth skills across Europeans", Health Policy and Technology, 6(3).

Gary Madden, Nicholas Apergis, Paul Rappoport & Aniruddha Banerjee (2018). "An application of nonparametric regression to missing data in large market surveys", *Journal of Applied Statistics*, 45(7).

Gary Madden & Hiroaki Suenaga (2018). "Understanding the equilibrium dynamics of European broadband diffusion", *Applied Economics*, 50(60).

Gary Madden, Walter Mayer & Chen Wu (2019). "Broadband and economic growth: A reassessment", *Information Technology for Development*, online.

INDEX

A
adoption, 67, 83, 96, 103–105, 107–111, 114–116, 118, 121–126, 128, 151–155, 157, 159, 161, 163, 285
adoption curve, 153, 154, 157
adversarial legalism, 290, 291, 296, 297
advertising, 11, 14, 155, 254, 271, 279, 304–306, 308–314, 317, 320
agriculture, 2–4, 6
airline, 152, 277, 278
algorithm, 195, 196, 201, 215, 270, 271, 316
Alta Vista, 11, 17
Amazon, 9–11, 13, 19, 270–272, 303, 305, 306, 309, 310, 317
American Express, 274–277, 305
American Revolutions, 3
Android, 10, 19
anticompetitive, 11, 17–20, 272, 274, 277, 279, 305

antitrust, 10–12, 14, 17–20, 272, 273, 275, 277, 304, 305, 312, 317–320
antitrust enforcement, 17
Antitrust Paradox, 305
artificial intelligence (AI), 122, 127
asymptotic variance-covariance (AVC), 212, 214, 215
Austria, 109, 157
automobile, 5

B
Babylon, 1, 3
banking, 156, 200, 201, 204, 304
basic telephone service, 210
Bass, A., 10
Bass diffusion model, 152
Bass, F.M., 151–157, 159–164
Bayesian, 203, 213
bidirectional causality, 64, 67, 68, 81, 96, 97
Big Data, 268

Bing, 11, 17
Bork, Robert, 272, 273, 305, 309
Britain, 3, 4
broadband causality, 68, 87, 91, 96, 97
broadband services, 72–74, 213, 251, 299
Bush v. Gore, 298

C
Cambridge Analytica, 304
capital, 5–8, 139, 319
Carnegie, Andrew, 5, 6, 13
cellular, 66
chemicals, 5
Chicago School, The, 20, 272, 309, 312
The Chicago School of antitrust, 20, 312, 317
China, 3, 11, 22
citizenship, 114
civic engagement, 181
competition, 12, 19, 195–198, 203, 209, 249–252, 255, 265, 267, 268, 270–276, 279–281, 292, 305, 312–318, 320
complementarity, perfect, 250, 256, 261
conjoint choice, 215, 216, 220
conservative, 197, 275, 278
consumer expenditures, 30, 38, 53
consumers' surplus, 243–245, 259, 260, 262, 268, 312, 314, 315
consumers' welfare, 267, 268, 270, 272, 273, 279, 280, 305, 320
content, 15, 104, 139, 194, 249–261, 264, 265, 267, 268, 271, 272
content providers (CPs), 249–252, 254, 255, 257, 261, 263, 264
content services, 252
cost complementarities, 257, 261
court, 268, 269, 274–281, 289, 290, 296–300

cross-country analysis, 63, 64
cross-price elasticities, 32, 38–40, 43, 45, 46, 48, 50, 52, 53, 209
cyber-civilization, 8–10, 13, 14, 20, 21, 23

D
data packets, 249
D.C. Circuit Court of Appeals, 287, 288, 296–300
D-efficient design, 213, 215, 220, 221, 223, 224
democracy(ies), 290, 293, 304, 316
Democratic party, 288, 291, 295, 296
Department of Justice (DOJ), 12, 17, 309, 317
deregulation, 209, 294, 305
D-error, 212–215, 220, 221, 223–225
developed countries, 6, 7, 22, 66, 67
development, 1, 7, 9, 10, 14, 23, 64, 65, 71, 73, 74, 77, 80, 81, 89, 93, 96–98, 126, 137, 152, 176, 188, 233, 249, 250, 286, 288, 300
Diebold-Mariano-West (DMW), 138
diffusion, 63, 66, 106, 107, 126, 151–153, 157, 163, 177
diffusion models, 155, 163
diffusion of innovations (DOI), 151, 152, 163
digital, 104, 106, 108, 109, 111, 127, 269, 271–274, 278, 280, 309–311
digital divide, 98, 188
digital economy, 107, 264
Digital Economy and Society Index (DESI), 103, 109
digital literacy, 179
digital skills, 104, 106, 111, 112, 115, 117–119, 121, 122, 124–126, 128, 129, 176, 179, 182, 186, 187
discrete choice, 115, 216

INDEX   337

discrimination, 249, 311
dominance, 4, 7, 11, 12, 14, 17, 267, 273, 279, 280, 304, 309, 314
Dupont, 5, 12
dynamic panel, 68, 74, 76, 95, 138, 139, 149

E
e-commerce, 73, 103–105, 107–112, 114–118, 121–128, 131, 304
econometrics, 66, 93, 95, 96, 98, 195
economic driver, 1, 3, 4, 6, 8, 23
economic growth, 63–68, 71–74, 77, 80, 81, 83, 85, 87, 89, 91, 93, 95–98, 137–140, 146–149, 202, 314
economics rent, 314
ecosystem, 273, 277, 278, 280
education, 20–22, 29, 31, 39, 40, 45–50, 52, 108, 111, 112, 115, 118, 122, 125, 126, 129, 177, 178, 184, 186, 203, 216, 291
Electronic Funds Transfer (ETF), 156, 217, 278
email, 153, 175, 180–182, 186, 187, 218
employment, 9, 10, 20–23, 108, 114, 121, 125, 177, 182, 186, 267, 291
England, 3, 289
entertainment, 31, 39, 40, 45–50, 52, 55
e-participation, 182, 184, 186
Euphrates, 1, 2
European, 159, 161, 163, 176, 179, 182
European Commission, 103, 109, 179, 180, 305
European Union (EU), 10–12, 15, 16, 19, 66, 83, 103, 104, 108, 109, 137, 180, 315
ex ante, 220–223, 268, 272
exclusive dealing, 231

exogenous variables, 156
expenditures, households, 30, 44
experimental design, 211, 213, 216, 220
ex post, 220–225, 234, 268, 272
externalities, 231, 257, 267, 269, 273, 280, 306, 315, 318, 319

F
Facebook, 9–11, 13, 18, 303, 306, 309, 317
Facebook, Amazon, Apple, Netflix, and Google (FAANG), 303
firm, 5, 6, 8–19, 65, 127, 234, 246, 256, 260, 279, 305, 309, 311, 313, 317–320
first wave, 3
Ford, 5, 21
forecasting, 193–204
foreclosure, 232, 235
fractional factorial, 210–212, 214, 215
French Revolutions, 3

G
Gamma/Shifted Gompertz model (G/SG), 155, 157, 161, 163
Germany, 9, 108, 109, 157
Global Competitiveness Index (GCI), 64, 68, 70, 71, 74, 95, 98
Global System for Mobile Communications Association (GSMA), 67, 87
Glorious Revolution, 3
Google, 9–12, 14, 15, 17, 19, 279, 303–306, 309–311, 313, 315–318, 320
Google, Apple, Facebook, Amazon, and Microsoft (GAFAM), 303, 314, 317–320
Granger-causality, 66, 68, 74–97
Great Britain, 4–6, 10

H

Heckman Selection Model (HSM), 105, 118, 121, 123, 128
Homogenous and Heterogeneous adoption, 155
hotel, 157, 159, 161, 163
households' consumer expenditures, 52
hypothesis tests of predictive accuracy, 138, 142

I

ICT Development Index (IDI), 64, 68, 70–74, 77–81, 83, 95, 98
income effects, 43
industrial production, 4–6
inequality, 7, 13, 14, 20, 23
information and communications technology (ICT), 64–69, 71–74, 85, 89, 93, 95–98, 105–108, 114, 153
information economy, 8–10, 14, 21–23
infrastructure, 66, 67, 71, 73, 85, 106, 137–140, 147–149, 249, 294, 318
innovation, 2, 8, 12, 70, 103, 151–157, 159, 163, 209, 267, 273, 280, 286, 312, 314
innovation adoption, 151
innovation diffusion, 107, 152, 156
Instagram, 9, 17, 18, 306, 310, 318
intermediary's platform, 267, 269, 271, 276, 280
International Telecommunication Union (ITU), 64, 71, 80, 89, 95
internet, 10, 11, 64, 66, 68, 69, 71, 73, 74, 83, 85, 87, 91, 95–97, 103, 104, 106–110, 114–117, 120, 122–125, 127–129, 152, 157, 159, 161, 162, 175, 177–182, 187, 188, 197, 210, 217, 218, 249–261, 264, 265, 267, 268, 272, 273, 279, 280, 285, 287, 296, 297, 299, 300, 304–306, 309, 310, 312, 314–320
internet service provider (ISP), 249–253, 255–265
investment, 2, 5, 7, 65–68, 71, 73, 77, 97, 249–254, 260, 263, 269

J

Japan, 1, 3–5, 7, 10, 15

K

Koch brothers, 294
Koch, Charles, 294, 295
Koch, David, 294, 295
Kodak, 9, 194

L

labor, 7, 14, 22, 23, 139, 186, 312
labor market, 186, 188
land, 3, 4, 6, 291
liberal, 22, 277, 293
licensing agreements, 232
Linear Probability Model (LPM), 117, 118, 121–123, 128
litigation, 272, 286, 288–291
logistic regression, 105, 120, 125, 128, 202, 214

M

machine learning, 194–204
Magna Carta, 3
Malaysia, 157, 159, 161
manufacturing, 4, 5, 8, 21, 194
market concentration, 268, 274, 281

marketplace distortions, 269
markets, 5, 9–12, 17–20, 73, 85, 96,
    152, 153, 161, 163, 209–211,
    213, 224, 232, 233, 247,
    249–252, 255, 258, 259, 264,
    265, 267–270, 272–281, 292,
    293, 295, 304–306, 308–315,
    317–320
mean absolute percentage error
    (MAPE), 198, 203
Meiji Restoration, 4, 5
merger, 17, 19, 314
Mesopotamia, 1–3, 23
Microsoft, 9, 10, 12, 13, 15, 17, 18,
    303
misinformation, 304
mobile internet penetration (PMI),
    74, 81–83, 85–87, 89–91
mobile services, 64, 69, 73, 210, 216,
    218, 220
modelling diffusion, 152, 163
monopolistic, 249–252, 264
monopoly, 14, 251, 252, 254, 305,
    310–317, 319, 320
monopoly rents, 12, 312, 314, 320
Morgan-Granger-Newbold (MGN)
    test, 138, 143–145, 148
Mozilla v. FCC, 287, 297–300
multisided markets (MSM), 309

N
*National Cable & Telecommunications
    Ass'n v. Brand X Internet Services*,
    299
network, 11, 18, 73, 115, 138, 186,
    196, 198, 201, 210, 232, 238,
    247, 249–252, 254, 265, 267,
    273, 280, 294, 295, 306, 309,
    311, 319
network neutrality, 272, 285–289,
    292, 295–297, 299, 300

network quality, 253
non-discriminatory, 250, 252,
    260–265

O
*Ohio et al. v. American Express Co.et
    al.*, 305
oil, 5, 32, 33, 45–52, 54, 294
online contacts, 178
online environment, 179, 186
online participation, 179–181, 184,
    186–188
online political activities, 180, 181,
    187
online tools, 176, 179, 187, 188
orthogonal design, 212, 214

P
per curiam, 297–300
platform, 15, 19, 126, 177, 209,
    231–238, 240–247, 267–281,
    305, 306, 309–312, 314–320
platform competition, 209
policy, 16, 98, 104, 105, 125, 126,
    176, 179–182, 186–188, 215,
    233, 242, 280, 285, 286,
    288–290, 292, 295–297, 300,
    309, 317, 319
policymakers, 17, 21, 24, 98, 125,
    126, 137, 247, 304
political activities, 175, 178, 187, 188,
    294
political agents, 186
political participation, 175–179, 181,
    182, 186–188
prediction, 138–140, 142–149, 194,
    195, 200, 201, 203, 280
Predictive Accuracy Tests, 137
price-caps, 319
price discrimination, 268

price elasticities, 29, 32, 41, 42, 53, 238
privacy, 11, 14–16, 19, 20, 196, 268, 274, 280, 281, 304, 309, 316
profit constraints, 319
profits, 238–243, 245, 246, 250, 252, 256–262, 270, 272, 273, 280, 281, 306, 311, 312, 314–316
publishing, 181, 304

Q

quality of service (QoS), 250–253, 255, 256, 259, 263, 264
quantitative forecasting, 105

R

railroads, 5
rate of return regulation, 319
regulating access, 268
regulation, 15, 16, 63, 73, 249–252, 260–265, 272, 287, 316, 318, 319
regulatory agencies, 268, 274, 277–279, 281
rent, 54
Republican Party, 287, 294
residential demand, 210
Restoring Internet Freedom Order, 285, 287, 288, 297
restrictive licensing agreements, 231
retail, 5, 10, 11, 19, 20, 194, 310
revealed preference (RP), 210
robot, 157
robotics, 13, 22

S

scarce resource, 3, 5, 6, 8, 23
search, 10, 11, 22, 197, 210, 279, 305, 306, 309, 311, 317
search engine, 11, 12, 17

Sears Roebuck, 5
Second wave, 5, 7, 21, 22
service quality, 251
Sherman Act, 12
Silicon Valley, 17
simulation, 38, 45–48, 51, 53, 56, 59, 60, 210, 212, 215
social costs, 314
social media, 11, 153, 179, 304, 306–308, 317
social networks, 161, 175–177, 179, 182, 184, 186, 187, 204, 267, 307
socioeconomic, 106, 107, 175, 176, 178–180, 187, 188
Spain, 103–105, 107–112, 114, 125
Spanish, 104, 108, 109, 114, 117, 121, 122, 125, 127, 129
Standard Oil, 5, 12
stated preference (SP), 210–212, 214, 215
statistical efficiencies, 212
substitution effects, 43, 53
supply chain, 201, 304
Supreme Court, 275, 276, 278, 305
surveillance capitalism, 268
survey, 29, 67, 105, 107–112, 114, 120, 124, 125, 127–130, 180, 194, 210, 211, 213–216, 218, 223
Swiss, 151, 157, 159, 161–163
Switzerland, 157

T

tax, 193, 294, 295, 304, 316
Tea Party, 294–296
technology adoption, 96, 105, 163
Telecommunications Act of 1996 (TA96), 292, 293, 295, 296, 300
third wave, 8, 21, 22
Tigris-Euphrates, 1–3

time series, 142, 143, 149, 193, 195, 196, 198, 201, 203
TopCoder, 233
tourism, 152, 153, 157, 159, 162, 163
transparency, 296, 318
Twitter, 303, 304
two-sided markets, 234, 268–270, 272, 273, 275, 277, 304, 305, 309, 310

U
United Nations (UN), 182, 184, 186
United States (US), 4–13, 15, 19–23, 31, 34, 36, 57, 58, 63, 67, 194, 210, 215, 216, 219, 253, 285–291, 295, 297, 300, 305–310, 316, 317
United States Supreme Court, 268, 274, 298–300
unregulated, 250, 252, 258, 259, 264, 265, 270
US Steel, 5, 12

US Telecom Association v. FCC, 287, 296
utility maximization, 116

V
Vapnik-Chervonenkis dimension, 200
variance-covariance, 212
Verizon v. FCC, 287, 299
vertical integration, 250, 251, 253, 257, 259, 262

W
wealth, 3, 6, 13
websites, 16, 126, 153, 157, 159, 161, 179, 180
WhatsApp, 18, 318
World Economic Forum (WEF), 64, 70, 95
World War II, 14, 291

Y
Yahoo!, 11, 17

Printed by Printforce, the Netherlands